J.H. Williams 10/90.

Reza Zamani 2001

Preparations of Vertebrate Central Nervous System
In Vitro

IBRO HANDBOOK SERIES
METHODS IN THE NEUROSCIENCES

General Editor: **A. D. Smith**
Department of Pharmacology,
University of Oxford

Volume 1: **Tracing Neural Connections with Horseradish Peroxidase**
Edited by M-Marsel Mesulam
0 471 10029 3 280 pages (paper) 1982

Volume 2: **Brain Microdissection Techniques**
Edited by A.C. Cuello
0 471 90019 2 186 pages (paper) 1983

Volume 3: **Immunohistochemistry**
Edited by A.C. Cuello
0 471 90052 4 544 pages (paper) 1983

Volume 4: **Methods for Neuronal Recording in Conscious Animals**
R. Lemon
0 471 90174 1 175 pages (cloth) 1983
0 471 90237 3 175 pages (paper) 1983

Volume 5: **Intracellular Perfusion of Excitable Cells**
Edited by P.G. Kostyuk and O.A. Krishtal
0 471 90379 5 152 pages (cloth) 1984
0 471 90392 2 152 pages (paper) 1984

Volume 6: **Measurement of Neurotransmitter Release *In Vivo***
Edited by C.A. Marsden
0 471 90445 7 233 pages (paper) 1984

Volume 7: **Molecular Biology Approach to the Neurosciences**
Edited by H. Soreq
0 471 90386 8 253 pages (cloth) 1985
0 471 90437 6 253 pages (paper) 1985

Volume 8: **Microiontophoresis and Pressure Ejection**
T. Stone
0 471 90607 7 222 pages (cloth) 1985
0 471 90608 5 222 pages (paper) 1985

Volume 9: **Advanced Micropipette Techniques for Cell Physiology**
K.T. Brown and D.G. Flaming
0 471 90952 1 296 pages (cloth) 1986

Volume 10: **Monoaminergic Neurons: Light Microscopy and Ultrastructure**
Edited by H.W.M. Steinbusch
0 471 91232 8 350 pages (cloth) 1987

Volume 11: **Neuropeptides: A Methodology**
Edited by G. Fink and A. Harmar
0 471 92072 X 268 pages (cloth) 1989

Volume 12: **Nerve Growth Factors**
Edited by R.A. Rush
0 471 92145 9 352 pages (cloth) 1989

Volume 13: **Preparations of Vertebrate Central Nervous System *In Vitro***
Edited by H. Jahnsen
0 471 92657 4 242 pages (cloth) 1990

Preparations of Vertebrate Central Nervous System In Vitro

Edited by

Henrik Jahnsen
Institute of Neurophysiology
University of Copenhagen
Denmark

A Wiley–Interscience Publication

JOHN WILEY & SONS
Chichester · New York · Brisbane · Toronto · Singapore

Published 1990 by John Wiley & Sons Ltd.
Baffins Lane, Chichester
West Sussex PO19 1UD, England

Other Wiley Editorial Offices

John Wiley & Sons, Inc., 605 Third Avenue,
New York, NY 10158-0012, USA

Jacaranda Wiley Ltd, G.P.O. Box 859, Brisbane,
Queensland 4001, Australia

John Wiley & Sons (Canada) Ltd, 22 Worcester Road,
Rexdale, Ontario M9W 1L1, Canada

John Wiley & Sons (SEA) Pte Ltd, 37 Jalan Pemimpin 05-04,
Block B, Union Industrial Building, Singapore 2057

Library of Congress Cataloging-in-Publication Data:

Preparations of vertebrate central nervous system in vitro / edited by
Henrik Jahnsen.
p. cm. — (IBRO handbook series ; v. 13)
'A Wiley-Interscience publication.'
Includes bibliographical references.
ISBN 0 471 92657 4
1. Central nervous system—Histochemistry—Technique.
2. Vertebrates—Histology. 3. Nervous system—Vertebrates.
I. Jahnsen, Henrik. II. Series.
[DNLM: 1. Central Nervous System—anatomy & histology.
2. Vertebrates—anatomy & histology. WL 101 P927]
QM575.P74 1990
596'.048'072—dc20
DNLM/DLC
for Library of Congress 90-11948
CIP

British Library Cataloguing in Publication Data:

Preparations from vertebrate central nervous systems in
vitro.
1. Vertebrates. Nervous system. Research. Laboratory
techniques
I. Jahnsen, Henrik II. Series
596.0188072

ISBN 0 471 92657 4

Printed and bound in Great Britain by Biddles Ltd., Guildford, Surrey.

Contents

Contributors

G.K. Aghajanian *Departments of Psychiatry and Pharmacology, Yale University School of Medicine, Connecticut Mental Health Center, 34 Park Street, New Haven, Connecticut 06508, USA*

A. Ames *Massachusetts General Hospital, Boston, Massachusetts 02114, USA*

C. Bourque *Center for Research in Neuroscience, McGill University and Division of Neurology, Montreal General Hospital, 1650 Cedar Avenue, Montreal, Quebec H3G 1A4, Canada*

L. Brodin *Nobel Institute for Neurophysiology, Karolinska Institute, Box 60400, S-10401 Stockholm, Sweden*

R. Fisher *Division of Neuroscience and Section of Neurophysiology, Department of Neurology, Baylor College of Medicine, Houston Medical Center, Houston, Texas 77030, USA*

B. Gähwiler *Brain Research Institute, University of Zurich, August-Forel-Strasse 1, CH-8029 Zurich, Switzerland*

R. Gray *Division of Neuroscience and Section of Neurophysiology, Department of Neurology, Baylor College of Medicine, Houston Medical Center, Houston, Texas 77030, USA*

S. Grillner *Nobel Institute for Neurophysiology, Karolinska Institute, Box 60400, S-10401 Stockholm, Sweden*

J. Hounsgaard *Institute of Neurophysiology, Blegdamsvej 3, DK-2200 Copenhagen, Denmark*

D. Johnston *Division of Neuroscience and Section of Neurophysiology, Department of Neurology, Baylor College of Medicine, Houston Medical Center, Houston, Texas 77030, USA*

T. Knöpfel *Brain Research Institute, University of Zurich, August-Forel-Strasse 1, CH-8029 Zurich, Switzerland*

R.H. Masland *Massachusetts General Hospital, Boston, Massachusetts 02114, USA*

C. Nicholson *Department of Physiology and Biophysics, New York University Medical Center, 550 First Avenue, New York, NY 10016, USA*

M. Simmonds *Department of Pharmacology, School of Pharmacy, University of London, 29/39 Brunswick Square, London WC1N 1AX, UK*

N. Spruston *Division of Neuroscience and Section of Neurophysiology, Department of Neurology, Baylor College of Medicine, Houston Medical Center, Houston, Texas 77030, USA*

Preface for IBRO Handbook Series

During the last 50 years there have been two changes in the way in which scientists have studied the nervous system. First of all, the traditional and largely independent major scientific disciplines of physics, chemistry, physiology, pharmacology and pathology gave rise to the more specialized subdisciplines of neurophysiology, neurochemistry, neuropharmacology, etc., and the science of experimental psychology was born. Then, after about another generation, it became clear that a deeper understanding of the brain could not be achieved by separate and unrelated studies in each of these subdisciplines. Rather, a unified approach was needed in which the specialized methods were applied in a coordinated way to solve a particular problem. Indeed, combinations of methods could often yield results not obtainable by the application of any individual technique. Thus scientists studying the nervous system began to call themselves neurobiologists or neuroscientists because they did not wish to be identified with any particular experimental discipline. Very soon meetings took place (e.g. in 1955 the First International Meeting of Neurobiologists) and organizations (Neuroscience Research Program, MIT) were founded to give formal recognition of this new approach to the study of the brain. In 1958 the decision was taken in Moscow to establish the International Brain Research Organization (IBRO), which became incorporated as an independent organization through a Bill in the Parliament of Canada in Ottawa in 1961.

In 1986 IBRO changed its status and became a world federation of neuroscientists, governed by representatives of all the corporate member societies. Individual membership now exceeds 20 000 from more than 50 countries of all political complexions. One of the programmes of IBRO, all of which aim to serve the international community of neuroscientists, is the publication programme. IBRO publications include *IBRO News*, *Neuroscience* and *IBRO Symposia Series*. With the present *Handbook Series*, IBRO aims to fill a major gap in the world literature. The neuroscientist needs to be able to turn to whichever specialized method is most suited for the problem he is currently studying. The series on *Methods in the Neurosciences* will help to provide expert advice on exactly how to carry out the experiments, on what difficulties can occur and on the limitations of the method. It is planned as a continuing series, so that new volumes can be published as and when new methods are developed, tested, and found useful. It is my hope that books in this series will have a significant impact on neuroscience throughout the world, by helping to provide the tools with which the scientist can tackle his problem.

A. David Smith
Director of Publications, IBRO,
University Department of Pharmacology,
South Parks Road,
Oxford OX1 3QT, UK

Preface

Scientists in general acknowledge that information on new methods should be spread throughout the scientific community quickly because awareness of techniques may facilitate the formulation of new and interesting hypotheses.

The purpose of this book is to present to a broad spectrum of investigators some *in vitro* techniques that can be used to study the vertebrate central nervous system. Earlier *in vivo* preparations dominated in the laboratories. Then, with the advent of the brain slice, people realized that tissue from the central nervous system is not as vulnerable to 'non-physiological' conditions as was thought previously. This opened up a new era in neuroscience and today a large number of *in vitro* preparations are available.

The brain slice has been described in many publications. It is essentially a section of the central nervous system in time and space, and this of course puts certain limits on the use of the technique. For instance, it is difficult to study properties of large three-dimensional neural networks. It is equally difficult to study processes lasting more than 18–24 hours. In this book the authors have sought to point out solutions to these problems as well as to indicate how it is possible to push *in vitro* techniques towards their limits.

The book has been divided into two parts. The first focuses on advanced use of the slice preparation. Gray, Fisher, Spruston and Johnston demonstrate how transmitter- and voltage-gated ionic channels can be studied in the hippocampal slice while some of the synaptic connections of the cells are preserved. Aghajanian shows how to analyse a classical problem of whether the activity of nerve cells in a particular area is generated by intrinsic membrane properties of the neurons or by properties of the neural network; he uses the brainstem slice as an example. Simmonds gives a guided tour through the problems and pitfalls that one encounters when brain slices are used for quantitative pharmacology – a method that enjoys increasing popularity especially in drug companies and related industries. Finally, Gähwiler and Knöpfel explain how it is possible to culture slices in order to greatly increase survival time. Thus, neurons in the slice culture form synapses with their neighbours and can easily be visualized.

The second part of the book is dedicated to methods alternative to the slice technique. These methods are useful when one wishes to study large assemblies of neurons, with most or all of the original synaptic connections intact. Brodin and Grillner describe the use of the central nervous system of lampreys and Hounsgaard and Nicholson present the isolated turtle brain and spinal cord. For people who need tissue from species evolved more recently there is a chapter by Ames and Masland on the mammalian retina *in vitro* and one by Bourque on the use of perfusion to maintain isolated parts of the mammalian brain; the hypothalamus is used as an example.

Some exciting and popular *in vitro* methods are not presented here. For instance, one will look in vain for chapters on primary cultures of neurons and glia and many of the things that slices can be used for are only briefly discussed. It is beyond the scope of this book to give a comprehensive survey of *in vitro* methods. Furthermore, some of the topics that might have been included will be discussed in forthcoming volumes in the IBRO Handbook Series 'Methods in the Neurosciences'. The authors and I hope that our book will serve as an inspiration for the reader to develop further the techniques. We hope that we have demonstrated that only imagination and sometimes money set limits.

Henrik Jahnsen
Copenhagen 1989

Part I: Slice Preparations

Preparations of Vertebrate Central Nervous System *In Vitro*
Edited by H. Jahnsen
Published by John Wiley & Sons Ltd

CHAPTER 1

Acutely Exposed Hippocampal Neurons: A Preparation for Patch Clamping Neurons from Adult Hippocampal Slices

RICHARD GRAY, RONALD FISHER, NELSON SPRUSTON
and DANIEL JOHNSTON
Division of Neuroscience and Section of Neurophysiology,
Department of Neurology, Baylor College of Medicine,
Houston, Texas 77030, USA

I INTRODUCTION

The hippocampal formation, made up of the hippocampus and the dentate gyrus, has been a popular preparation with cellular neurophysiologists since the initial *in vivo* intracellular investigations by Spencer and Kandel (Kandel *et al.*, 1961; Kandel and Spencer, 1961; Spencer and Kandel, 1961a,b) and the pioneering use of the *in vitro* hippocampal slice for intracellular recording by Yamamoto (1972a,b) and Schwartzkroin (1975).

The widespread use of the *in vitro* hippocampal slice preparation is at least partially due to its unique structure in which the cell bodies of the primary cell types and their major synaptic connections are arranged in layers transverse to the long axis of the hippocampus (Andersen *et al*., 1971; Andersen, 1983). The major intrinsic synaptic pathways (perforant path, mossy fibers, and Schaffer collaterals), therefore, are largely preserved in transverse slices. This has resulted in the hippocampal slice preparation being used extensively for studies of membrane and synaptic physiology and pharmacology.

Until about 1980, the main electrophysiological techniques used for studying the hippocampus involved either measuring the activity of groups of neurons with extracellular field recordings, or measuring voltage changes within an individual neuron with intracellular current-clamp recordings. These techniques have provided a wealth of information on the properties of hippocampal neurons. However, as knowledge has been gained about the ionic mechanisms underlying electrical activity in more accessible neurons (e.g. those in molluscan ganglia, peripheral ganglia, and tissue-cultured neurons), questions have naturally arisen as to whether results obtained from these isolated and/or immature neurons can be generalized to the adult mammalian central nervous system. It is only recently that quantitative, biophysical techniques have been applied to mammalian cortical neurons. This was aided by the advent of the single-electrode voltage clamp (Wilson and Goldner, 1975) and its use to study the properties of hippocampal neurons (Johnston *et al*., 1980; Johnston, 1981; Johnston and Brown, 1981, 1983; Brown and Johnston, 1983).

There are three generally recognized limitations of voltage clamping a cortical neuron with an intracellular microelectrode. The first limitation is the lack of adequate space clamp for most cellular events of interest. A careful analysis of the limitations of the single-electrode voltage clamp for clamping cortical neurons with complex geometries has yielded the conclusion that synaptic conductances arising within an electrotonic distance of less than 0.1 can be adequately clamped (Johnston and Brown, 1983). More distal synaptic conductances or large and fast voltage-dependent conductances (that may be distributed over broad electrotonic distances) would be expected to be very poorly clamped.

The second limitation of voltage clamping with a microelectrode is the lack of fidelity or frequency response of the voltage-clamp system. Generally, the properties of the neuron itself are rate-limiting (Johnston and Brown, 1983), but the low-pass filtering by the microelectrode in the bath and tissue and the speed limitations inherent in a time-share clamp system limit the frequency response of the voltage clamp to well under 1 kHz and usually closer to 200–500 Hz (Johnston and Brown, 1983). This frequency response is sufficient for most synaptic and slow voltage-dependent conductance changes but is inadequate for studying the voltage-dependent conductances underlying the faster electrical activities of cortical neurons.

The third limitation of voltage clamping with a microelectrode is the lack of adequate current separation. The complete separation of currents carried by different ionic species is a formidable task in membrane biophysics, particularly when using intracellular microelectrodes. Although in a slice preparation one usually has free access to the extracellular environment for ionic substitutions and the addition of channel-blocking drugs, one typically has only limited access to the intracellular side of the membrane. The injection of channel blockers through intracellular microelectrodes rarely results in the complete elimination of the unwanted currents.

In an attempt to overcome some of these difficulties, we have developed an adult hippocampal slice preparation in which the neurons, after treatment with proteolytic enzymes, are physically exposed and amenable to the use of patch-clamp recording techniques (Figure 1). By making recordings from small patches of membrane, or from whole-cell configurations, the problem of inadequate space clamp is either eliminated or reduced significantly in this preparation. Furthermore, the frequency response of the current measurements is roughly five-fold greater than that using microelectrodes and, with single-channel recordings or intracellular dialysis through whole-cell electrodes, ionic current separation is excellent.

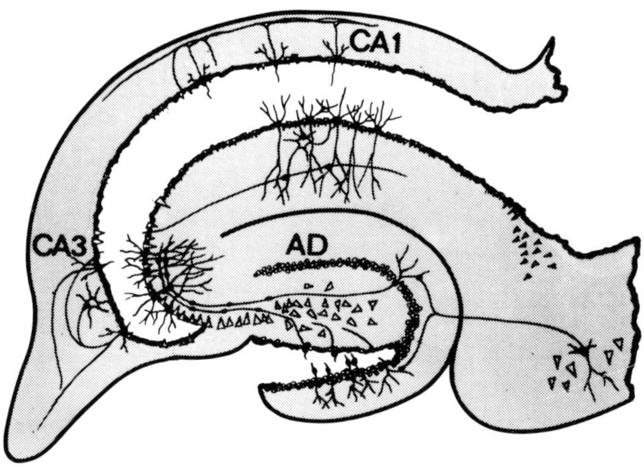

Figure 1 Schematic drawing showing the hippocampal slice split along the major cell body layers (pyramidal, CA1–CA3; area dentata, AD)

This preparation has been utilized for studies of single ligand-gated (γ-aminobutyric acid or GABA) channels using excised-patch configurations, and voltage-dependent Ca^{2+} channels, using both whole-cell and attached-patch configurations. Many of these results have been presented previously elsewhere as abstracts (Gray *et al*., 1982, 1983; Griffith *et al*., 1983; Gray and Johnston, 1985a,b, 1986a,b; Fisher *et al*., 1988), as full publications (Gray and Johnston, 1985c, 1987); or as parts of reviews (Johnston *et al*., 1987, 1988a,b, 1989a,b). In this chapter we will present the detailed methods for preparing exposed hippocampal neurons. These methods have only been published previously as part of a thesis (Gray,

1988). We will also briefly review results of several investigations that have used this preparation in various ways.

II METHODS

Adult guinea pigs of either sex and weighing 200–600 g are used. Animals are anesthetized with ethyl ether and the brain is removed after cervical dislocation or decapitation. The hippocampus is carefully dissected from the brain and slices 350 μm thick are cut with a McIlwain tissue chopper. The slices are cut approximately 20° (clockwise) off of a line perpendicular to the septotemporal axis of the hippocampus.

II.1 Preparation of Exposed Cells

The first attempts at producing isolated hippocampal neurons used techniques very similar to those used to isolate horizontal cells from catfish retinae (Johnston and Lam, 1981). The yield of healthy neurons from this approach, however, was generally low, and the ability to obtain neurons at all was quite variable.

Initial attempts at improving the yield centered around choosing the proper enzymes and solutions for dispersal of the cells. During this time the following enzymes were tried, all at a wide range of concentrations: papain (Worthington and Sigma), chymopapain, trypsin, chymotrypsin, hyaluronidase, Dispase, Subtilisin BPN′, Subtilopeptidase A, collagenase types I, IA, IV and V, pronase E (Sigma), pronase (Calbiochem) and thermolysin (Calbiochem and Sigma).

Different dispersing salines were also tried with the different enzymes. Initial salines were based on normal, Ca^{2+}-containing artificial cerebrospinal fluid (aCSF) with bovine serum albumin added to provide a saturating concentration of substrate for any remaining proteases. During the course of these studies, it became apparent that a higher yield of cells was obtained when Ca^{2+} was not present during the dissociation step. Subsequent dispersing salines contained either no added Ca^{2+} or EGTA to buffer the free Ca^{2+} concentration to low levels.

It was also observed that, while it was possible to obtain totally isolated neurons, those that remained as part of a slice fragment appeared much healthier. During the early stage of development of the preparation, 'health' was judged by the appearance of the cells – those whose surfaces appeared shiny under Hoffman optics and had many intact processes were judged healthy, whereas cells with grainy surfaces and rounded-up processes were considered unhealthy. Since our goal was to develop a preparation amenable to single-channel recording, it was not essential to have totally isolated cells. In fact, we quickly realized that a preparation with exposed neural somata and proximal processes might have some advantages over totally isolated cells, especially if synaptic transmission remained intact. Although this property of the preparation has not been exploited, pilot experiments have shown that synaptic transmission is conserved (Griffith *et al*., 1983).

To maximize the success rate of obtaining acutely exposed, as opposed to totally isolated, neurons, enzyme concentrations and treatment times were reduced, and the mechanical components of exposure were made more gentle. Many of the enzymes listed above were re-tried under these conditions, and we ultimately arrived at the exposure protocol described below. This protocol has remained essentially unchanged over the last five years, although some individuals have observed an increased yield of healthy cells by raising magnesium to 6–10 mM in the dissection and exposure salines (see Table 2).

II.1.a Detailed protocol for acutely exposed hippocampal neurons

The procedures for the entire preparation are given in chronological order. Supplies (and our usual sources) are listed in Table 1, and salines commonly used are listed in Table 2.

(1) (a) Make up 200 ml Ca-aCSF and 100 ml EGTA-aCSF. (Recently, R.F. and N.S. have been using a high-Mg^{2+} (6–10 mM, with appropriate adjustment of NaCl to maintain proper osmolality) aCSF for slicing and incubating, while others still use a more standard aCSF. The high-Mg^{2+} aCSF may improve the

Table 1 Supplies used

Item	Vendor	Part no.
Plastic 2-ml analyzer cups	Fisher	02-544
Tygon tubing for superfusing Petri dish with O_2/CO_2 (i.d. 1/32″, o.d. 3/32″)	Fisher	14-169-1A
Plastic transfer pipettes	Fisher	13-711-7
35 mm Falcon 1008 Petri dishes	Fisher	08-757-100A
12 × 75-mm disposable borosilicate glass culture tubes	Fisher	14-959-17
Thermolysin (Protease Type X)	Sigma	P-1512
Pronase	Calbiochem	53702

Table 2 Salines for dissection, incubation, and patch clamping

	NaCl	KCl	$MgSO_4$	$NaHCO_3$	Dextrose	NaH_2PO_4	$MgCl_2$	HEPES	$CaCl_2$
aCSF	124.00	2.50	2.00	26.00	10.00	1.25			*
Mg^{2+}-aCSF	114.00	2.50		30.00	10.00		8.00		*
Exp.bath	149.00	2.50			10.00		1.00	10.00†	2.00

* Add $CaCl_2$ to 2 mM for Ca-aCSF or EGTA to 2 mM for EGTA–aCSF.
† Adjust pH to 7.35 with NaOH.

yield of healthy cells, presumably by reducing Ca^{2+} influx through NMDA channels during the dissection.) Put the Ca^{2+}-containing saline on ice and bubble with 95% O_2/5% CO_2 (cover with Parafilm). Air stones with a small bubble size are used for oxygenating bulk solutions. Put the EGTA-containing saline on a stir plate to mix at room temperature.

(b) Turn on the water bath. A variety of types of water bath are used in our laboratory – they all have a metal shelf (stainless steel or homemade) in contact with the top of the water, and a cover. One type is a small Fisher Versa-Bath. Make sure the water level is just up to the bottom of the shelf. The temperature should be set for 31°C (so the temperature in the Petri dish is approximately 30°C).

(c) Let the enzyme containers warm up for ~30 min before opening to reduce condensation. The enzymes are stored at −20°C with a desiccant.

(2) (a) Weigh out 2.0 mg of pronase, put it into a glass culture tube, cap it, and put it on ice. The amount of enzymes needed will change with their age and potency. The quantities mentioned here are good starting values; you may have to use slightly more or less. Use less if the slices disintegrate very easily or if you see lots of dead cells; use more if the slice does not split apart with gentle shaking. Once you find an optimum amount, it will generally stay constant for the life of the vial of enzyme. You will probably need a slightly higher enzyme concentration if you use the high-Mg^{2+} saline for incubation.

(b) Weigh out 1.5 mg of thermolysin and put it into a glass culture tube, cap it, and put it on ice.

(3) Use a soldering iron to melt a hole in the cover of a 35-mm plastic Petri dish for inserting a tube to superfuse with 95% O_2/5% CO_2. The gas mixture is warmed and humidified by bubbling it through distilled water in a bottle placed in the water bath.

(4) (a) Add 3 ml cold oxygenated Ca^{2+}-containing saline to the culture tube containing pronase. Shake to mix it. Cap the tube, and put it back on ice.

(b) Dissect out the hippocampus. Immediately before slicing pour the pronase-containing saline into the 35-mm Petri dish. Cut the slices 350 μm thick. We have found that it is important to remove the brain quickly (in less than 1 min) into ice-cold saline to minimize ischemic damage.

(c) Transfer the slices into the Petri dish, cover it, insert the air tube, and put the Petri dish in the incubator.

(d) The slices should be superfused with 95% O_2/5% CO_2 during the incubation. Adjust the flow of gas so that you can see the solution moving around in the dish, but not so high that the slices are moving around. Do not bubble the gas directly into the saline.

(e) Set a timer for 30 min.

(5) Continue oxygenating the remaining normal saline, but take it off ice so it warms to room temperature. By now the EGTA-aCSF should be mixed and ready to be oxygenated. Do this at room temperature.

(6) (a) After 30 min in the pronase solution, add 3 ml oxygenated normal saline to the culture tube containing thermolysin. Shake until the enzyme is dissolved. The tube is usually held in the hand for a minute or so to warm it so that there is not a large temperature change when it is added to the slices.
(b) Using a plastic pipette, gently draw off the pronase solution, and replace it with the thermolysin solution. Try to be fairly gentle during these solution changes. Sometimes the slices are rinsed a few times with normal saline before the enzyme change, but this does not seem to be essential.
(c) Put the Petri dish back in the incubator. Adjust the air flow so that there is just a dimple on the surface of the solution. This is a slightly lower flow rate than that used for the first enzyme treatment.
(d) Set the timer for 15 min.

(7) After 15 min in the thermolysin solution, take a look at the slices. They should be beginning to look a little opaque. If not, wait 5–10 min longer before removing slices.

(8) (a) Rinse and then fill a 2-ml plastic analyzer cup with EGTA-aCSF.
(b) Transfer one slice into the analyzer cup. We use a Pasteur pipette cut and fire-polished to a 3–5-mm opening for transferring individual slices. Rinse the slice five times with the EGTA-aCSF saline. Be gentle so that the slice is not unnecessarily agitated.
(c) With the analyzer cup about one-half full of EGTA-aCSF, cover with a piece of Parafilm and shake a few times (gently, at first). If the slice does not begin to split apart, you need to shake a little more vigorously. If a lot of debris is floating around, rinse with EGTA-aCSF before shaking any more. The key here is to use just enough force to split the slice – too much and the cells will be sheared off the edge. You may wish to transfer the slice to the microscope and look for exposed cells to decide if you need to shake more. You will still need to transfer the slice back to the analyzer cup to rinse it with EGTA-aCSF and then with whatever experimental saline you will be using. For granule cells and CA1 pyramidal cells, the optimal situation is when the slice splits after a few shakes. For CA3 cells, more shaking is necessary (see below). Once the cells are exposed, however, further shaking quickly kills them. If the final experimental saline contains Ca^{2+}, it must be added to the analyzer cup slowly – fill the cup with EGTA-aCSF, then add a few drops of the Ca^{2+}-containing saline; suck out a little of the mixture, add a little more Ca^{2+} saline, and repeat. The cells do not tolerate going from 0 to 2 mM Ca^{2+} rapidly.

(9) The slices will survive quite a while (3–5 h) sitting in the enzyme solution, although they usually are not as healthy after the first 3 h. It seems that there is an asymptotic limit to either the lifetime of the enzyme or the amount of digestion. Sometimes oxygenated Ca-aCSF is added to the Petri dish (slowly, so that the temperature does not rapidly drop) to dilute the enzyme.

(10) A slight variation of this protocol is used to expose CA3 cells:
(a) Follow the original protocol until the slice has been rinsed in EGTA-aCSF. Then shake very gently in EGTA-aCSF, stopping before the slice begins to split apart.
(b) Remove the slice from the analyzer cup and place in a small Petri dish filled with EGTA-aCSF. Make a cut in the slice using a scalpel with a small blade. The cut begins in the subiculum and extends to the cell body layer of CA1, about halfway to CA3. The purpose of this cut is simply to facilitate the splitting of the slice along the CA1 region. This increases the likelihood that, upon shaking, the CA3 cells will be the last cells exposed. This is important, since they are easily damaged by shaking after they are exposed.
(c) Replace the slice in the analyzer cup and proceed with step 8(c). Stop shaking as soon as the CA3 border begins to be exposed. If cells are not exposed enough, simply replace the slice in the cup and gently shake again.
(d) If the experimental bath saline you will be using contains very little Ca^{2+} (e.g. 140 mM potassium aspartate with 10 mM EGTA), you can change the solution to this immediately following step 10(a). All subsequent steps are then carried out in this saline. With the potassium aspartate saline, we find it useful to conduct the saline exchange rapidly; the combination of high potassium aspartate and Na^+ in the external saline seems to be harmful to the cells.

(11) We sometimes have problems with slices drifting very slowly during patch recordings. This can be controlled by placing a small glass coverslip coated with Cell Tak (a natural glue isolated from marine mussels) in the recording chamber before each slice is put in. The slices stick to the Cell Tak within 2–10 min. Cell Tak is commercially available from Collaborative Research Inc. (Two Oak Park, Bedford MA 01730).

II.2 Patch Clamping

The patch-clamp system used for our studies is conventional, having been derived from the Yale Mark V design, and constructed in our laboratory from an unpopulated printed circuit board purchased from the Physiology Electronics Shop, Yale University School of Medicine. Headstages (current to voltage converters) with 10 and 1 GΩ feedback resistors were built here from the design of Sigworth (1983) with modifications suggested by David Corey and Rick Levis. This patch-clamp amplifier design uses a high-frequency boost circuit that raises its intrinsic frequency response to >50 kHz (step rise time $< 20\ \mu s$). The frequency response of the clamp system is decreased slightly under experimental conditions because of the filtering characteristics of the patch electrode and holder. These effects, however, can be reduced by proper coating of the electrodes and by keeping the bath saline level low, both of which act to reduce the capacitive load. When these precautions are followed, the frequency response during experiments is flat to 10–15 kHz, allowing reliable measurements of currents with rise times slower than ~70 μs. The main frequency limitation during single-channel measurements is the level of low-pass filtering necessary to resolve small-amplitude single-channel currents.

Patch pipettes were pulled from 1.7-mm borosilicate glass (Boralex, Rochester Scientific) and coated with Sylgard (Dow Corning) to within 30–50 μm of the tip. For single-channel recordings, the fire-polished pipettes had 'bubble numbers' (Corey and Stevens, 1983) of 1.5–3.0 and d.c. resistances of 5–15 MΩ, depending on the specific pipette saline used. Seal resistances of the order of 50 to greater than 100 GΩ were routine.

Patch pipettes for whole-cell recordings had 'bubble numbers' of 3.0–4.5 and d.c. resistances of 3–10 MΩ, depending on the pipette saline. Whole-cell electrodes were also coated with Sylgard, but usually not as heavily as those used for single-channel recordings. Capacitive transient cancellation and series resistance compensation as described by Sigworth (1983) were used for all experiments.

All data were low-pass filtered by an 8-pole Bessel filter (Frequency Devices, Inc.) prior to digitization. Programs for data acquisition and analysis were written in either BASIC-23 (Indec, Inc.) or BASIC-68 (developed in our own laboratory for use on the Masscomp/Concurrent line of computers).

III RESULTS

III.1 Possible Enzyme Effects on Exposed Neurons

The possibility that the proteolytic enzymes have undesirable effects on neurons and/or synapses is an extremely important issue. For example, pronase has been shown in squid axon to remove Na^+ channel inactivation when applied internally (Armstrong *et al*., 1973), trypsin has been shown to alter tetrodotoxin binding to Na^+ channels in invertebrate neurons (Lee *et al*., 1977), and a number of enzymes have been reported to alter *N*-methyl-D-aspartate (NMDA) responses in dissociated neurons. This latter finding, however, has been inconsistent. Although Akaike and colleagues (Akaike *et al*., 1988) reported a lack of NMDA responses from dissociated neurons, others (Connor *et al*., 1988; Mody *et al*., 1989) have reported NMDA responses using similar but not identical methods. From our own studies, we have found no striking differences between intact and enzyme-treated slices, at least with respect to the electrophysiological responses we have measured.

The neurons exposed at the edges of different slice fragments are within a continuum of morphological conditions, ranging from cells that look very similar to intact tissue (Figure 2) (Johnston and Brown, 1984; Turner and Schwartzkroin, 1983), to those that are rounded with obvious structural abnormalities. These cells generally continue to round up and deteriorate over the next few minutes. Healthy cells, however, remain fairly unchanged for 1–2 h.

Exposed neurons have been impaled with conventional KCl-containing intracellular microelectrodes and judged healthy by normal electrophysiological

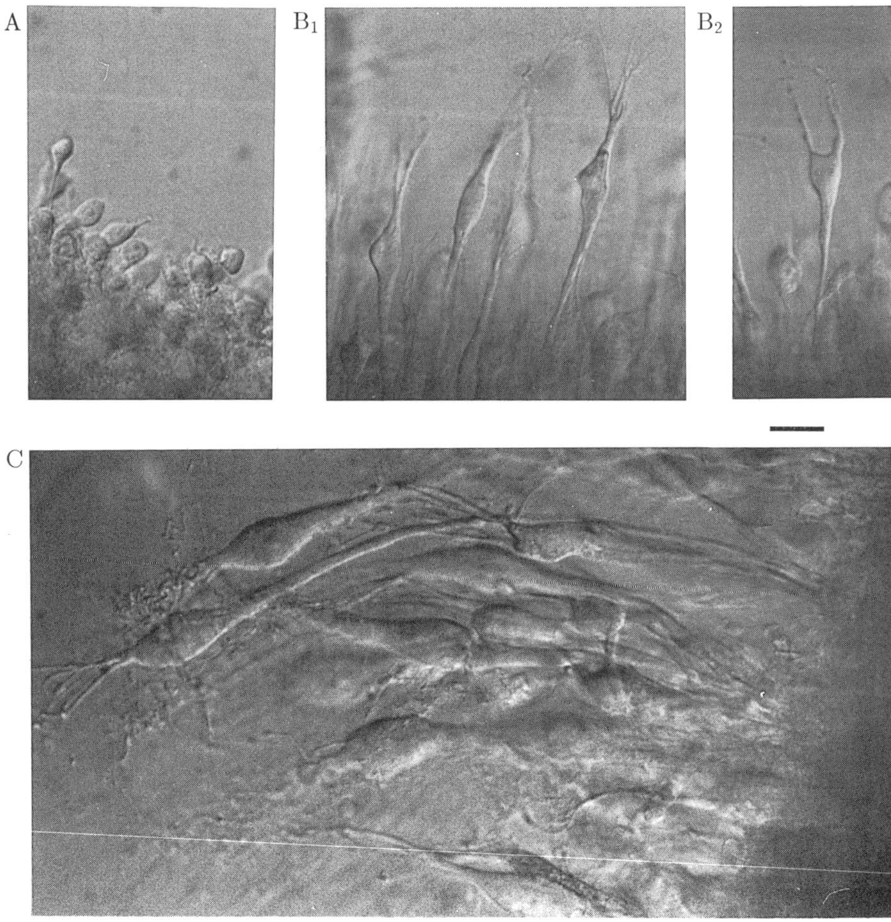

Figure 2 Photomicrographs of exposed neurons from hippocampal slices. (A) Granule cells from area dentata. (B) CA1 pyramidal cells. The most common type appears fusiform shaped (B_1); a less common subtype appears more pyramidal shaped (B_2). (C) CA3 pyramidal cells. Bar = 20 μm

criteria: i.e., stable resting potentials more negative than −55 mV; membrane time constants greater than 10 ms; overshooting action potentials; and spontaneous and evoked synaptic input (Figure 3) (Gray *et al*., 1982, 1983; Griffith *et al*., 1983). Similar observations have been made in more recent studies in which the passive membrane properties are being investigated with whole-cell patch recordings (Figure 4). Morphological and standard physiological criteria suggest that the relatively brief enzymatic treatment required to expose neurons with normal-appearing morphology probably has minimal effects on the properties

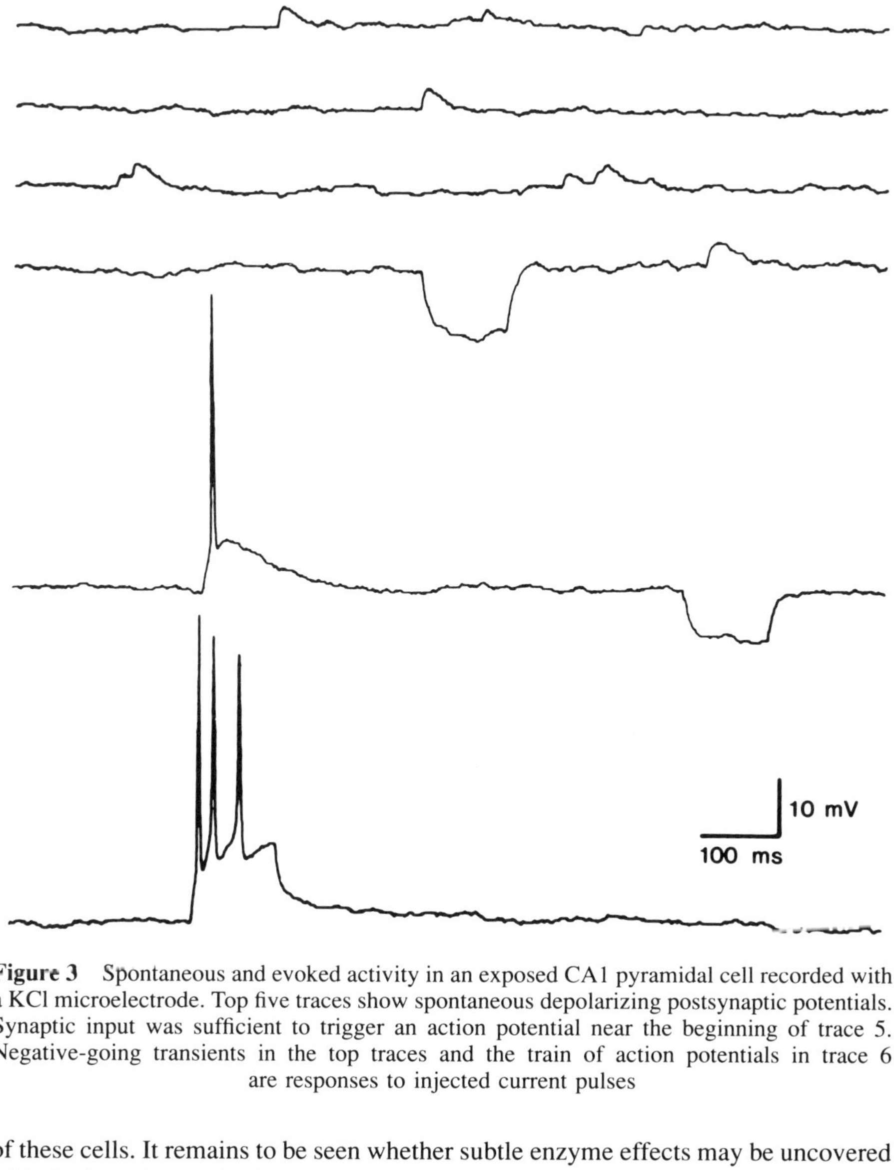

Figure 3 Spontaneous and evoked activity in an exposed CA1 pyramidal cell recorded with a KCl microelectrode. Top five traces show spontaneous depolarizing postsynaptic potentials. Synaptic input was sufficient to trigger an action potential near the beginning of trace 5. Negative-going transients in the top traces and the train of action potentials in trace 6 are responses to injected current pulses

of these cells. It remains to be seen whether subtle enzyme effects may be uncovered with further electrophysiological and pharmacological investigations.

III.2 Ligand-gated Channels

GABA-gated Cl^- channels have been investigated in inside-out and outside-out patches from acutely exposed hippocampal neurons (Figure 5) (Gray and Johnston, 1985c). Although our main goal in those experiments was to characterize the

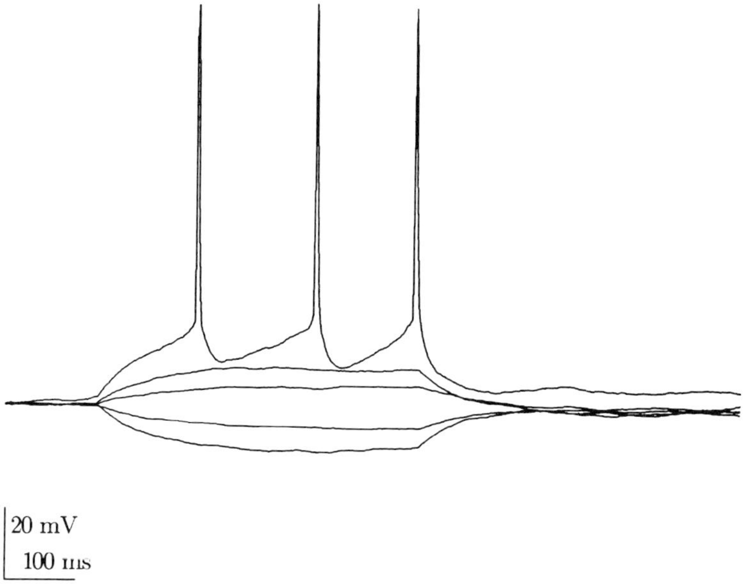

Figure 4 Passive membrane responses and action potentials recorded from a dentate granule cell with a whole-cell current clamp. Charging curves are averages of 5–9 voltage responses to current injections of −3, −1, +1, and +3 pA. Action potentials are a single trace in response to a current injection of +4 pA. The input resistance of this cell was 4.2 GΩ and the membrane time constant was 75 ms. The patch electrode contained (in mM): 10 KCl, 125 KAsp, 11 EGTA, 1 $CaCl_2$, 10 phosphocreatine (di-Tris salt), 4 MgATP, and 50 units/ml creatine phosphokinase (pH adjusted to 7.35 with KOH). Extracellular saline consisted of (in mM): 150 NaCl, 2 KCl, 2 $MgCl_2$, 2 $CaCl_2$, 10 HEPES, and 10 dextrose (pH adjusted to 7.35 with NaOH)

conductance of Cl^- through the GABA-gated channel in adult neurons, we also made some initial measurements on the opening and closing kinetics of GABA-gated channels. Recordings from patches of membrane containing a single channel were analyzed for the time that the channel spent in the open and closed states. The open and closed times were then collected into histograms and fitted with the best sum of exponentials. Examples of open- and closed-time distributions are shown in Figure 6. In this experiment both the open- and closed-time distributions were best fitted with the sum of two exponential components with mean open times of 1.5 and 4.4 ms, and mean closed times of 0.8 and 4.4 ms. In some experiments the closed-time distribution was better fitted with the sum of three exponentials.

Observations regarding the single-channel conductance of GABA-gated channels have been reported previously (Gray and Johnston, 1985c). Briefly, we found that the single-channel current displayed outward rectification due either to rectification of single channels or to rapid and undetectable transitions to a larger conductance state of the channel.

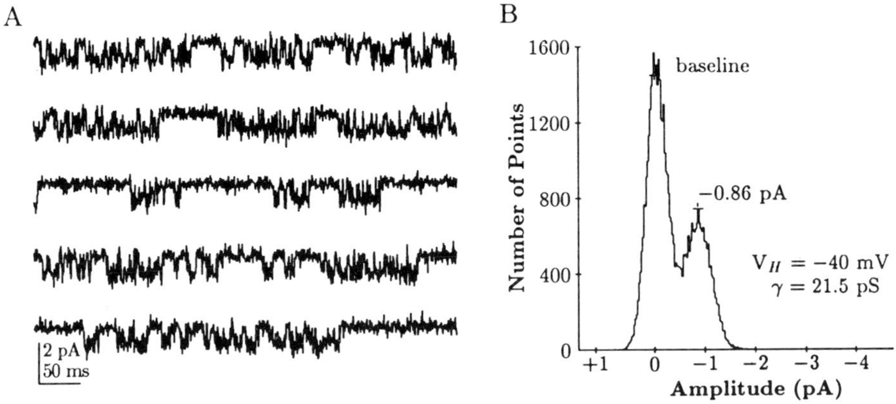

Figure 5 GABA-gated single-channel currents from a CA1 pyramidal neuron. (A) Consecutive GABA-gated current records from an inside-out patch from a CA1 pyramidal cell. The patch was held at −40 mV; channel openings are illustrated as downward deflections. (B) Total-amplitude histogram. The two peaks correspond to the closed-channel (baseline) and open-channel current levels. The mean current level at −40 mV was −0.86 pA, corresponding to a single-channel chord conductance (γ) of 21.5 pS

III.3 Voltage-dependent Calcium Channels

Norepinephrine (NE) (through β-adrenoceptors) has been shown to increase Ca^{2+} currents in cardiac cells (Reuter, 1983). After the demonstration that NE enhances long-term potentiation (LTP) at mossy fiber synapses (Hopkins and Johnston, 1984; Johnston *et al.*, 1987), it was of interest to determine whether NE also increases Ca^{2+} currents in hippocampal neurons. This led to an investigation into the properties of voltage-dependent Ca^{2+} channels in acutely exposed hippocampal neurons and their possible modulation by β-adrenergic agonists.

III.3.a Whole-cell recordings

We chose to investigate Ca^{2+} currents in hippocampal granule cells for several reasons. First, the spherical geometry of the granule cell, compared to the more complex geometry of the pyramidal cells, was expected to be much more favorable for obtaining an adequate space clamp. Second, granule cells give rise to the mossy fibers, and we felt that any effect of β-adrenergic agonists on Ca^{2+} currents in the cell body might provide information relevant to its possible action at the mossy fiber synaptic ending. Third, although macroscopic Ca^{2+} currents had been studied in a variety of cell types, virtually no quantitative data were available on their properties in granule cells from the adult hippocampus.

Like most excitable cells, granule cells contain many parallel voltage-dependent conductance pathways, which in summation are responsible for the active properties of the cell. Examples of whole-cell ionic currents seen in a typical granule cell are shown in Figure 7A. To measure current flow through Ca^{2+} channels, it is

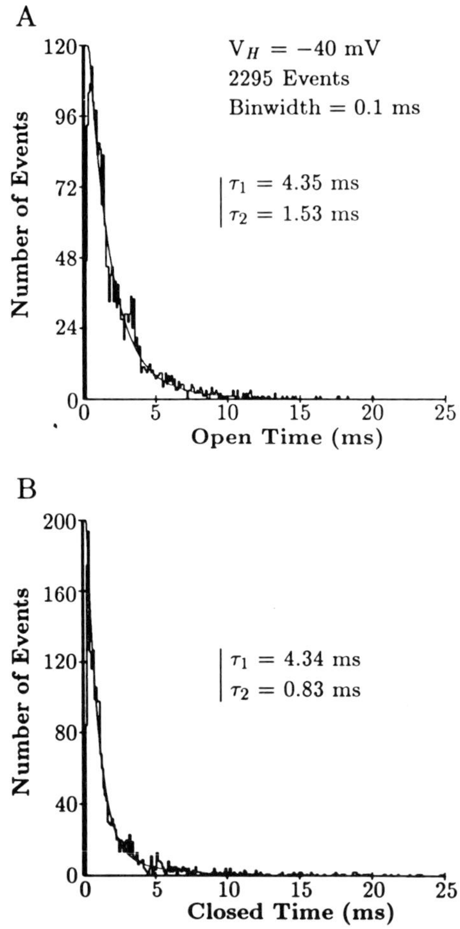

Figure 6 Open- and closed-time distributions of GABA-gated channels. (A) Open-time distribution. The best fit to the distribution is a double exponential curve with time constants of 4.35 and 1.53 ms (solid line). (B) Closed-time distribution. The best fit had time constants of 4.34 and 0.83 ms. The binwidth for both distributions was 0.1 ms. Durations less than 300 μs were not included in the fitting process. Raw data were sampled at 12.5 kHz and low-pass filtered at 2.5 kHz (−3 dB)

obviously necessary to block these other currents using drugs or ion substitution. Several experiments were performed to determine adequate levels of K^+ channel blockers necessary to eliminate contamination of the Ca^{2+} current. It was found that an internal saline containing (in mM) 50 CsCl, 90 potassium acetate, 5 EGTA, 10 HEPES, with 100 μM 3,4-diaminopyridine (3,4-DAP) and 5–10 mM TEA added to both internal and external salines, was sufficient to block all outward currents at potentials as depolarized as +40 mV during commands as long as 1 s duration. The inward Na^+ current could be totally blocked by the addition of 1 μM TTX to the bath saline. The remaining whole-cell Ca^{2+} currents from a granule cell using these blockers and salines are shown in Figure 7.

Because of their nearly spherical shape and small processes, granule cells might be expected to provide a better space clamp than pyramidal neurons. For each cell, the quality of space clamp was evaluated by two methods. First, current flow across unclamped membrane can produce a sharp threshold for activation of inward

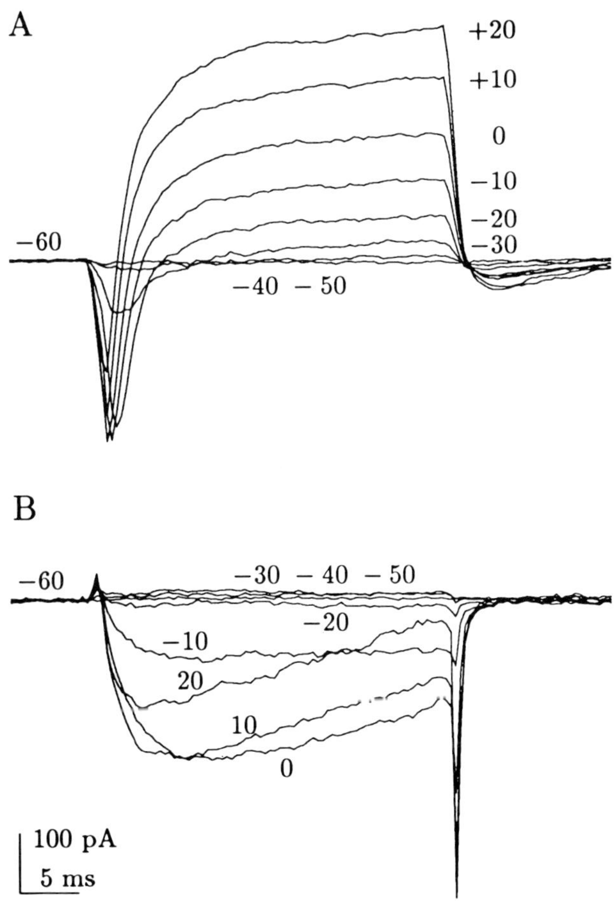

Figure 7 Whole-cell currents recorded with and without K^+ and Na^+ channel blockers. (A) Rapid inward Na^+ currents and outward K^+ currents from a granule cell with no channel blockers present. Bath saline: HEPES-buffered aCSF. Intracellular saline (mM): 140 potassium acetate, 10 HEPES, 10 EGTA. (B) Inward Ca^{2+} currents (from a different granule cell) were observed when K^+ and Na^+ currents were blocked. Bath saline: same as in (A), but with addition of 100 μM 3,4-diaminopyridine, 1 μM TTX, and 6 mM TEA. Intracellular saline: 50 CsCl, 90 potassium acetate, 10 HEPES, 10 EGTA, 0.1 3,4-diaminopyridine, 15 TEA. Command potentials are shown for each trace

currents and a variable delay in their activation. Cells accepted for analysis were required to show a smooth voltage-dependent activation of the Ca^{2+} current with no variable delays. Second, the capacity charging transient of neurons with a complex electrotonic structure will contain multiple exponential components, reflecting the charging of membrane area remote from the site of current injection. Exponential fits were made to the relaxation of the capacity charging current and only cells that displayed a single exponential component were candidates for further analysis. The frequency response of the clamp system was also measured under experimental conditions to ensure that currents with rise times of the order of 100 μs could be measured reliably.

Step depolarizations from a holding potential of −60 mV to potentials of −35 mV and more positive activated an inward current whose peak amplitude was a function of the command potential (Figure 8A). This inward current was inferred to be carried by Ca^{2+}, since it could be totally eliminated by adding Cd^{2+} (0.1 mM) or Ni^{2+} (0.5 mM) to the bath. A plot of the peak inward Ca^{2+} current versus command potential for one experiment is shown in Figure 8B. The maximum inward current in this and other experiments was usually reached with commands to around 0 mV.

NE, isoproterenol, forskolin, and 8-bromo-cAMP were found to enhance the peak amplitude of the whole-cell Ca^{2+} currents (Johnston *et al*., 1987). With the possibility that there are multiple types of Ca^{2+} channels in hippocampal neurons (Gray and Johnston, 1986b) and that β-adrenergic agonists might have differential effects on these channels, it was necessary to investigate Ca^{2+} currents and their modulation at the single-channel level.

III.3.b Single calcium channels

A number of reports have described the existence of multiple types of Ca^{2+} channels in a variety of cell types (Carbone and Lux, 1984; Nowycky *et al*.,

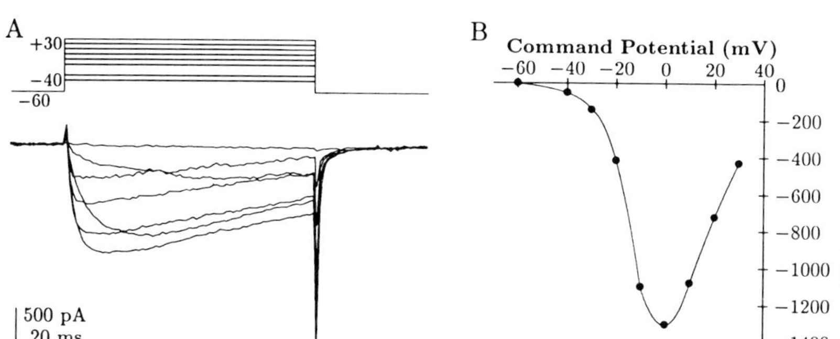

Figure 8 Ca^{2+} currents after blockade of other currents. (A) Voltage commands (top) and resultant inward currents (bottom). (B) Plot of peak inward current versus command potential

1985; Armstrong and Matteson, 1985; McCleskey *et al.*, 1986). These observations raised the following questions relevant to our investigations: do multiple types of Ca^{2+} channels exist in our preparation, and, if so, which of them are modulated by β-adrenergic agonists? We addressed this question by performing cell-attached single-channel measurements of Ba^{2+} current through Ca^{2+} channels (see Gray, 1988).

Based on single-channel conductance measurements and voltage ranges of activation and inactivation, at least three types of Ca^{2+} channels appear to be present in adult hippocampal neurons (Gray and Johnston, 1986a). The three types correspond reasonably well to the T, N, and L types described by Tsien and colleagues in cultured dorsal root ganglion neurons (Fox *et al.*, 1987). A more complete description of Ca^{2+} channel subtypes in hippocampal neurons will appear (Fisher *et al.*, in press).

In granule cells β-adrenoceptor activation enhances the activity of the N-type Ca^{2+} channel (Figure 9), with no effect on the T type (Gray, 1988). The L type is rarely observed in attached patches from granule cells, and so we were unable to assess possible effects of β-adrenergic agonists on L channels. Using CA3 neurons, Fisher *et al.* (1990, submitted) have recently demonstrated an enhancement of the L type after application of isoproterenol. β-Adrenoceptor activation has no effect on single-channel conductance of the N type, but merely increases the total time the channel stays in the open state during a depolarizing command (see Gray and Johnston, 1987). The increase in the whole-cell Ca^{2+} current in granule cells following application of NE, isoproterenol, and 8-bromo-cAMP appears to be accounted for by an increase in activity of one type of voltage-dependent Ca^{2+} channel. This increase in channel activity results from an increase in the fractional open time of single channels or from an increase in the number of channels available for opening.

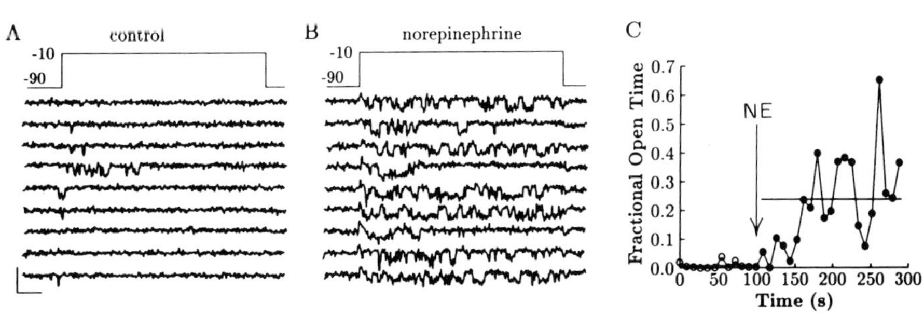

Figure 9 Effect of NE on Ca^{+} channel activity. (A) Consecutive current traces in control saline in response to step commands from −90 to −10 mV. Calibration bar: 2 pA × 10 ms. (B) Channel activity in the same patch 95 s after application of 10 μM NE. (C) Fraction of time during the command when the current level was more negative than a threshold level (−0.3 pA) before and after drug application

IV DISCUSSION

This chapter was intended to present the detailed methods for preparing acutely exposed adult hippocampal neurons suitable for patch clamping. These methods have only been published previously as part of a Ph.D. dissertation (Gray, 1988) or in abbreviated form as part of original research findings (Gray and Johnston, 1985c; Gray and Johnston, 1987). The reader is referred to the latter research publications for more information regarding experimental results obtained using this preparation. As mentioned in section I, our goal was to develop a preparation of adult hippocampal neurons that would permit a thorough investigation of their biophysical properties. We feel that the acutely exposed neuron preparation at least partly meets this goal. It should be noted, however, that there are two other preparations that offer some of the same advantages for such studies: tissue-cultured hippocampal neurons and totally isolated adult hippocampal neurons.

Tissue-cultured hippocampal neurons from embryonic or newborn animals are suitable for patch clamping and have been used extensively for biophysical investigations during the past few years (Segal and Barker, 1984a,b; Barker *et al*., 1984; Rogawski, 1986; Llano *et al*., 1988; Mayer *et al*., 1988). There are several limitations of tissue-culture systems, particularly when one is interested in properties of adult neurons and their synaptic connections. These include difficulty in identification of specific cell types, differences between embryonic and adult cells, and the influence of culture conditions on neuronal properties.

The use of totally dissociated neurons for neurochemical investigations was pioneered by Lam and colleagues (Lam, 1972, 1975) in the retina with subsequent application of neurophysiological techniques (Kaneko *et al*., 1976; Johnston and Lam, 1981). Totally dissociated heart muscle has also been a useful preparation for electrophysiologists (Giles and Ginneken, 1985; Brown *et al*., 1986). Wong and colleagues developed a preparation of totally isolated hippocampal neurons (Numann and Wong, 1984; Kay and Wong, 1986) that has been used for a number of studies of ion channels. This preparation offers the possibility of excellent space clamp and ionic current separation. The main limitation is that most of the dendrites and all synaptic inputs are lost during the dissociation, and this limits the scope of investigation to the somatic or proximal dendritic membrane.

The preparation of acutely exposed hippocampal neurons described here has been used successfully in our laboratory for the past five years for several investigations of ligand- and voltage-gated ion channels from neurons in the adult brain. The limitations of this preparation are primarily those related to the unknown effects of proteolytic enzymes on neurons. Furthermore, although synaptic transmission appears to be intact, there is some disruption of neuronal circuitry in the slice when it is split apart to expose the cells. Nevertheless, this preparation permits the use of patch-clamp techniques with adult cortical tissue, which was not possible previously, and, with suitable further refinements, should be a useful preparation for many future studies.

NOTE ADDED IN PROOF

Methods for patch-clamping neurons in brain slices have been recently described using thin (100–140μm) slices and an upright compound microscope (Edwards *et al.*, 1989) or thicker slices and a dissecting microscope (Blanton *et al.*, 1989). Neurophysiologists who require patch-clamp approaches now have available preparations ranging from totally isolated cells to intact slices, with each preparation having its own strengths and weaknesses.

ACKNOWLEDGEMENTS

This work was supported by NIH Grants NS11535 and HL31164 and AFOSR 88-0142.

REFERENCES

Akaike, N., Kaneda, M., Hori, N., and Krishtal, O.A. (1988). 'Blockade of *N*-methyl-D-aspartate response in enzyme-treated rat hippocampal neurons', *Neurosci. Lett.*, **87**, 75–79.

Andersen, P. (1983). 'Operational principles of hippocampal neurons', in *Neurobiology of Hippocampus* (Ed. W. Seifert), pp. 81–86, Academic Press, New York.

Andersen, P., Bliss, T.V.P., and Skrede, K.K. (1971). 'Lamellar organization of hippocampal excitatory pathways', *Exp. Brain Res.*, **13**, 222–238.

Armstrong, C.M., and Matteson, D.R. (1985). 'Two distinct populations of calcium channels in a clonal line of pituitary cells', *Science*, **227**, 65–67.

Armstrong, C.M., Bezanilla, F., and Rojas, E. (1973). 'Destruction of sodium conductance inactivation in squid axons perfused with pronase', *J. Gen. Physiol.*, **62**, 375–391.

Barker, J.L., Owen, D.G., and Segal, M. (1984). 'GABA actions on the excitability of cultured CNS neurons', *Neurosci. Lett.*, **47**, 313–318.

Blanton, M.G., Lo Tureo, J.-J., and Kriegstein, A.R. (1989). 'Whole cell recording from neurons in slices of reptilian and mammalian cerebral cortex', *F. Neurosci. Methods*, **30**, 203–210.

Brown, A.M., Kunze, D.L., and Yatani, A. (1986). 'Dual effects of dihydropyridines on whole cell and unitary calcium currents in single ventricular cells of guinea-pig', *J. Physiol. (Lond.)*, **379**, 495–514.

Brown, T.H., and Johnson, D. (1983). 'Voltage-clamp analysis of mossy fiber synaptic input to hippocampal neurons', *J. Neurophysiol.*, **50**, 487–507.

Carbone, E., and Lux, H.D. (1984). 'A low voltage-activated calcium conductance in embryonic chick sensory neurons', *Biophys. J.*, **46**, 413–418.

Connor, J.A., Wadman, W.J., Hockberger, P.E., and Wong, R.K.S. (1988). 'Sustained dendritic gradients of Ca^{2+} induced by excitatory amino acids in CA1 hippocampal neurons', *Science*, **240**, 649–653.

Corey, D.P., and Stevens, C.F. (1983). 'Science and technology of patch-recording electrodes', in *Single-Channel Recording* (Eds B. Sakmann and E. Neher), pp. 53–68. Plenum Press, New York.

Edwards, F.A., Konnerth, A., Sakmann, B., and Takahashi, T. (1989). 'A thin slice preparation for patch clamp recordings from neurones of the mammalian central nervous system', *Pflügers Arch.*, **414**, 600–612.

Fisher, R.E., Gray, R., and Johnston, D. (1988). 'β-adrenoceptor modulation of calcium channels in acutely exposed CA3 pyramidal neurons of adult guinea pig hippocampus', *Soc. Neurosci. Abstr.*, **14**, 645.

Fisher, R.E., Gray, R., and Johnston, D. (1990). 'Properties and distribution of single voltage-gated calcium channels in adult hippocampal neurons', *J. Neurophysiol.* (in press).

Fisher, R., and Johnston, D. (1990). 'Differential modulation of single voltage-gated calcium channels by cholinergic and adrenergic agonists in adult hippocampal neurons', *submitted*.

Fox, A.P., Nowycky, M.C., and Tsien, R.W. (1987). 'Single-channel recordings of three types of calcium channels in chick sensory neurones', *J. Physiol. (Lond.)*, **394**, 173–200.

Giles, W.R., and Ginneken, A.C.G.V. (1985). 'A transient outward current in isolated cells from the crista terminalis of rabbit heart', *J. Physiol. (Lond.)*, **368**, 243–264.

Gray, R. (1988). *Patch Clamp Analysis of Ligand and Voltage Gated Ion Channels in Adult Hippocampal Neurons*, Baylor College of Medicine, Ph.D. Thesis, Houston, TX.

Gray, R., and Johnston, D. (1985a). 'Macroscopic calcium currents in acutely-exposed granule cells from adult hippocampus', *Soc. Neurosci. Abstr.*, **11**, 792.

Gray, R., and Johnston, D. (1985b). 'Macroscopic calcium currents in acutely-exposed neurons from adult hippocampal slices', *Biophys. J.*, **47**, 66a.

Gray, R., and Johnston, D. (1985c). 'Rectification of single GABA-gated chloride channels in adult hippocampal neurons', *J. Neurophysiol.*, **54**, 134–141.

Gray, R., and Johnston, D. (1986a). 'Multiple types of calcium channels in acutely exposed neurons from the adult guinea pig hippocampus', *J. Gen. Physiol.*, **88**, 25a–26a.

Gray, R., and Johnston, D. (1986b). 'Multiple types of calcium channels in acutely-exposed neurons from adult hippocampus', *Biophys. J.*, **49**, 432a.

Gray, R., and Johnston, D. (1987). 'Noradrenaline and β-adrenoceptor agonists increase activity of voltage-dependent calcium channels in hippocampal neurons', *Nature*, **327**, 620–622.

Gray, R., Kellaway, J., and Johnston, D. (1982). 'Electrical properties of acutely isolated hippocampal neurons', *Physiologist*, **25**, 221.

Gray, R., Kellaway, J., and Johnson, D. (1983). 'Recordings of single GABA-activated channels from acutely-isolated hippocampal neurons', *Soc. Neurosci. Abstr.*, **9**, 1189.

Griffith, W.H., Gray, R., and Johnson, D. (1983). 'Patch-clamp recordings from hippocampal neurons in synaptically-intact slice fragments', *Soc. Neurosci. Abstr.*, **9**, 1190.

Hopkins, W.F., and Johnston, D. (1984). 'Frequency-dependent noradrenergic modulation of long-term potentiation in the hippocampus', *Science*, **226**, 350–352.

Johnston, D. (1981). 'Passive cable properties of hippocampal CA3 pyramidal neurons', *Cell Mol. Neurobiol.*, **1**, 41–55.

Johnston, D., and Brown, T.H. (1981). 'Giant synaptic potential hypothesis for epileptiform activity', *Science*, **211**, 294–297.

Johnston, D., and Brown, T.H. (1983). 'Interpretation of voltage-clamp measurements in hippocampal neurons', *J. Neurophysiol.*, **50**, 464–486.

Johnston, D., and Brown, T.H. (1984). 'Biophysics and microphysiology of synaptic transmission in hippocampus', in *Brain Slices* (Ed. R. Dingledine), pp. 51–86, Plenum Press, New York.

Johnston, D., and Lam, D.M.K. (1981). 'Regenerative and passive membrane properties of isolated horizontal cells from a teleost retina', *Nature*, **292**, 451–454.

Johnston, D., Hablitz, J.J., and Wilson, W.A. (1980). 'Voltage clamp discloses slow inward current in hippocampal burst-firing neurones', *Nature*, **286**, 391–393.

Johnston, D., Hopkins, W.F., and Gray, R. (1987). 'Cellular mechanisms of noradrenergic enhancement of long-term synaptic potentiation in hippocampus', in *The Role of Neuroplasticity in the Response to Drugs*, Vol. 78 (Eds D.P. Friedman and D.H. Clouet), pp. 95–107, NIDA Research Monograph Series, Rockville, MD.

Johnston, D., Hopkins, W.F., and Gray, R. (1988a). 'Noradrenergic enhancement of long-term synaptic potentiation', in *Long-Term Potentiation: From Biophysics to Behavior* (Eds P.W. Landfield and S. Deadwyler), pp. 355–376, Alan R. Liss, New York.

Johnston, D., Hopkins, W.F., and Gray, R. (1988b). 'Norepinephrine enhances long-term potentiation at hippocampal mossy fiber synapses', in *Synaptic Plasticity in the Hippocampus* (Eds H.L. Haas and G. Buzsàki), pp. 57–60, Springer-Verlag, Berlin.
Johnston, D., Hopkins, W.F., and Gray, R. (1989a). 'The role of norepinephrine in long-term potentiation at mossy fiber synapses in the hippocampus', in *Neural Models of Plasticity* (Eds J.H. Byrne and W.O. Berry), pp. 307–328, Academic Press, San Diego.
Johnston, D., Williams, S.H., Gray, R., and Fisher, R.E. (1989b). 'Cholinergic and noradrenergic modulation of long-term potentiation in hippocampal CA3 neurons', in *Brain Signal Transduction and Memory* (Eds M. Ito and Y. Nishizuka), pp. 171–183, Academic Press, San Diego.
Kandel, E.R., and Spencer, W.A. (1961). 'Electrophysiology of hippocampal neurons. II. After-potentials and repetitive firing', *J. Neurophysiol.*, **24**, 243–259.
Kandel, E.R., Spencer, W.A., and Brinley, F.G. (1961). 'Electrophysiology of hippocampal neurons. I. Sequential invasion and synaptic organization', *J. Neurophysiol.*, **24**, 225–242.
Kaneko, A., Lam, D.M.-K., and Wiesel, T.N. (1976). 'Isolated horizontal cells of elasmobranch retinae', *Brain Res.*, **105**, 567–572.
Kay, A.R., and Wong, R.K.S. (1986). 'Isolation of neurons suitable for patch-clamping from adult mammalian central nervous systems', *J. Neurosci. Meth.*, **16**, 227–238.
Lam, D.M.-K. (1972). 'Biosynthesis of acetylcholine in turtle photoreceptors', *Proc. Natl. Acad. Sci. USA*, **69**, 1987–1991.
Lam, D.M.-K. (1975). 'Biosynthesis of γ-aminobutyric acid by isolated axons of cone horizontal cells in the goldfish retina', *Nature*, **254**, 345–347.
Lee, K.S., Akaide, N., and Brown, A.M. (1977). 'Trypsin inhibits the action of tetrodotoxin on neurones', *Nature*, **265**, 751–753.
Llano, I., Marty, A., Johnson, J.W., Ascher, P., and Gähwiler, B.H. (1988). 'Patch-clamp recording of amino acid-activated responses in "organotypic" slice cultures', *Proc. Natl. Acad. Sci. USA*, **85**, 3221–3225.
Mayer, M.L., Westbrook, G.L., and Vyklický, Jr, L. (1988). 'Sites of antagonist action on *N*-methyl-D-aspartic acid receptors studied using fluctuation analysis and a rapid perfusion technique', *J. Neurophysiol.*, **60**, 645–663.
McCleskey, E.W., Fox, A.P., Feldman, D., and Tsien, R.W. (1986). 'Different types of calcium channels', *J. Exp. Biol.*, **124**, 177–190.
Mody, I., Salter, M.W., and MacDonald, J.F. (1989). 'Whole-cell voltage clamp recordings in granule cells acutely isolated from hippocampal slices of adult or aged rats', *Neurosci. Lett.*, **96**, 70–75.
Nowycky, M.C., Fox, A.P., and Tsien, R.W. (1985). 'Three types of neuronal calcium channel with different calcium agonist sensitivity', *Nature*, **316**, 440–443.
Numann, R.E., and Wong, R.K.S. (1984). 'Voltage-clamp study on GABA response desensitization in single pyramidal cells dissociated from the hippocampus of adult guinea pigs', *Neurosci. Lett.*, **47**, 289–294.
Reuter, H. (1983). 'Calcium channel modulation by neurotransmitters, enzymes and drugs', *Nature*, **301**, 569–573.
Rogawski, M.A. (1986). 'Single voltage-dependent potassium channels in cultured rat hippocampal neurons', *J. Neurophysiol.*, **56**, 481–493.
Schwartzkroin, P.A. (1975). 'Characteristics of CA1 neurons recorded intracellularly in the hippocampal *in vitro* slice preparation', *Brain Res.*, **85**, 423–436.
Segal, M., and Barker, J.L. (1984a). 'Rat hippocampal neurons in culture: Voltage-clamp analysis of inhibitory synaptic connections', *J. Neurophysiol.*, **52**, 469–487.
Segal, M., and Barker, J.L. (1984b). 'Rat hippocampal neurons in culture: Potassium conductances', *J. Neurophysiol.* **51**, 1409–1433.

Sigworth, F.J. (1983). 'Electronic design of the patch clamp', in *Single-Channel Recording* (Eds B. Sakmann and E. Neher), pp. 3–35, Plenum Press, New York.
Spencer, W.A., and Kandel, E.R. (1961a). 'Electrophysiology of hippocampal neurons: III. Firing level and time constant', *J. Neurophysiol.*, **24**, 260–285.
Spencer, W.A., and Kandel, E.R. (1961b). 'Electrophysiology of hippocampal neurons: IV. Fast prepotentials', *J. Neurophysiol.*, **24**, 272–285.
Turner, D.A., and Schwartzkroin, P.A. (1983). 'Electrical characteristics of dendrites and dendritic spines in intracellularly stained CA1 and dentate hippocampal neurons', *J. Neurosci.*, **3**, 2381–2394.
Wilson, W.A., and Goldner, M.M. (1975). 'Voltage clamping with a single microelectrode', *J. Neurobiol.*, **6**, 411–422.
Yamamoto, C. (1972a). 'Activation of hippocampal neurons by mossy fiber stimulation in thin brain sections *in vitro*', *Exp. Brain Res.*, **14**, 423–435.
Yamamoto, C. (1972b). 'Intracellular study of seizure-like afterdischarges elicited in thin hippocampal sections *in vitro*', *Exp. Neurol.*, **35**, 154–164.

Preparations of Vertebrate Central Nervous System *In Vitro*
Edited by H. Jahnsen
Published by John Wiley & Sons Ltd

CHAPTER 2

Use of Brain Slices in the Study of Serotonergic Pacemaker Neurons of the Brainstem Raphe Nuclei

GEORGE K. AGHAJANIAN
Departments of Psychiatry and Pharmacology,
Yale University School of Medicine,
and the Abraham Ribicoff Research Facilities,
Connecticut Mental Health Center, New Haven,
Connecticut 06508, USA

I INTRODUCTION

The serotonergic neurons of the brainstem raphe nuclei provide one of the most striking examples of a homogeneous population of tonic pacemaker cells in the central nervous system. From the time they were first recorded *in vivo*, these cells were observed to have a slow (~1–2 spikes/s), regular, metronome-like spontaneous firing pattern (Aghajanian *et al.*, 1968, 1970). Subsequent studies by autocorrelation techniques further emphasized the remarkable regularity in the firing of these cells *in vivo* (Wang and Aghajanian, 1982). However, in brain slices, most serotonergic raphe cells fall silent and no longer display their characteristic spontaneous, rhythmic firing pattern (VanderMaelen and Aghajanian, 1983). This observation suggests that the impetus for pacemaker activity in dorsal raphe neurons is not entirely endogenous and that important inputs necessary for sustaining pacemaker function may have been disconnected in the slice-making process. The purpose of this chapter is to show the usefulness of the brain slice preparation in sorting out extrinsic factors from intrinsic membrane mechanisms that underlie the pacemaker cycle of serotonergic raphe cells.

II SEROTONERGIC NEURONS *IN VIVO*

II.1 Anatomical Features

The existence of serotonergic neurons in the mammalian CNS was discovered by means of the formaldehyde-condensation histochemical method applied to freeze-dried brain tissue (Carlsson *et al.*, 1962). By this method a mapping of the location of neuronal perikarya (Dahlstrom and Fuxe, 1965), terminals (Fuxe, 1965), and pathways (Anden *et al.*, 1966; Ungerstedt, 1971) containing serotonin was first accomplished. More recently, serotonergic neurons and terminals have been localized and mapped by immunocytochemical methods (Steinbusch, 1981). Both methods reveal that the great majority of cell bodies of serotonergic neurons are located in the brainstem raphe nuclei. Arising from cells in the raphe region is a vast network of ascending and descending serotonergic nerve fibers that terminate at all levels of the brain and spinal cord.

II.2 Pacemaker Characteristics

In view of its widespread anatomical distribution, the serotonergic system potentially can have an influence on every aspect of brain and spinal cord function. Thus, it is significant that presumed serotonergic neurons in the dorsal and other raphe nuclei are characterized under virtually all conditions by a slow, regular firing pattern with spontaneous rates of approximately 1–2 spikes/s. This pattern has been found in the anesthetized rat (Aghajanian *et al.*, 1970; Bramwell, 1974; Mosko and Jacobs, 1974) and cat (Nakahama *et al.*, 1981), the cerveau isole rat (Wang and Aghajanian, 1982), and the awake unrestrained cat (McGinty and Harper, 1976; Trulson and Jacobs, 1979). The serotonergic identity of rhythmically firing neurons of the rat dorsal raphe nucleus has been confirmed by intracellular double labeling techniques (Aghajanian and VanderMaelen, 1982b). Although dorsal raphe neurons tend to have a higher mean rate of firing during periods of behavioral arousal than during slow-wave sleep, a regular rhythm is maintained unless the animal is perturbed by sensory stimuli (Heym *et al.*, 1982; Trulson and Jacobs, 1979). Dorsal raphe neurons are completely silent only during periods of rapid eye movement (REM) or paradoxical sleep (Cespuglio *et al.*, 1981; McGinty and Harper, 1976; Puizillout *et al.*, 1979; Trulson and Jacobs, 1979). The fact that serotonergic neurons of the dorsal raphe nucleus maintain their slow, rhythmic firing pattern under a wide variety of conditions *in vivo* has suggested that they possess intrinsic pacemaker properties (Aghajanian and Wang, 1978).

II.3 Serotonin Autoreceptors and Collateral Inhibition

Serotonin applied locally by microiontophoresis has a powerful inhibitory effect on the firing of serotonergic neurons in the dorsal raphe nucleus (Aghajanian *et al.*, 1972; Bramwell and Gonye, 1976; Haigler and Aghajanian, 1974b). It has been suggested that serotonin may exert its inhibitory effect by acting on 'autoreceptors' in the somatodendritic region of serotonergic neurons (Aghajanian and Wang, 1978). There is evidence that serotonergic autoreceptors may serve to mediate collateral inhibition in the raphe nuclei. When serotonergic neurons are driven antidromically by stimulating the serotonergic pathway at the level of the ventromedial tegmentum, they exhibit a pronounced period of poststimulus depression in firing (Wang and Aghajanian, 1977). Poststimulus suppression can be demonstrated even where the recorded neuron itself is subthreshold for antidromic activation, indicating that the Ca^{2+}-activated K^+ conductance (see below) cannot be the sole mechanism for the inhibition. A mechanism of recurrent or collateral inhibition is suggested by the observation that the suppression of firing induced by antidromic stimulation of the serotonergic pathway depends on the integrity of serotonergic fibers (Wang and Aghajanian, 1977).

Theoretically, recurrent inhibition could be mediated via axon collaterals and/or collateral dendrodendritic junctions. The possibility of interactions between dendrite bundles in the dorsal raphe nucleus has been suggested on the basis of Golgi, histofluorescence, and electron-microscopic studies (Felten and Harrigan, 1980; Mollgard and Moller, 1973). Electron-microscopic autoradiographic studies have shown that there is [^{3}H]serotonin labeling of cell bodies and dendrites but not nerve terminals within the neuropil of the dorsal raphe nucleus (Baraban and Aghajanian, 1981; Descarries *et al.*, 1979). Together these observations seem to favor a dendrodendritic rather than an axon–collateral basis for interactions between serotonergic neurons.

In the dorsal raphe nucleus, cross-correlation analysis of simultaneously recorded neuronal pairs shows a high probability of correlated discharges (i.e., firings within 10 ms) between adjacent serotonergic cells (Wang and Aghajanian, 1982). In addition to this synchronous firing pattern, adjacent serotonergic neurons often display direct inhibition or mutual inhibition (defined as reduced probability of firing of one neuron before and after firing of another). There are several possible interpretations for synchronous firing among adjacent serotonergic neurons. Synchronous firing could result from the direct activation of one cell by another or from sharing a direct activation or inhibition from a third neuron (Moore *et al.*, 1970; Perkel *et al.*, 1967). The interactions among serotonergic neurons that have been detected by correlational techniques are largely restricted to adjacent neurons (i.e. within 100 μm). Cross-correlograms generated from non-adjacent serotonergic neuronal pairs rarely show any interaction, indicating a very low probability that a correlated interaction between the two adjacent serotonergic cells is due to chance alone (Wang and Aghajanian, 1982). These studies further suggest that physiological coupling in the dorsal raphe nucleus takes place predominantly within small clusters of serotonergic cells rather than between serotonergic cells in distant parts of the nucleus.

Although the precise anatomical basis for local physiological interactions in the dorsal raphe nucleus remains to be determined, collateral interactions mediated by serotonin autoreceptors are likely to exert a negative feedback control over serotonergic pacemaker activity.

II.4 The Pharmacology of Serotonin Autoreceptors

Serotonin autoreceptors are of obvious interest in relation to pacemaker activity because of their possible role in mediating recurrent or collateral inhibition in the dorsal raphe nucleus. The development of selective ligands for various serotonin receptor subtypes (e.g. 5-HT_{1A}, 5-HT_{1B}, 5-HT_2, and 5-HT_3) has proven useful in the pharmacological characterization of serotonin autoreceptors. One such compound is a novel agonist with serotonin-like actions but devoid of an indole nucleus, 8-hydroxy-2-(di-n-propylamino)tetralin (8-OH-DPAT; Arvidsson *et al.*, 1981) which possesses an almost 1000-fold selectivity for the 5-HT_{1A} binding site (Middlemiss and Fozard, 1983). A marked reduction in the binding of 8-OH-DPAT in the dorsal raphe nucleus occurs following 5,7-dihydroxytryptamine-induced degeneration of serotonergic cell bodies, suggesting that the somatic autoreceptor of neurons in the dorsal raphe is of the 5-HT_{1A} subtype (Verge *et al.*, 1985). Moreover, small intravenous doses of 8-OH-DPAT inhibit the firing rate of serotonin neurons in the dorsal raphe of chloral hydrate-anesthetized rats (Fallon *et al.*, 1983). This effect is also seen in microiontophoretic studies in which 8-OH-DPAT exerts a depressant effect comparable to LSD (deMontigny *et al.*, 1984). Furthermore, buspirone, a non-benzodiazepine anxiolytic which displays 5-HT_{1A} binding properties (Peroutka, 1985), has been shown to potently inhibit the firing of serotonergic dorsal raphe neurons when administered systemically or microiontophoretically, or added to media bathing brain slices (VanderMaelen *et al.*, 1986). Buspirone has also been shown to slow dorsal raphe cell firing when it is administered to awake, freely moving cats (Wilkinson *et al.*, 1987).

A direct comparison of the effects of 5-HT_{1A}- and 5-HT_{1B}-selective compounds on the spontaneous firing of dorsal raphe neurons has disclosed striking differences between the two ligand subtypes (Sprouse and Aghajanian, 1987). The intravenous administration of ipsapirone or *p*-aminophenylethyl-*m*-

trifluoromethylphenylpiperazine (PAPP), two long-chain substituted piperazines chosen for their 5-HT_{1A} binding affinities (Asarch *et al*., 1985; Dompert *et al*., 1985; Glaser *et al*., 1985; Ransom *et al*., 1986), potently inhibits raphe cell firing in a dose-dependent manner. The ED_{50} for ipsapirone is 9 μg/kg (i.v.), a dose which compares favorably with the raphe-inhibiting potency of LSD (Aghajanian *et al*., 1970). In contrast, trifluoromethylphenylpiperazine (TFMPP) and *m*-chlorophenylpiperazine (mCPP), two short-chain substituted piperazines chosen for their relatively low 5-HT_{1A} binding affinities as well as some degree of selectivity for the 5-HT_{1B} site (Hamon *et al*., 1986; Sills *et al*., 1984), display much weaker or irregular actions. Raphe cell firing is also suppressed by the microiontophoretic application of ipsapirone and PAPP, indicating a direct action on serotonergic neurons. Dose–response curves for these two 5-HT_{1A}-selective compounds are virtually indistinguishable from that of 5-HT. However, the two 5-HT_{1B}-selective compounds, TFMPP and mCPP, when applied by microiontophoresis, again appear to be only weak autoreceptor agonists.

In general, the classical serotonin antagonists have proven ineffective in blocking the effect of serotonin at the somatodendritic autoreceptor, or, for that matter, at most inhibitory postsynaptic sites (Haigler and Aghajanian, 1974a). However, behavioral studies have suggested that (−)-propranolol may possess 5-HT_{1A} antagonistic properties (Green and Grahame-Smith, 1976; Tricklebank *et al*., 1984). Electrophysiological studies indicate that low microiontophoretic currents of (−)- but not (+)-propranolol effectively block the suppressant effects of ipsapirone and 8-OH-DPAT on raphe cell firing (Sprouse and Aghajanian, 1986). Surprisingly, (−)-propranolol is only a relatively weak antagonist of serotonin itself, suggesting the possibility that the endogenous transmitter may interact with 5-HT_{1A} receptors in a manner which is not identical with that of the selective 5-HT_{1A} ligands.

Taken together, these studies show that serotonin autoreceptors have distinctive pharmacological characteristics and drugs that act as agonists at these receptors can suppress the normal pacemaker activity of dorsal raphe neurons.

II.5 Noradrenergic Input and Pacemaker Activity

The dorsal raphe nucleus receives one of the heaviest noradrenergic innervations in the brain. This input has been demonstrated by histofluorescence (Fuxe, 1965), biochemical (Levitt and Moore, 1979; Saavedra *et al*., 1976; Versteeg *et al*., 1976), and immunocytochemical (Grzanna and Molliver, 1980) techniques. Physiological studies *in vivo* in the anesthetized rat suggest that the spontaneous firing of serotonergic cells in the dorsal raphe nucleus depends on a tonically active adrenergic system. Serotonergic cell firing is suppressed by the administration of either a low dose of clonidine (Svensson *et al*., 1975), which acts presynaptically to inhibit noradrenergic cell firing, or reserpine, which impairs noradrenergic transmission (Baraban *et al*., 1978). Furthermore amphetamine, which releases catecholamines, restores firing after either of these treatments. Several lines of evidence indicate that the adrenergic influence on serotonergic cells is mediated by α_1-adrenoceptors: (1) α-adrenoceptor antagonists, administered either systemically or locally, suppress the 'spontaneous' firing of serotonergic neurons (Baraban and Aghajanian, 1980a,b); (2) norepinephrine reverses the suppression produced by α-adrenoceptor antagonists; (3) phenylephrine, a selective α_1-adrenoceptor agonist, shares the ability of norepinephrine to reverse the suppression produced by α-adrenoceptor antagonists; and (4) α-adrenoceptor antagonists block the activation of serotonergic cells by norepinephrine after the administration of reserpine. In the non-anesthetized cat, clonidine or α-adrenoceptor antagonists also appear to suppress basal activity of the dorsal raphe neurons, but phasic auditory or visual stimulation may serve as an alternative driving force to maintain cell firing (Heym *et al*., 1981).

III SEROTONERGIC NEURONS IN BRAIN SLICES

III.1 Methodological Considerations

Although the above *in vivo* studies suggest an important role for norepinephrine inputs in maintaining normal pacemaker activity, the complexities inherent in the *in vivo* situation (e.g. anesthesia, sensory inputs) are difficult to control experimentally. Thus, the brain slice, in addition to making intracellular recording more feasible, offers clear advantages for distinguishing noradrenergic and other extrinsic factors from intrinsic factors which may contribute to raphe pacemaker activity. However, certain caveats are necessary in the interpretation of data pertaining to the issue of pacemaker activity in brain slices. For example, artifactual 'spontaneous' activity may occur in a damaged slice due to the inability of cells to recover fully to their normal resting potential. For these reasons, it is essential that the method used in slice preparation be optimal for obtaining cells in a healthy state.

Recently we have been using a method that was originally developed for the difficult problem of preserving of motoneuron viability in brain slices (Aghajanian and Rasmussen, 1989); we also find that this method gives excellent results in raphe slices. The essence of the method is to substitute sucrose (at 252 mM to maintain constant calculated osmolarity) temporarily for NaCl (126 mM) during slice preparation and to have a 1-h recovery period; the actual recordings are made only after switching back to a normal medium. A variant of the method is to make the transition back to a normal medium gradually in several stages. Other aspects of the procedure are conventional (for details see Aghajanian and Rasmussen, 1989). The use of the modified artificial cerebrospinal fluid is based on the following considerations: (1) neurons are prone to depolarization during anoxia, a condition which is virtually unavoidable during the early phases of the slice procedure (e.g. decapitation, blocking, and slicing); (2) *acute* neurotoxicity induced by depolarization (in contrast to *delayed* neurotoxicity, which is Ca^{2+}-dependent) results primarily from passive Cl^- influx, followed by cation and water entry, which can lead ultimately to swelling and cell lysis (Rothman, 1985). The replacement of NaCl by sucrose, a relatively impermeant molecule, appears to prevent this type of cellular damage.

III.2 Loss of Pacemaker Activity in Brain Slices

In almost all respects, serotonergic neurons in the brain slice resemble serotonergic neurons recorded *in vivo* from anesthetized rats (Mosko and Jacobs, 1976; VanderMaelen and Aghajanian, 1983). However, while the great majority of serotonergic dorsal raphe neurons *in vivo* are spontaneously active, most dorsal raphe neurons in the slice are silent (VanderMaelen and Aghajanian, 1983). The existence in raphe slices of a large population of silent serotonergic neurons was demonstrated by a sampling procedure in which glutamate is applied microiontophoretically while advancing an electrode through multiple tracts in the dorsal raphe nucleus. By this procedure, almost 80% of otherwise typical serotonergic neurons were found to lack spontaneous activity. One obvious explanation for the paucity of spontaneous activity is that essential excitatory afferents to the nucleus which originate from outside of the slice have been severed. More recent studies, using an improved preparatory method (see above), indicate that the figure of 80% may be an underestimate of the proportion of serotonergic neurons that are silent in the brain slice preparation.

In support of the hypothesis that noradrenergic deafferentation accounts for the loss of spontaneous activity in the raphe slice is the fact that adding norepinephrine or the α_1-adrenoceptor agonist phenylephrine to the perfusate results in a reversible restoration of normal tonic firing (VanderMaelen and Aghajanian, 1983). These activations can be blocked by bath application of the non-selective α-antagonist phentolamine or the selective α_1-antagonist prazosin in exact parallel with the ability of these drugs to suppress serotonergic cell firing *in vivo* in the anesthetized rat.

III.3 Pacemaker Potentials: Interactions with Norepinephrine

Intracellular recordings both *in vivo* (Aghajanian and VanderMaelen, 1982a) and *in vitro* (VanderMaelen and Aghajanian, 1983) show that spikes of serotonergic neurons arise from gradual interspike depolarizations ('pacemaker potentials') rather than discrete synaptic potentials. The interspike behavior of a spontaneously active serotonergic neuron is characterized by a large afterhyperpolarization, followed by a short period of relatively rapid depolarization, and finally a plateau phase lasting several hundred milliseconds which precedes spike initiation (Figure 1A). The magnitude of the afterhyperpolarization is on the order of 15–20 mV, which by itself establishes a minimum relative refractory period of approximately 200 ms. Acceleration of pacemaker activity by α_1-adrenoceptor stimulation markedly shortens the plateau phase but does not reduce the initial amplitude of the afterhyperpolarization (Figure 1B). Similarly, in voltage-clamp experiments, the amplitude of the Ca^{2+}-activated K^+ current ($I_{K(Ca)}$), which underlies the afterhyperpolarization, is not reduced by the α_1-agonist phenylephrine (Freedman and Aghajanian, 1987). This contrasts with β-adrenoceptor effects on hippocampal pyramidal neurons, where there is an inhibition of Ca^{2+}-dependent afterhyperpolarization, resulting in increased excitability due to reduced spike frequency accommodation (Haas and Konnerth, 1983; Madison and Nicoll, 1982).

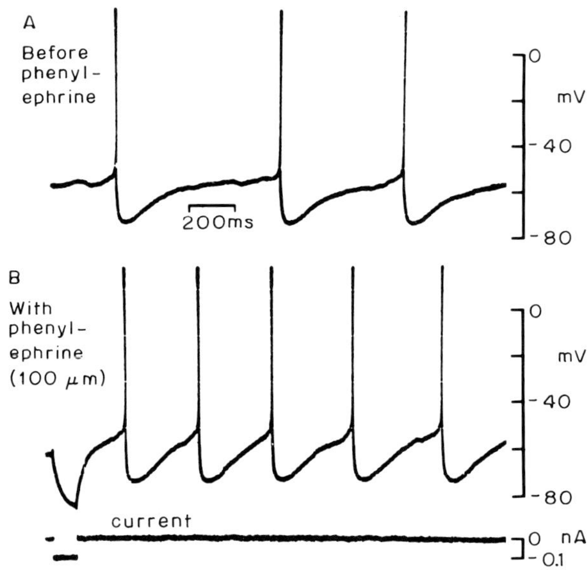

Figure 1 Intracellular recording from a serotonergic dorsal raphe neuron before (A) and during (B) the bath application of phenylephrine (100 μM). The spontaneous discharge rate in (A) was about 1.5 Hz. This was approximately doubled by the phenylephrine. The rate of interspike depolarization was accelerated during the response to phenylephrine, but the initial amplitude of the afterhyperpolarization was undiminished. Intracellularly injected current pulse in (B) (0.1 nA) indicated resistance of about 210 MΩ (but this varied as a function of time since the last spike). Note that this cell is somewhat atypical in that most serotonergic neurons in the slice are silent. (From VanderMaelen and Aghajanian, 1983, with permission)

Figure 2 Intracellular recording from a serotonergic neuron in a dorsal raphe slice preparation showing the effect of the α_1-agonist phenylephrine (PE) on membrane potential, input resistance, and firing rate (A–C). Simultaneous chart recordings of low-pass-filtered membrane potential, input resistance, and integrated firing rate. (A) The introduction of PE (25 μM) into the slice perfusate induces a rapid depolarization. (B) The depolarization is accompanied by a sharp increase in input resistance as measured by the voltage deflections to a repeated constant current (0.08 nA). (C) The depolarization leads to spiking in this previously quiescent cell. All of these changes are rapidly terminated by the introduction of serotonin (5-HT), which is known to inhibit serotonergic neurons in the dorsal raphe slice (cf. VanderMaelen and Aghajanian, 1983). (D) Oscilloscope traces of raw data for current–voltage (I–V) curves that are plotted in (E) for the PE and 5-HT periods. (From Aghajanian, 1984, with permission)

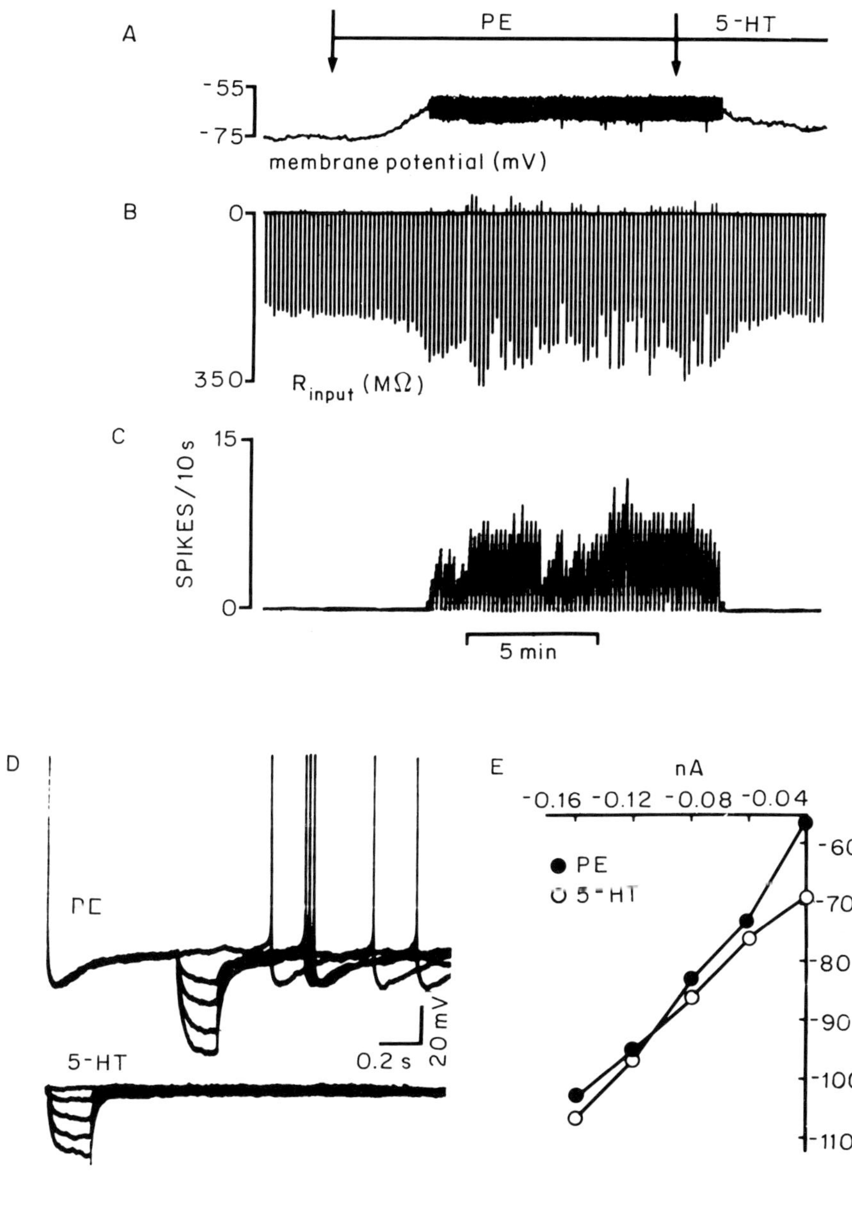
A
PE
5-HT
-55
-75
membrane potential (mV)
B
0
350
R input (MΩ)
C
15
0
SPIKES/10 s
5 min
D
PE
5-HT
20 mV
0.2 s
E
nA
-0.16 -0.12 -0.08 -0.04
● PE
○ 5-HT
-60
-70
-80
-90
-100
-110
mV

Figure 2

The means by which α-adrenoceptor stimulation shortens the plateau period of the interspike interval in dorsal raphe neurons has been explored by current-clamp and voltage-clamp methods in brain slices. In quiescent dorsal raphe cells, norepinephrine or phenylephrine produces a slow depolarization accompanied by an increased input resistance, leading to an initiation of tonic firing (Figure 2). The apparent reversal potential for this slow depolarization, even when Cl^--containing electrodes are used, is in the deeply hyperpolarized direction (e.g. −90 to −100 mV), suggestive of a change in K^+ conductance. Single-electrode voltage-clamp studies show that serotonergic dorsal raphe neurons possess an early transient outward current (I_A) which can be elicited by depolarizations from potentials below resting levels. This outward current, which tends to delay the rate of depolarization toward the end of the interspike interval, is suppressed by norepinephrine or phenylephrine (Figure 3; Aghajanian, 1985). Since serotonergic

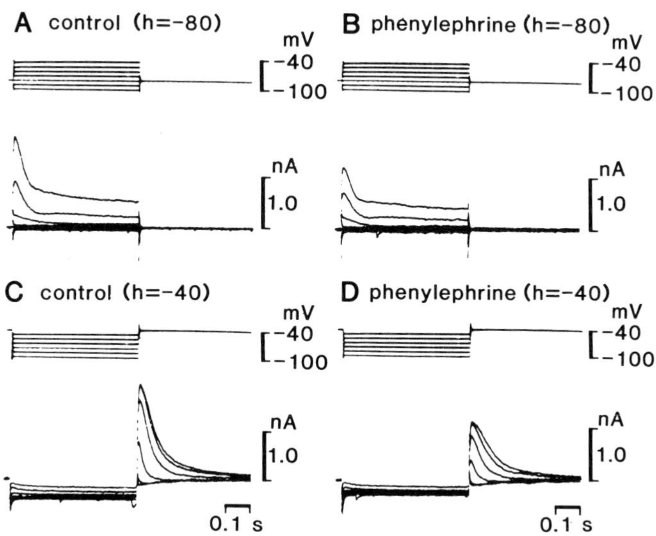

Figure 3 The effect of phenylephrine on the activation of I_A in a serotonergic neuron. In (A) and B, holding potential (*h*) is at −80 mV (holding current in (A), −0.07 nA); current responses (lower traces) are to voltage steps (upper traces) ranging from −40 to −100 mV. Note that I_A first becomes activated at −60 mV and increases progressively up to the −40-mV step. In (B), the administration of phenylephrine (25 μM; 18 min) results in a decrease in amplitude at each step. In (C) and (D), the holding potential is at −40 mV (holding current in (C), +0.25 nA) with hyperpolarizing prepulse steps ranging from −50 to −100 mV; note that I_A is de-inactivated only by hyperpolarizing steps below −60 mV. In (D), following the administration of phenylephrine (15 min), the holding current shifts to +0.16 nA and there is a reduction in the amplitude of I_A elicited by a return to the holding potential (−40 mV). Note the decrease in leakage current in the prepulse series (−50 to −100 mV) during phenylephrine perfusion as compared with the control series in (C). (Reprinted by permission from *Nature*, Vol. 315, pp. 501–503. Copyright ©1985 Macmillan Magazines Ltd)

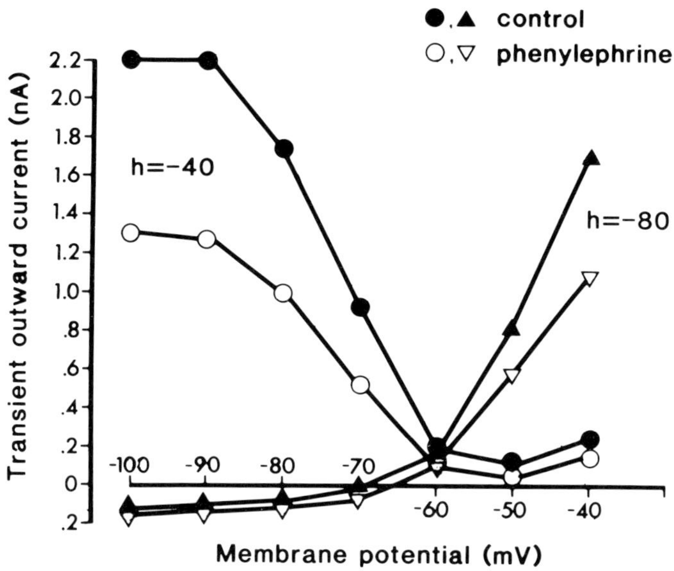

Figure 4 Activation–inactivation curves for I_A in a serotonergic dorsal raphe neuron: the effect of phenylephrine. Control and phenylephrine traces for holding potentials of $h = -40$ mV and $h = 80$ mV are plotted as direct current–voltage (I–V) curves (no leakage current subtraction); as can be seen, at all voltage steps where I_A is activated, phenylephrine reduces the maximum amplitude. Also note that phenylephrine induces a small increase in 'resting' inward current. The slice was perfused with artificial CSF containing TTX (1 μM) throughout the experiment to block Na^+ spikes. (Reprinted by permission from *Nature*, Vol. 315, pp. 501–503. Copyright ©1985 Macmillan Magazines Ltd)

neurons in the dorsal raphe nucleus receive a direct noradrenergic synaptic input, these results suggest that the suppression of I_A by norepinephrine represents a true physiological modulation of a voltage-dependent current. The reversal potential of I_A in dorsal raphe cells is approximately -90 mV, indicating that it is primarily a K^+ current. The plateau period of the pacemaker potential in serotonergic neurons occurs between -65 and -60 mV (see above) which is precisely the range in which I_A starts to become activated (Figure 4). Thus, by suppressing I_A, stimulation of α_1-adrenoceptors would shorten the plateau period and, together with a reduction in resting K^+ conductance, accelerate pacemaker activity.

III.4 A Low Threshold Calcium Conductance and Raphe Pacemaker Potentials

In brain slices, several distinct anodal break phenomena can be observed in silent or slowly firing dorsal raphe neurons (Burlhis and Aghajanian, 1987). In the cell illustrated in Figure 5A, upon anodal break an early hyperpolarizing response occurs (a). In other studies this type of response has been termed an anodal break

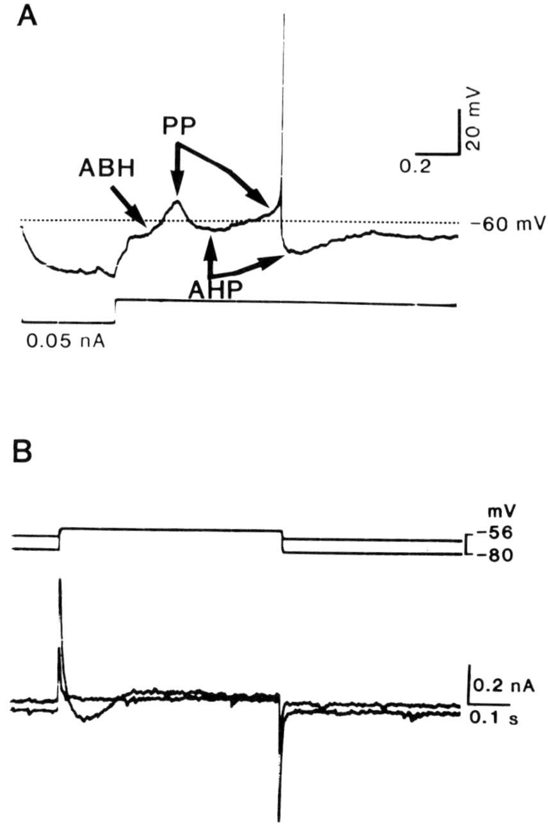

Figure 5 Anodal break phenomena elicited in the dorsal raphe neuron (A). Intracellular current-clamp experiment showing (a) an anodal break hyperpolarization (ABH), (b) an anodal break depolarizing prepotential (PP), and (c) an afterhyperpolarization (AHP), which follows a subsequent full spike. Note that the cell did not exhibit spontaneous activity and that both the PP and the full spike were induced by the anodal break. (B) Single-electrode voltage-clamp experiment analogous to (A). A depolarizing step from −80 to −56 mV shows a transient outward current (I_A; see text), followed by a transient inward current and a delayed outward current. In contrast, a depolarizing step from −65 to −56 mV has little effect. In (B), current traces are beneath the voltage traces. Outward currents are upward going. (From Burlhis and Aghajanian, 1987, with permission)

hyperpolarization (ABH) and has been shown to be a manifestation of I_A, which decreases the rate of return of membrane potential to spike threshold (Yarom *et al*., 1985). The occurrence of ABHs in dorsal raphe neurons is consistent with the finding that these cells possess a prominent I_A (see above; Aghajanian, 1985; Segal, 1985). The end of the hyperpolarizing pulse is also followed by a *depolarizing prepotential* (PP) (b) which in this sweep does not lead to an action potential. Subsequently, a full spike occurs in the sequence of anodal break events. As observed in previous studies (Crunelli *et al.*, 1983; VanderMaelen and Aghajanian, 1983), a large afterhyperpolarization (c) follows the action potential. An experiment analogous to that of Figure 5A, carried out using the single-electrode voltage-clamp technique, is shown in Figure 5B. The membrane potential of the cell was maintained at either −65 or −80 mV. Upon a depolarizing step from −80 to −56 mV, a fast transient outward current, which corresponds in time to the ABH, occurs (a). The fast outward current (I_A) is followed by a transient inward current which corresponds to the depolarizing prepotential seen in current clamp (b). Finally, there is a more delayed outward current (c). In contrast, upon a depolarizing step from −65 to −56 mV, I_A is barely detectable, and no inward current occurs (d). Thus, the occurrence of both I_A and the subsequent inward current depends on a sufficient prior hyperpolarization (i.e. 'priming').

Ni^{2+} as well as other divalent cations (e.g. Mn^{2+}, Co^{2+}, Cd^{2+}) that block Ca^{2+} conductances reversibly block the depolarizing prepotential, indicating that the inward movement of Ca^{2+} underlies the depolarizing prepotential (Burlhis and Aghajanian, 1987). High-threshold (approximately −40 mV) Ca^{2+} spikes as well as AHPs can be elicited by a depolarizing current under control conditions. However, the high-threshold spikes are not diminished after perfusion with Ni^{2+}, showing a specificity of Ni^{2+} for inhibiting the low-threshold Ca^{2+} conductance. Other divalent cations tested (Mn^{2+}, Co^{2+}, and Cd^{2+}) do not share this specificity.

The depolarizing prepotential in dorsal raphe neurons appears to represent a type of low-threshold Ca^{2+} conductance which could contribute to the pacemaker potential of serotonergic neurons by triggering a spike toward the end of the pacemaker cycle. This conductance is similar to the low-threshold Ca^{2+} conductance underlying rebound excitation in inferior olivary and thalamic neurons, enabling them to display oscillatory behavior in the absence of synaptic interactions (Jahnsen and Llinás, 1984a,b; Llinás, 1984; Llinás and Yarom, 1981a,b). In general, these conductances, which are inactive at rest, are removed from inactivation (or de-inactivated) with membrane hyperpolarization, the de-inactivation being time- and voltage-dependent. Subsequent to an action potential, a large AHP generated by $I_{K(Ca)}$ enables the de-inactivation of low-threshold Ca^{2+} channels. Other low-threshold Ca^{2+} conductances have been reported in the substantia nigra (Llinás *et al.*, 1984), and the pontine reticular formation (Greene *et al.*, 1986). The time and voltage dependence of de-inactivation and activation of the low-threshold inward Ca^{2+} current in serotonergic neurons are virtually identical to those of the outward I_A (Burlhis and Aghajanian, 1987). Thus, these two voltage-activated currents have directly competing influences on later phases of the pacemaker potential of dorsal raphe neurons.

Under voltage clamp neither norepinephrine nor serotonin appear to directly alter low-threshold Ca^{2+} currents in dorsal raphe serotonergic neurons. However, under current clamp, the depolarizing prepotentials are reduced in amplitude by serotonin, perhaps because of a drop in input resistance as well as a hyperpolarizing shift in membrane potential away from threshold for the prepotential.

III.5 Serotonergic Suppression of Pacemaker Potentials: Ionic and G Protein Coupling Mechanisms

As previously described, extracellular studies *in vivo* have pointed to the importance of serotonin autoreceptors in mediating collateral inhibition in the dorsal raphe nucleus. Intracellular studies *in vivo* have shown that the inhibition of serotonergic dorsal raphe neurons by systemically administered LSD (an agonist at serotonin 5-HT_{1A} autoreceptors), is associated with a hyperpolarization and a decrease in input resistance (i.e. increased conductance). Intracellular recordings *in vivo* show that the hyperpolarization is not reversed when Cl^--containing electrodes are used, suggesting that an increase in K^+ conductance is the probable ionic mechanism (Aghajanian and VanderMaelen, 1982a). However, as 5-HT does not cross the blood–brain barrier readily, it is not feasible to compare the intracellular effects of this putative transmitter (or modulator) directly with that of systemically administered LSD *in vivo*. Thus, the brain slice preparation becomes essential for the study of ionic and other membrane mechanisms for the action of serotonin and autoreceptor agonists on dorsal raphe cells

Both 5-HT and LSD, administered in known concentration through the perfusing medium, induce a hyperpolarization and a decrease in input resistance of serotonergic dorsal raphe neurons in the brain slices (Aghajanian and Lakoski, 1984). Reversal potentials for the 5-HT- and LSD-induced hyperpolarizations in artificial cerebrospinal fluid (aCSF) containing either 5 or 10 mM KCl show a shift according to the Nernst equation for a K^+-dependent change in conductance (viz. an 18.3-mV shift in reversal potential for a two-fold change in K^+ concentration at 34°C). As can be seen for the cell shown in Figure 6, the reversal potential for 5-HT is −90 mV in 5 mM KCl and −72 mV in 10 mM KCl, the difference of 18 mV being almost exactly that predicted by the Nernst equation for a pure K^+ conductance. There is a similar shift in the reversal potentials for LSD at 5 and 10 mM KCl. These results indicate that serotonin and LSD hyperpolarize serotonergic neurons predominantly if not exclusively through increasing K^+ conductance. When serotonin is applied at high concentrations, the magnitude of the hyperpolarization is sufficient to abort pacemaker activity entirely.

There is now evidence that the autoreceptor-mediated opening of K^+ channels in dorsal raphe neurons is mediated by guanosine 5′-triphosphate (GTP)-binding proteins (Innis *et al*., 1988; Williams *et al*., 1988). The effects of GTP-binding proteins (or G proteins) on transmembrane signal transduction were initially thought to be limited to modulation of the levels of the second messenger cyclic adenosine monophosphate (cAMP) (for review see Gilman, 1987; Stryer and Bourne, 1986). However, recent research has shown in heart membranes, for example, that a G protein directly mediates inhibition induced by the opening of K^+ channels in a cAMP-independent manner (Yatani *et al*., 1987). Similar cAMP-independent effects have been reported for serotonin (5-HT)- and $GABA_B$-receptor-mediated hyperpolarization of hippocampal pyramidal neurons (Andrade *et al*., 1986).

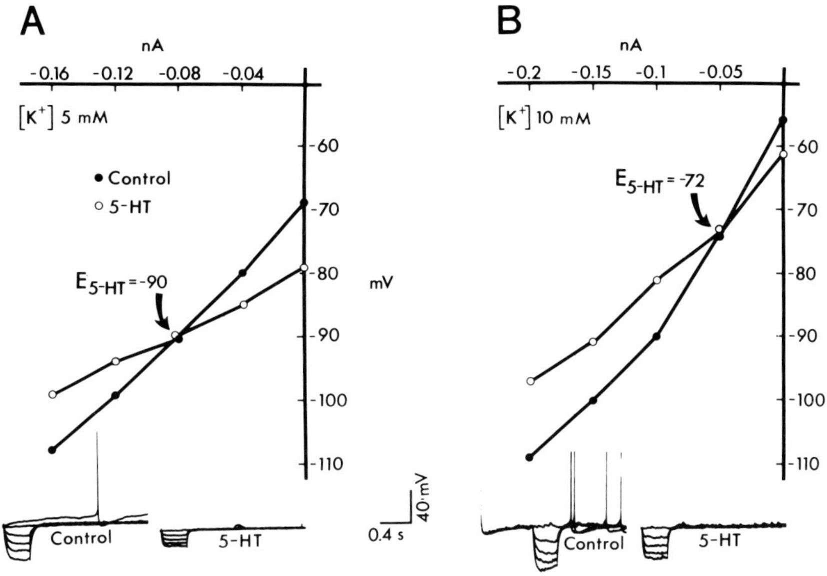

Figure 6 Current–voltage (I–V) relationships before and during perfusion (15–20 min) of serotonin (80 μM) in artificial cerebrospinal fluid solutions containing either 5 or 10 mM KCl. In (A), in the presence of 5 mM KCl, the 'control' and serotonin I–V curves intersect at -90 mV; insets beneath graph show superimposed voltage responses to incremental constant-current pulses from which the graphs were constructed. In (B), in the presence of 10 mM KCl, the intersection of control and serotonin I–V curves occurs at -72 mV; note in the inset that there is spontaneous activity which results from the depolarizing effect of the high K^+ concentration. The points of intersection represent the reversal potentials since, by definition, at the reversal potential for a given ion the diffusional potential is equal and opposite to the transmembrane potential; thus, at the reversal potential, since there would be no net current flow with a change in conductance, no change in membrane potential would occur. (From Aghajanian and Lakoski, 1984, with permission)

Pertussis toxin (islet-activating protein) has proven to be a useful tool in investigating the role of G proteins, because it leads to the ADP-ribosylation of the α-subunit of G_i and G_o and the uncoupling of the G protein from the receptor (Katada and Ui, 1980; Sternweis and Robishaw, 1984). Thus, blockade of a receptor-mediated event by pertussis toxin suggests that a G protein (either G_i or G_o) is involved in mediation of the receptor's action. Experiments in the dorsal raphe nucleus show that local preinjection of pertussis toxin causes a virtually complete blockade of serotonin- and baclofen-induced hyperpolarization (Figure 7; Innis *et al*., 1988). Preinjections with pertussis toxin by the intracerebroventricular route are also able to block hyperpolarizing responses to serotonin (Innis *et al*., 1988; Williams *et al*., 1988). The intracellular injection of GTPγS, a non-hydrolyzable analog of GTP which induces an irreversible activation of G proteins, mimics the action of serotonin by causing hyperpolarization and a decrease in R_{in}

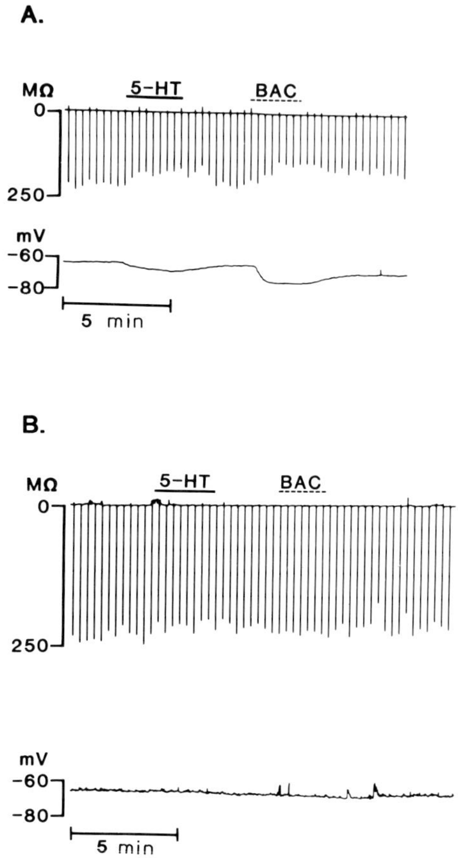

Figure 7 Blockade by pertussis toxin of the hyperpolarizing effects of serotonin and baclofen on serotonergic dorsal raphe neurons. (A) In a brain slice prepared from a control animal, serotonin (100 μM) and baclofen (50 μM) applied in the perfusing artificial CSF for the times indicated induced hyperpolarization of membrane potential and a decrease in R_{in}; (B) In a brain slice prepared from an animal injected 2 days earlier with pertussis toxin (0.5 μg), the effects of serotonin and baclofen were nearly completely blocked. For both (A) and (B), the top trace of the record is the measurement of apparent R_{in}. For ease of visualization, each downward vertical line represents the isolated hyperpolarizing voltage deflection from a constant baseline in response to the application of a constant negative current (−0.5 nA, 350 ms duration, 20 s interval). Input resistance was calculated from Ohm's law ($R = V/I$). The bottom trace in each record represents the membrane potential of the cell. (From Innis *et al*., 1988, with permission)

(Innis *et al*., 1988). The hyperpolarizing action of GTPγS is non-additive with that of serotonin, suggesting that they share a common effector system (i.e. a G protein). Concentrations of 8-Br-cAMP (1–2 mM) which are electrophysiologically effective in certain other regions have no effect on membrane potential, R_{in}, or the AHP, and do not block serotonin- or baclofen-induced hyperpolarizations in dorsal raphe neurons. Thus, inhibition mediated by serotonin appears to be cAMP-independent. Finally, a comparison between two methods of pertussis toxin injection (local versus intracerebroventricular) shows that the sensitivities to serotonin and baclofen correlate with the levels of remaining non-ADP-ribosylated G protein (both G_i and G_o separately, and the combination of G_i plus G_o) (Innis *et al*., 1988).

Taken together, these results suggest a major role for G proteins as mediators of the autoreceptor-mediated inhibition of serotonergic neurons in rat dorsal raphe nucleus. As the inhibition appears to be independent of cAMP, these findings are consistent with a direct action of a G protein(s) in opening resting K^+ channels in the dorsal raphe nucleus, although it is possible that an unidentified second messenger may be involved.

IV SUMMARY AND CONCLUSIONS

IV.1 The Pacemaker Cycle in Dorsal Raphe Cells

The pacemaker potential (or 'automaticity') has been defined as the gradual depolarization of a cell during the interspike interval (Jung and Moore, 1966; Katzung, 1978). The rhythmically firing neurons of the dorsal raphe nucleus have several characteristics that fit this definition. For example, they exhibit a repetitive cycle of postspike hyperpolarizations followed by gradual depolarizations during the interspike interval. Based primarily on studies in brain slices, the following sequence of events can be proposed to account for the pacemaker cycle of serotonergic neurons: (1) during a spike, Ca^{2+} enters the cell through voltage-dependent Ca^{2+} channels leading to a large afterhyperpolarization (15–20 mV) due to the Ca^{2+}-activated K^+ conductance; (2) the hyperpolarization not only serves to suppress firing but it also removes the low-threshold Ca^{2+} and A-channels from inactivation; (3) the cell gradually depolarizes during the interspike interval as the afterhyperpolarization fades (secondary to the sequestration and extrusion of Ca^{2+}); (4) when the cell's membrane potential enters the appropriate voltage range (i.e. −60 to −58 mV) the low-threshold Ca^{2+} current triggers a spike; (5) Ca^{2+}, which enters during the spike, once again induces a large afterhyperpolarization to reinitiate the pacemaker cycle.

This scheme points up the general importance of various Ca^{2+} and K^+ conductances in regulating tonic pacemaker activity. Although, to varying degrees, many other types of cells possess similar voltage-activated currents, dorsal raphe cells are somewhat unusual in having such a large afterhyperpolarization (15–20 mV). In addition to rendering cells refractory in the early phase of the interspike interval, the large afterhyperpolarization serves to remove from inactivation two *overlapping* and *competing* voltage-activated currents: I_A, which slows depolarization, and the low-threshold Ca^{2+} current, which accelerates depolarization in the late phases of the interspike interval. By suppressing I_A, the tonic noradrenergic input tips the balance of forces in favor of the low-threshold Ca^{2+} depolarizing prepotential.

IV.2 Is the Dorsal Raphe Pacemaker Purely Endogenous?

In vivo, a reduction or blockade of the noradrenergic input to serotonergic dorsal raphe neurons in the anesthetized animal results in a marked decrease in spontaneous activity (Baraban and Aghajanian, 1980a,b; Baraban *et al*., 1978; Svensson *et al*., 1975). By inference, therefore, it appears likely that the reduced level of spontaneous activity of serotonergic dorsal raphe neurons in brain slices results, at least in part, from a dysfacilitation due to the removal of functional tonic noradrenergic input to these cells. The dysfacilitation occurs despite the fact that serotonin autoreceptor-mediated collateral inhibition is probably reduced in the brain slice in the absence of the initial precursor of serotonin, L-tryptophan. Experimental support for the dysfacilitation hypothesis comes from finding that the addition of norepinephrine or the α_1-agonist phenylephrine restores pacemaker activity to silent dorsal raphe neurons in the brain slice preparation.

If most serotonergic dorsal raphe neurons are dependent upon extrinsic excitatory noradrenergic inputs for their characteristic spontaneous activity to be expressed, does this contradict previous studies suggesting that these neurons may function as 'autonomous pacemakers' (Mosko and Jacobs, 1976) with an 'endogenous rhythm' (Trulson *et al*., 1982) due to the presence of 'pacemaker potentials' (Aghajanian and VanderMaelen, 1982a)? Such a contradiction exists only if one insists that 'endogenous rhythms' and 'pacemaker potentials' must, by definition, be *totally* autonomous (i.e. completely independent of all extrinsic synaptic or neurohumoral influences). However, such a definition would seem excessively restrictive in view of the fact that some invertebrate neurons display prominent and well-described 'pacemaker potentials' only when certain afferent fibers are stimulated (Kupfermann and Kandel, 1970) or only when exposed to certain neurohumoral substances (Barker and Smith, 1976; Heller *et al*., 1980).

For most serotonergic dorsal raphe neurons the manifestation of pacemaker potentials is dependent upon the presence of adequate tonic excitatory influences. This excitation may be provided by noradrenergic synaptic inputs (as occurs *in vivo*), by exogenously applied α-adrenoceptor agonists in brain slices, or possibly by other means not as yet identified. In awake freely moving cats, some spontaneous activity is present in the dorsal raphe nucleus even after administration of high doses of α-adrenoceptor blocking drugs, especially when the animal is aroused by sensory stimulation (Heym *et al*., 1981). It would therefore appear that in awake animals there are additional excitatory or facilitatory influences operating which can maintain spontaneous activity, or at least temporarily restore it, even after pharmacological blockade of the noradrenergic input to serotonergic dorsal raphe neurons.

By isolating the dorsal raphe nucleus from all but its local afferent inputs, the brain slice preparation has given new insights into both the exogenous and endogenous determinants of the raphe neuron pacemaker cycle. Clearly, serotonergic neurons are finely programmed, given sufficient extrinsic excitatory drive, to fire as tonic pacemakers. It remains a matter of speculation as to what physiological significance this highly regulated tonic pacemaker rhythm may have for a wide range of brain functions which have been proposed to be modulated by serotonin (e.g. sleep–waking cycle, affective states, analgesia, motoneuron excitability, sensory processing, and cognition).

REFERENCES

Aghajanian, G.K. (1984) 'The physiology of central a- and B-adrenoceptors', in *Catecholamines: Neuropharmacology and Central Nervous System–Theoretical Aspects* (Eds E. Usdin, A. Carlsson, A. Dahlstrom and J. Engel), pp. 85–92, Alan R. Liss, Inc., New York.

Aghajanian, G.K. (1985). 'Modulation of a transient outward current in serotonergic neurones by α_1-adrenoceptors', *Nature*, **315**, 501–503.

Aghajanian, G.K., and Lakoski, J.M. (1984). 'Hyperpolarization of serotonergic neurons by serotonin and LSD: studies in brain slices showing increased K^+ conductance', *Brain Res.*, **305**, 181–185.

Aghajanian, G.K., and Rasmussen, K. (1989). 'Intracellular studies in the facial motor nucleus illustrating a simple new method for obtaining viable motoneurons in adult rat brain slices', *Synapse*, **3**, 331–338.

Aghajanian, G.K., and VanderMaelen, C.P. (1982a). 'Intracellular recordings from serotonergic dorsal raphe neurons: pacemaker potentials and the effect of LSD', *Brain Res.*, **238**, 463–469.

Aghajanian, G.K., and VanderMaelen, C.P. (1982b). 'Intracellular identification of central noradrenergic and serotonergic neurons by a new double labeling procedure', *J. Neurosci.*, **2**, 1786–1792.

Aghajanian, G.K., and Wang, R.Y. (1978). 'Physiology and pharmacology of central serotonergic neurons', in *Psychopharmacology: A Generation of Progress* (Eds M.A. Lipton, A. DiMascio and K.F. Killam), pp. 171–183, Raven Press, New York.

Aghajanian, G.K., Foote, W.E., and Sheard, M.H. (1968). 'Lysergic acid diethylamide: sensitive neuronal units in the midbrain raphe', *Science*, **161**, 706–708.

Aghajanian, G.K., Foote, W.E., and Sheard, M.H. (1970). 'Action of psychotogenic drugs on single midbrain raphe neurons', *J. Pharmacol. Exp. Ther.*, **171**, 178–187.

Aghajanian, G.K., Haigler, H.J., and Bloom, F.E. (1972). 'Lysergic acid diethylamide and serotonin: direct actions on serotonin-containing neurons', *Life Sci.*, **11**, 615–622.

Anden, N.E., Dahlstrom, A., Fuxe, K., Larsson, K., Olson, L., and Ungerstedt, U. (1966). 'Ascending monoamine neurons to the telencephalon and diencephalon', *Acta Physiol. Scand.*, **67**, 313–326.

Andrade, R., Malenka, R.C., and Nicoll, R.A. (1986). 'A G protein couples serotonin and $GABA_B$ receptors to the same channels in hippocampus', *Science*, **234**, 1261–1265.

Arvidsson, L.-E., Hacksell, U., Nilsson, J.L.G., Hjorth, S., Carlsson, A., Lindberg, P., Sanchez, D., and Wikstrom, H. (1981). '8-hydroxy-2-(di-n-propylamino)tetralin, a new centrally acting 5-hydroxytryptamine receptor agonist', *J. Med. Chem.*, **24**, 921–923.

Asarch, K.B., Ransom, R.W., and Shih, J.C. (1985). '$5\text{-}HT_{1A}$ and $5\text{-}HT_{1B}$ selectivity of two phenylpiperazine derivatives: evidence for $5\text{-}HT_{1B}$ heterogeneity', *Life Sci.*, **36**, 1265–1273.

Baraban, J.M., and Aghajanian, G.K. (1980a). 'Suppression of serotonergic neuronal firing by α-adrenoceptor antagonists: Evidence against GABA mediation', *Eur. J. Pharmacol.*, **66**, 287–294.

Baraban, J.M., and Aghajanian, G.K. (1980b). 'Suppression of firing activity of 5-HT neurons in the dorsal raphe by α-adrenoreceptor antagonists', *Neuropharmacology*, **19**, 355–363.

Baraban, J.M., and Aghajanian, G.K. (1981). 'Noradrenergic innervation of serotonergic neurons in the dorsal raphe: demonstration by electron microscopic autoradiography', *Brain Res.*, **204**, 1–11.

Baraban, J.M., Wang, R.Y., and Aghajanian, G.K. (1978). 'Reserpine suppression of dorsal raphe neuronal firing: mediation by adrenergic system', *Eur. J. Pharmacol.*, **52**, 27–36.

Barker, J.L., and Smith, T.G., Jr (1976). 'Peptide regulation of neuronal membrane properties', *Brain Res.*, **103**, 167–170.

Bramwell, G.J. (1974). 'Factors affecting the activity of 5-HT-containing neurones', *Brain Res.*, **79**, 515–519.

Bramwell, G.J., and Gonye, T. (1976). 'Responses of midbrain neurones to microiontophoretically applied 5-hydroxytryptamine: comparison with the response to intravenously administered lysergic acid diethylamide', *Neuropharmacology*, **15**, 456–461.

Burlhis, T.M., and Aghajanian, G.K. (1987). 'Pacemaker potentials of serotonergic dorsal raphe neurons: contribution of a low threshold Ca^{2+} conductance', *Synapse*, **1**, 582–588.

Carlsson, A., Falck, B., and Hillarp, N.A. (1962). 'Cellular localization of brain monoamines', *Acta Physiol. Scand.*, **196**, 1–28.

Cespuglio, R., Faradji, H., Gomez, M.E., and Jouvet, M. (1981). 'Single unit recordings in the nuclei raphe dorsalis and magnus during the sleep–waking cycle of semi-chronic prepared cats', *Neurosci. Lett.*, **24**, 133–138.

Crunelli, V., Forda, S., Brooks, P.A., Wilson, K.C.P., Wise, J.C.M., and Kelly, J.S. (1983). 'Passive membrane properties of neurones in the dorsal raphe and periaqueductal grey recorded in vitro', *Neurosci. Lett.*, **40**, 263–268.

Dahlstrom, A., and Fuxe, K. (1965). 'Evidence for the existence of monoamine-containing neurons in the central nervous system. I. Demonstration of monoamines in the cell bodies of brainstem neurons', *Acta Physiol. Scand*. Suppl., **232**, 1–55.

deMontigny, C., Blier, P., and Chaput, Y. (1984). 'Electrophysiologically-identified serotonin receptors in the rat CNS', *Neuropharmacology*, **23**, 1511–1520.

Descarries, L., Beaudet, A., Watkins, K.C., and Garcia, S. (1979). 'The serotonin neurons in nucleus raphe dorsalis of adult rat', *Anat. Rec.*, **193**, 520.

Dompert, W.U., Glaser, T., and Traber, J. (1985). '3H-TVX Q 7821: identification of 5-HT_1 binding sites as target for a novel putative anxiolytic', *Naunyn-Schmied. Arch. Pharmacol.*, **328**, 467–470.

Fallon, S.L., Kim, H.S., and Welch, J.J. (1983). 'Electrophysiological evidence for modulation of dopamine neurons by 8-hydroxy-2-(di-n-propylamino)tetralin', *Soc. Neurosci. Abstr.*, **9**, 716.

Felten, D.L., and Harrigan, P. (1980). 'Dendrite bundles in nuclei raphe dorsalis and centralis superior of the rabbit: a possible substrate for local control of serotonergic neurons', *Neurosci. Lett.*, **16**, 275–280.

Freedman, J.E., and Aghajanian, G.K. (1987). 'Role of phosphoinositide metabolites in the prolongation of afterhyperpolarizations by α_1-adrenoceptors in rat dorsal raphe', *J. Neurosci.*, **7**, 3897–3906.

Fuxe, K. (1965). 'Evidence for the existence of monoamine neurons in the central nervous system. IV. Distribution of monoamine nerve terminals in the central nervous system', *Acta Physiol. Scand*. Suppl., **247**, 37–85.

Gilman, A.G. (1987). 'G proteins: transducers of receptor-generated signals', *Ann. Rev. Biochem.*, **56**, 615–649.

Glaser, T., Rath, M., Traber, J., Zilles, K., and Schleicher, A. (1985). 'Autoradiographic identification and topographical analyses of high affinity serotonin receptor subtypes as a target for the novel putative anxiolytic TVX Q 7821', *Brain Res.*, **358**, 129–136.

Green, A.R., and Grahame-Smith, D.G. (1976). '(−)-Propranolol inhibits the behavioural responses of rats to increased 5-hydroxytryptamine in the central nervous system', *Nature*, **262**, 594–596.

Greene, R.W., Haas, H.L., and McCarley, R.W. (1986). 'A low threshold calcium spike mediates firing pattern alterations in pontine reticular neurons', *Science*, **234**, 738–740.

Grzanna, R., and Molliver, M.E. (1980). 'The locus coeruleus in the rat: an immunohistochemical delineation', *Neuroscience*, **5**, 21–40.

Haas, H.L., and Konnerth, A. (1983). 'Histamine and noradrenaline decrease calcium-activated potassium conductance in hippocampal pyramidal cells', *Nature (Lond.)*, **302**, 432–450.

Haigler, H.J., and Aghajanian, G.K. (1974a). 'Peripheral serotonin antagonists: failure to antagonize serotonin in brain areas receiving a prominent serotonergic input', *J. Neural Trans.*, **35**, 257–273.

Haigler, H.J., and Aghajanian, G.K. (1974b). 'Lysergic acid diethylamide and serotonin: a comparison of effects on serotonergic neurons and neurons receiving a serotonergic input', *J. Pharmacol. Exp. Ther.*, **188**, 688–699.

Hamon, M., Cossery, J.M., Spampinato, U., and Gozlan, H. (1986). 'Are there selective ligands for 5-HT_{1A} and 5-H_{1B} receptor binding sites in brain?', *TIPS*, **7**, 336–338.

Heller, E., Kaczmarek, L.K., Hunkapiller, M.W., Hood, L.E., and Strumwasser, F. (1980). 'Purification and primary structure of two neuroactive peptides that cause bag cell afterdischarge and egg laying in Aplysia', *Proc. Nat. Acad. Sci. USA*, **77**, 2328–2332.

Heym, J., Trulson, M.E., and Jacobs, B.L. (1981). 'Effects of adrenergic drugs on raphe unit activity in freely moving cats', *Eur. J. Pharmacol.*, **74**, 117–125.

Heym, J., Trulson, M.E., and Jacobs, B.L. (1982). 'Raphe unit activity in freely moving cats: effects of phasic auditory and visual stimulation', *Brain Res.*, **232**, 29–39.

Innis, R.B., Nestler, E.J., and Aghajanian, G.K. (1988). 'Evidence for G protein mediation of serotonin- and $GABA_B$-induced hyperpolarization of rat dorsal raphe neurons', *Brain Res.*, **459**, 27–36.

Jahnsen, H., and Llinás, R. (1984a). 'Ionic basis for the electroresponsiveness and oscillatory properties of guinea-pig thalamic neurons in vitro', *J. Physiol. (Lond)*, **349**, 227–247.

Jahnsen, H., and Llinás, R. (1984b). 'Electrophysiological properties of guinea-pig thalamic neurones: An in vitro study', *J. Physiol. (Lond)*, **349**, 205–226.

Jung, D., and Moore, G.P. (1966). 'Interspike interval fluctuations in Aplysia pacemaker neurons', *Biophys. J.*, **6**, 411–434.

Katada, T., and Ui, M. (1980). 'Slow interaction of islet-activating protein with pancreatic islets during primary culture to cause reversal of α-adrenergic inhibition of insulin secretion', *J. Biol. Chem.*, **255**, 9580–9588.

Katzung, B.G. (1978). 'Automaticity in cardiac cells', *Life Sci.*, **23**, 1309–1316.

Kupfermann, I., and Kandel, E.R. (1970). 'Electrophysiological properties and functional interconnections of two symmetrical neurosecretory clusters (bag cells) in abdominal ganglion of Aplysia', *J. Neurophysiol.*, **33**, 865–876.

Levitt, P., and Moore, R.Y. (1979). 'Origin and organization of brain stem catecholamine innervation in the rat', *J. Comp. Neurol.*, **186**, 505–528.

Llinás, R.R. (1984). 'Rebound excitation as the physiological basis for tremor: A biophysical study of the oscillatory properties of mammalian central neurones in vitro' in *Movement Disorders: Tremor* (Eds L.J. Findley and R. Capildeo), pp. 165–182, Oxford University Press, Oxford.

Llinás, R.R., and Yarom, Y. (1981a). 'Properties and distribution of ionic conductances generating electroresponsiveness of mammalian inferior olivary neurones in vitro', *J. Physiol. (Lond)*, **315**, 569–584.

Llinás, R.R., and Yarom, Y. (1981b). 'Electrophysiology of mammalian inferior olivary neurones in vitro. Different types of voltage-dependent ionic conductances', *J. Physiol. (Lond)*, **315**, 549–567.

Llinás, R.R., Greenfield, S.A., and Jahnsen, H. (1984). 'Electrophysiology of pars compacta cells in the in vitro substantia nigra – a possible mechanism for dendritic release', *Brain Res.*, **294**, 127–132.

Madison, D.V., and Nicoll, R.A. (1982). 'Noradrenaline blocks accommodation of pyramidal cell discharge in the hippocampus', *Nature (Lond.)*, **299,** 636–638.

McGinty, D.J., and Harper, R.M. (1976). 'Dorsal raphe neurons: depression of firing during sleep in cats', *Brain Res.*, **101**, 569–575.

Middlemiss, D.N., and Fozard, J.R. (1983). '8-hydroxy-2-(di-n-propylamino)-tetralin discriminates between subtypes of the 5-HT_1 recognition site', *Eur. J. Pharmacol.*, **90**, 151–153.

Mollgard, K., and Moller, M. (1975). 'Dendrodendritic gap junctions: a developmental approach', in *Advances in Neurology. Physiology and Pathology of Dendrites* (Ed. G.W. Kreutzberg), pp. 79–89, Raven Press, New York.

Moore, G.P., Segundo, J.P., Perkel, D.H., and Levitan, H. (1970). 'Statistical sign of synaptic interaction in neurones', *Biophys. J.*, **10**, 876–900

Mosko, S.S., and Jacobs, B.L. (1974). 'Midbrain raphe neurons: spontaneous activity and response to light', *Physiol. Behav.*, **13**, 589–593.

Mosko, S.S., and Jacobs, B.L. (1976). 'Recording of dorsal raphe unit activity in vitro', *Neurosci. Lett.*, **2**, 195–200.

Nakahama, H., Shima, K., Yamamoto, M., and Kojiro, A. (1981). 'Regularity of the spontaneous discharge of neurons in the nucleus raphe dorsalis of the cat', *Neurosci. Lett.*, **23**, 161–165.

Perkel, D.H., Gerstein, G.L., and Moore, C.P. (1967). 'Neuronal spike trains and stochastic point processes. II. Simultaneous spike trains', *Biophys. J.*, **7**, 419–440.

Peroutka, S.J. (1985). 'Selective interaction of novel anxiolytics with 5-hydroxytryptamine$_{1A}$ receptors', *Biol. Psychiatry*, **20**, 971–979.

Puizillout, J.J., Gaudin-Chazal, G., Daszuta, A., Seyfritz, N., and Ternaux, J.P. (1979). 'Release of endogenous serotonin from "encephale isole" cats. II. Correlations with raphe neuronal activity and sleep and wakefulness', *J. Physiol. Paris*, **75**, 531–537.

Ransom, R.W., Asarch, K.B., and Shih, J.C. (1986). [^{3}H]1-[2-(4-aminophenyl)ethyl]-4-(3-trifluoromethylphenyl) piperazine: a selective radioligand for 5-HT_{1A} receptors in rat brain', *J. Neurochem.*, **46**, 68–75.

Rothman, S.M. (1985). The neurotoxicity of excitatory amino acids is produced by passive chloride influx. *J. Neurosci.*, **5**, 1483–1489.

Saavedra, J.M., Grobecker, H., and Zivin, J. (1976). 'Catecholamines in the raphe nuclei of the rat', *Brain Res.*, **114**, 339–345.

Segal, M. (1985). 'A potent transient outward current regulates excitability of dorsal raphe neurons', *Brain Res.*, **359**, 347–350.

Sills, M.A., Wolfe, B.B., and Frazer, A. (1984). 'Determination of selective and nonselective compounds for the 5-HT_{1A} and 5-HT_{1B} receptor subtypes in rat frontal cortex', *J. Pharmacol. Exp. Ther.*, **231**, 480–487.

Sprouse, J.S., and Aghajanian, G.K. (1986). '(−)-Propranolol blocks the inhibition of serotonergic dorsal raphe cell firing by 5-HT_{1A} selective agonists', *Eur. J. Pharmacol.*, **128**, 295–298.

Sprouse, J.S., and Aghajanian, G.K. (1987). 'Electrophysiological responses of serotonergic dorsal raphe neurons to 5-HT_{1A} and 5-HT_{1B} agonists', *Synapse*, **1**, 3–9.

Steinbusch, H.W.M. (1981). 'Distribution of serotonin-immunoreactivity in the central nervous system of the rat-cell bodies and terminals', *Neuroscience*, **6**, 557–618.

Sternweis, P.C., and Robishaw, J.D. (1984). 'Isolation of two proteins with high affinity for guanine nucleotides from membranes of bovine brain', *J. Biol. Chem.*, **259,** 13806–13813.

Stryer, L., and Bourne, B.H.R. (1986). 'G proteins: a family of signal transducers', *Ann. Rev. Biochem.*, **2**, 391–425.

Svensson, T.H., Bunney, B.S., and Aghajanian, G.K. (1975). 'Inhibition of both noradrenergic and serotonergic neurons in brain by the α-adrenergic agonist, clonidine', *Brain Res.*, **92**, 291–306.

Tricklebank, M.D., Forler, C., and Fozard, J.R. (1984). 'The involvement of subtypes of the 5-HT_1 receptor and of catecholaminergic systems in the behavioural response to 8-hydroxy-(2-di-n-propylamino) tetralin in the rat', *Eur. J. Pharmacol.*, **106**, 271–282.

Trulson, M.E., and Jacobs, B.L. (1979). 'Raphe unit activity in freely moving cats: correlation with level of behavioral arousal', *Brain Res.*, **163**, 135–150.

Trulson, M.E., Howell, G.A., Brandstetter, J.W., Frederickson, M.H., and Frederickson, C.J. (1982). 'In vitro recording of raphe unit activity: evidence for endogenous rhythms in presumed serotonergic neurons', *Life Sci.*, **31**, 785–790.

Ungerstedt, U. (1971). 'Stereotaxic mapping of the monoamine pathways in the rat brain', *Acta Physiol. Scand.*, **367**, 1–48.

VanderMaelen, C.P., and Aghajanian, G.K. (1983). 'Electrophysiological and pharmacological characterization of serotonergic dorsal raphe neurons recorded extracellularly and intracellularly in rat brain slices', *Brain Res.*, **289**, 109–119.

VanderMaelen, C.P., Matheson, G.K., Wilderman, R.C., and Patterson, L.A. (1986). 'Inhibition of serotonergic dorsal raphe neurons by systemic and iontophoretic administration of buspirone, a non-benzodiazepine anxiolytic drug', *Eur. J. Pharmacol.*, **129**, 123–130.

Verge, D., Daval, G., Patey, A., Gozlan, H., El Mestikawy, S., and Hamon, M. (1985). 'Presynaptic 5-HT autoreceptors on serotonergic cell bodies and/or dendrites but not terminals are of the 5-HT_{1A} subtype', *Eur. J. Pharmacol.*, **113**, 463–464.

Versteeg, D.H.G., VanDerGugten, J., DeJong, W., and Palkovits, M. (1976). 'Regional concentrations of noradrenaline and dopamine in rat brain', *Brain Res.*, **113**, 563–574.

Wang, R.Y., and Aghajanian, G.K. (1977). 'Antidromically identified serotonergic neurons in the rat midbrain raphe: evidence for collateral inhibition', *Brain Res.*, **132**, 186–193.

Wang, R.Y., and Aghajanian, G.K. (1982). 'Correlative firing patterns of serotonergic neurons in rat dorsal raphe nucleus', *J. Neurosci.*, **2**, 11–16.

Wilkinson, L.O., Abercrombie, E.D., Rasmussen, K., and Jacobs, B.L. (1987). 'Effect of buspirone on single unit activity in locus coeruleus and dorsal raphe nucleus in behaving cats', *Eur. J. Pharmacol.*, **136**, 123 127.

Williams, J.T., Colmers, W.F., and Pan, Z.Z. (1988). 'Voltage- and ligand-activated inwardly rectifying currents in dorsal raphe neurons in vitro', *J. Neurosci.*, **8**, 3499–3506.

Yarom, Y., Sugimori, M., and Llinás, R. (1985). 'Ionic currents and firing patterns of mammalian vagal motoneurons *in vitro*', *Neuroscience*, **16**, 719–737.

Yatani, A., Codina, J., Brown, A.M., and Birnbaumer, L. (1987). 'Direct activation of mammalian atrial muscarinic potassium channels by GTP regulatory protein G_k, *Science*, **235**, 207–211.

Preparations of Vertebrate Central Nervous System *In Vitro*
Edited by H. Jahnsen
Published by John Wiley & Sons Ltd

CHAPTER 3

Use of Slices for Quantitative Pharmacology

M.A. SIMMONDS
Department of Pharmacology, School of Pharmacy, University of London, 29/39 Brunswick Square, London WC1N 1AX, UK

Quantitative pharmacology is fundamentally concerned with the measurement of cellular responses evoked by interactions of receptor systems with drugs and physiological mediators. Progressively, molecular biology is giving us much clearer ideas of the structures of membrane-spanning proteins that contain drug/transmitter recognition sites. However, this has not yet reached the stage where the immediate effect of a drug on the receptor protein structure can be determined directly. Perhaps the nearest approach is to monitor the opening and closing of single ion channels for which the molecular structure is also known, e.g. channel proteins that incorporate $GABA_A$ or glycine or acetylcholine nicotinic receptors (Barnard *et al.*, 1987; Bormann, 1988; Levitan *et al.*, 1988). Nevertheless, for most pharmacological studies, the chemistry that transduces drug–receptor interactions into observable responses remains insufficiently defined and we can only represent such interactions in the form of theoretical models.

The 'occupancy theory' of drug–receptor interaction was developed by Gaddum and Schild more than 40 years ago from the earlier work of Langley, Hill and Clark (see Colquhoun, 1973). Out of this arose the traditional procedures of quantitative pharmacology based on an assumed relationship between the response of a tissue to a drug and receptor occupancy by the drug. These procedures still remain useful, even though we now know much more about the complexity of the multiple stages involved between receptor occupation and the measured response. Clearly, there is advantage in reducing these multiple stages to a minimum but, generally speaking, this involves an increase in the technical difficulty of the measurements. There is thus a trade-off between technical convenience and minimizing the problems of data interpretation when the end-responses of complex systems are measured. I shall explore the nature of this trade-off in the use of slices of CNS tissue for quantitative studies.

I INTERPRETATION OF DOSE–RESPONSE RELATIONSHIPS

The occupancy theory of drug–receptor interactions provides a basis for the use of log dose–response curves for agonists and predictions of how those curves will be affected in the presence of different types of antagonist. It is based on some important assumptions that are difficult to achieve in practice and to which due consideration must be given when data are interpreted.

Among the initial assumptions are: (1) that the response of a tissue to an agonist results from an interaction with a uniform population of receptors; (2) that the agonist has equal access to all receptors; (3) that the maximum response is achieved only when all receptors are occupied by the agonist; (4) that the agonist concentration in the vicinity of the receptors is known. In practice, these conditions are rarely met. Administration of an agonist for short periods of time to a slice preparation can be done either by superfusion of the whole slice or by more localized application within the slice by ionophoresis or pressure-ejection from a micropipette. In the first case, there will be a concentration gradient of agonist into the slice and it is unlikely that a uniform concentration will be achieved throughout the extracellular space of the slice during the usual agonist contact time of not more than 2 min, especially if active removal processes such as uptake into neurons or glial cells and metabolism are operative (Garthwaite, 1985). Likewise, local application of agonist from a micropipette will set up a concentration gradient from the point of release which will be subject to similar influences. Despite this, individual responses can appear to reach a plateau, suggesting that an equilibrium is attained. Such an equilibrium is likely to have a complicated mixture of factors contributing to it. Thus, receptors exposed to the high end of the agonist concentration gradient will contribute more to the response than receptors exposed to lower concentrations; however, at the higher concentrations, there is a greater chance that desensitization will diminish contributions to the response.

When an increased dose of agonist is applied, either by increasing the superfusion concentration or by increasing the rate of ejection from a micropipette, it will diffuse a greater distance in the tissue and achieve suprathreshold concentrations in the vicinity of a larger population of receptors. The increased response will, therefore, be due to an increase in the receptor population reached by the agonist as well as to an increased likelihood of individual receptor occupation by the agonist. At the same time, the contribution to the response by receptors close to the source of the agonist would be more likely to diminish through desensitization. It is possible that the latter phenomenon could sufficiently limit the increase in response to an increased dose of agonist that the log dose–response curve appears to reach a maximum.

Caution will, therefore, be required in the interpretation of the maxima of log dose–response curves obtained from tissue slices. This, in turn, affects the interpretation of the slopes of the curves. The relationship between dose and effect, as derived from occupancy theory, may be expressed as

$$r = [D]^n/([D]^n + K_D)$$

where r is the response expressed as a fraction of the maximum response, $[D]$ is the molar concentration of agonist, n is the number of molecules of agonist required to combine with a receptor to activate it, and K_D is the equilibrium dissociation constant. In the simplest situation when $n = 1$, it can be calculated that r should increase from 0.1 to 0.9 over an 81-fold increase in agonist concentration. If $n = 2$, the same increase in response should occur over a nine-fold increase in agonist concentration. The value of n can be determined from the so-called Hill plot of

$$\log \{r/(1 - r)\} \text{ versus log concentration}$$

in which the slope = n. This plot requires a reliable estimate of maximum response that represents full occupation of the available receptors, a condition that it is very difficult to be sure of achieving in practice.

When a value of $n = 2$ is obtained for an agonist, the agonist–receptor interaction may be written as

$$\begin{array}{ccccc} D + D + R & \rightleftharpoons & DR + D & \rightleftharpoons & D_2R \\ & & \updownarrow & & \updownarrow \\ & & DR^* & & D_2R^* \end{array}$$

where D = agonist, R = receptor in the resting state and R^* = receptor isomerized into the active state. The presumption for $n = 2$ is that DR^* is a much less frequently occurring form than D_2R^*.

II DOSE–RESPONSE RELATIONSHIPS IN PRACTICE

In recent years, the grease-gap arrangement has proved to be quite useful for recording population responses of neurons to drugs that alter membrane potential. It has been used with the roots of hemisected spinal cord from the immature rat

(Evans, 1978), the dorsal funiculus in slices of rat cuneate nucleus (Simmonds, 1980a), segments of rat optic nerve (Simmonds, 1983a), pyramidal cell projections in slices of cerebral cortex from the rat (Harrison and Simmonds, 1985) and the mouse (Burton *et al*., 1988), Purkinje cell and granule cell parallel fibre projections in rat cerebellar slices (Garthwaite *et al*., 1986) and the alvear projection of CA1 neurons in slices of rat hippocampus (Blake *et al*., 1988).

A typical arrangement is shown in Figure 1 for the cuneate nucleus slice on which extensive quantitative studies of the $GABA_A$ receptor complex have been performed (e.g. Pickles and Simmonds, 1980; Simmonds, 1980a, 1981, 1982; Harrison and Simmonds, 1983, 1984; Turner and Simmonds, 1989). The principal requirement is for a reasonably dense projection of axons within the slice which is arranged in the perfusion bath so that the axons pass through a grease-filled gap in a barrier separating the two compartments of the bath. Each compartment is separately perfused and the potential difference between the compartments is recorded continuously on a chart recorder via Ag–AgCl electrodes connected to the high-impedance probe of a preamplifier. Drugs are normally perfused through one compartment only to access the appropriate end of the axonal projection.

Drugs that depolarize neurons cause increased negativity in the drug-containing compartment. The probable explanation for this is that membrane depolarization results in a flow of current into the neurons and along the inside of the axons and then out again at the points of least resistance. Some of the outflow points will be on the other side of the barrier from the drug-induced current inflow so that the return current must pass through the restricted interstitial space between the axons in the grease-gap. This represents an electrical resistance of several KΩ and the passage of current sets up a potential difference across the barrier. Thus, the potential difference recorded in response to a drug will depend upon a number of factors. These include: (1) drug concentration in the perfusate; (2) the density of receptors exposed to the drug on each neuronal element projecting through the grease-gap; (3) the difference between resting membrane potential and the reversal potential for the drug effect; (4) the extent of membrane conductances tending to stabilize membrane potential locally; (5) the space constant of the axons which will determine what proportion of the current flowing into the axons will flow out of them on the other side of the barrier; (6) the number of axons projecting through the barrier; (7) the electrical resistance of the extracellular path through the grease seal. Apart from drug concentration, which will be deliberately varied, all the other factors should, ideally, remain constant during an experiment. This requires good mechanical stability of the tissue projecting through the grease seal and perfusion conditions that allow membrane properties and ionic distributions to be fully recovered following each application of agonist. For optimal signal amplitude, attention should be given to the plane and thickness of the slice so that the number of viable axons projecting through the grease-gap is maximized.

The projecting axon may be the terminal projection of a pathway, as in the case of the dorsal funiculus projecting into the cuneate nucleus. It can also be the

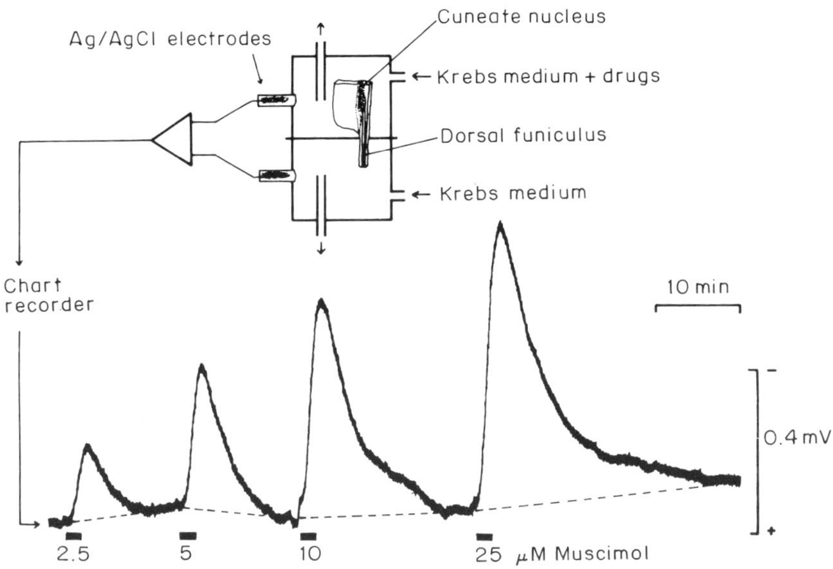

Figure 1 Diagram (not to scale) of a two-compartment Perspex bath system used for grease-gap recordings. The compartments are separated by a removable barrier which is made in two pieces; the bottom piece contains a 1-mm square slot cut into the centre of its upper edge and the top piece has a straight lower edge. The 1 mm thickness of the barrier is machined down to 0.5 mm in the region of the slot in both the top and bottom pieces. The tissue is placed through the barrier as follows: the bottom piece of the barrier is put in place, the slot having been packed with silicone grease and its upper edge lightly greased; the meniscus of the medium is raised just above the barrier so that the slice can be floated into position over the slot; the top piece of the barrier is then brought down to ease the tissue into the grease in the slot and make a watertight contact with the bottom piece of the barrier. The meniscus is then raised on both sides to submerge the tissue. Each compartment can be perfused by pumping medium into the bottom and sucking off at the meniscus; alternatively, the medium can be dripped onto the meniscus and removed by a filter paper syphon arrangement. The latter perfusion method has been adopted by some workers to avoid the use of a roller pump, which can induce a fluctuating charge on the medium and, thus, interfere with the recordings. Such interference can be avoided with the former perfusion system by incorporating sealed air-breaks into the perfusion lines to isolate electrically the medium passing through the pump from the medium in the bath. A convenient size of bath will contain 1.5–2.0 ml of medium and a flow rate of about 2 ml/min has been found adequate for most purposes. The potential difference between the two compartments is recorded continuously with Ag–AgCl electrodes embedded in agar/saline via a high-impedance, low-gain amplifier onto a chart recorder. The typical trace shown is from a rat cuneate nucleus slice exposed to 2-min superfusions of muscimol

proximal axonal projection from a dense grouping of cell bodies, as in the cortical slice. It may even be a mid-segment of nerve, as in the optic nerve preparation. In the latter case, it is presumed that the cut ends of the myelinated axons reseal and that the functional glycine receptor–ion channel complexes that are present on the exposed resealed ends of these particular axons were either originally present in the membrane beneath the myelination or were inserted from an intracellular location during the resealing process. A similar situation is found with the caudal cut ends of the dorsal funiculus in the cuneate nucleus slice. An intriguing difference between these two preparations, however, is that the substantial population of glycine receptors on the optic nerve is accompanied by a rather sparse population of $GABA_A$ receptors whereas, on the dorsal funiculus, there is a substantial population of $GABA_A$ receptors accompanied by only a sparse population of glycine receptors (Simmonds, 1983a).

If a nondescript piece of CNS tissue is placed in the grease-gap, no convincing responses to amino acids are obtained, presumably due to the lack of sufficient neurons that bear amino acid receptors projecting through the gap. However, responses to elevations of K^+ by 3 or 6 mM are almost always obtained. This probably represents glial depolarizations and a degree of electrotonic coupling between glial cells to form a conducting pathway through the grease seal (Cohen, 1970). Thus, even when a neuronal projection through the grease seal is present, responses to K^+ may substantially arise from glial cells.

So far, then, there appears to be little difficulty arising from the recording technique that would complicate the interpretation of log dose–response curves. But, because of the anticipated difficulties discussed earlier that relate to the distribution of applied agonist within a slice preparation, little detailed analysis has been made of the slopes of agonist log dose–response curves. One notable exception is the recent publication by Williams *et al.* (1988) in which an attempt has been made to fit responses of mouse neocortical slices to excitatory amino acids into the theoretical framework of occupancy theory described above. Responses to a range of concentrations of the excitatory amino acids *N*-methyl-D-aspartate (NMDA) and quinolinate, which were stated to extend up to a clear maximal level, yielded a Hill slope of about 2 for each agonist. This was interpreted as being consistent with a receptor–agonist interaction in which two molecules of agonist are required to be bound simultaneously for receptor activation. I have, therefore, now taken a closer look at some of our own log dose–response curves for inhibitory as well as excitatory amino acids to determine whether convincing interpretations of Hill plots can be made.

Figure 2a shows typical log dose–response curves to the $GABA_A$ receptor agonists muscimol and GABA obtained from the dorsal funiculus terminals in slices of rat cuneate nucleus with the grease-gap apparatus shown in Figure 1. It is obvious that the curve for muscimol is steeper than that for GABA. The data are replotted in Figure 2b as Hill plots from which Hill numbers are obtained by calculating the slopes of the steepest portions of the plots, i.e. within the range ± 1.0 on the vertical axis (see Podleski, 1973). The Hill number for muscimol is 1.80 and that for GABA is 0.88. Referring back to the agonist–receptor interaction scheme set out earlier, it is not impossible that, with appropriate rate constants, D_2R^* could occur much more frequently than DR^* for one agonist, muscimol in this case, whilst for another agonist, GABA, the reverse could be the case. Nevertheless,

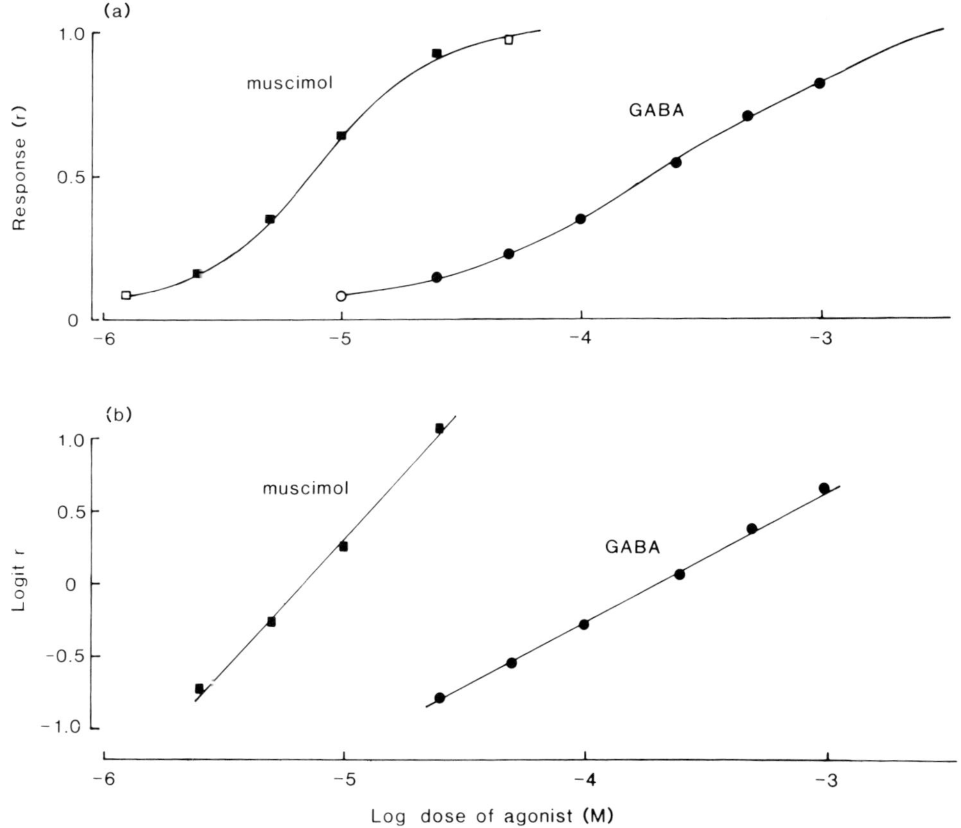

Figure 2 (a) Typical log dose–response curves to GABA and muscimol obtained from dorsal funiculus terminals in slices of rat cuneate nucleus. Each curve was obtained on a different slice and the responses are plotted as fractions of the maximum response. The maxima were estimated from Hofstee plots of the data. (b) Hill plots of logit r versus log dose of GABA or muscimol for the data shown as filled symbols in (a), where r = fractional response and logit $r = \log(r/1 - r)$. The slopes are 0.88 for the GABA example and 1.80 for the muscimol example

before such a proposition can be seriously considered, alternative explanations for the different Hill numbers should be examined.

The first question to be addressed is which, if either, of these Hill numbers can be reliably interpreted. In one respect, the muscimol data should be more reliable, since GABA is much more susceptible to removal by uptake processes than is muscimol and this might account for the shallower slope with GABA. On the other hand, desensitization of responses may truncate the log dose–response curves for both GABA and muscimol so that they extend over shortened dose ranges and thereby yield artificially high Hill numbers. However, using the 'concentration-clamp' technique, which minimizes the influence of uptake and desensitization, Akaike *et al*. (1986) found a Hill number of 2 for GABA on frog sensory neurons. The same considerations that apply to GABA might also apply to glycine. The Hill plot derived from a previously published glycine log dose–response curve from the rat optic nerve (Simmonds, 1983a) yields a Hill number of 1.94. In this case, the value closely matches the Hill number of 1.8 found for glycine-induced currents in rat isolated ventromedial hypothalamic neurons by Tokutomi *et al*. (1989) with the 'concentration-clamp' technique.

For excitatory amino acids, log dose–response curves have been obtained on slices of rat cerebral cortex. Responses to NMDA (Figure 3a) have been obtained in Krebs medium in the absence of Mg^{2+}, which is a non-competitive antagonist of this response. These data yield a Hill number of 1.09. This is clearly at variance with the Hill number of 1.97 reported by Williams *et al*. (1988) for similar experiments on neocortical slices from the mouse. Furthermore, in the rat preparation, quisqualate and AMPA (2-amino-3-hydroxy-5-methyl-4-isoxazole-propionic acid), which are both agonists at the 'quisqualate receptor', produce log dose–response curves that have clearly different slopes from each other (Horne and Simmonds, 1989). Typical examples from these data are presented as Hill plots in Figure 3b. In a series of four experiments with each agonist, Hill numbers of 0.84 ± 0.05 for quisqualate and 1.90 ± 0.14 for AMPA (mean $\pm$ SEM) were obtained (Horne and Simmonds, 1989). There is some indirect evidence that quisqualate may be susceptible to removal by uptake (Lodge *et al*., 1980) and this might, therefore, affect the Hill number obtained. Since the absolute values of maximal response to the two agonists on the same preparation were only slightly less for quisqualate than for AMPA, it would appear that any differences in response desensitization were insufficient to explain the Hill numbers.

Perhaps the parameter that lies behind the distinctly different Hill numbers for pairs of amino acids that act at the same receptor is agonist 'efficacy'. If it is assumed that a maximal response is obtained only when all the available receptors are occupied, an agonist of low efficacy should give a low maximal response. On the other hand, if the observed maximal response is limited by other factors, such as desensitization of receptors on superficial neurons, no difference may be apparent between the depressed maximal response to agonists of differing efficacy. However, the slope of the log dose–response curve for an agonist of high efficacy

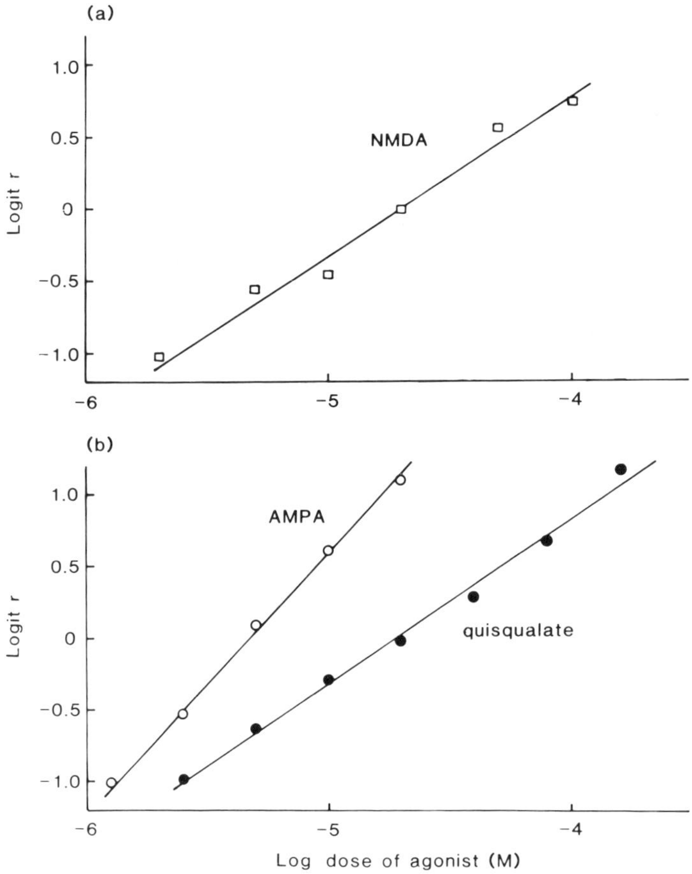

Figure 3 (a) Hill plot of the averaged data from two dose response curves obtained with NMDA on slices of rat cerebral cortex in Mg^{2+}-free Krebs medium. The maxima were estimated from Hofstee plots of the data. The slope of the Hill plot is 1.09. (b) Hill plots of the data from typical dose–response curves to quisqualate and AMPA on slices of rat cerebral cortex. In both cases, convincing maxima were obtained experimentally. The slopes of the Hill plots are 1.14 for quisqualate and 1.83 for AMPA

in this situation might appear steeper than that for an agonist of lower efficacy. This would yield a higher Hill number for the former than for the latter and both Hill numbers would be greater than the true stoichiometry of the agonist–receptor combination required for receptor activation.

There still remains, however, the possibility mentioned earlier of a true difference in stoichiometry for different agonists acting at the same receptor. Furthermore, the model would allow intermediate values between 1 and 2, depending on the ratio of DR^* to D_2R^*, together with the possibility that these values could be subject to modulatory influences acting on the protein that carries the agonist recognition site.

Unfortunately, the difficulty of distinguishing between all the different factors that might affect the slopes of dose–response relationships in slices leads me to conclude that, in this type of study, the Hill numbers obtained cannot be reliably interpreted in terms of the stoichiometry of the agonist–receptor interaction. Despite this, agonist log dose–response relationships in slices can still be used for studies of the effects of antagonists and potentiators, provided that the quantitative interpretations take into account the uncertainties just discussed.

III STUDIES WITH ANTAGONISTS

It is not difficult to conceive of several different mechanisms by which antagonist drugs might have their effects. The classical concept of competitive antagonism requires that the antagonist and the agonist mutually displace each other either from the same binding site or from allosteric sites. The theoretical model of such antagonism in terms of the occupancy theory predicts a quantitative relationship between antagonist concentration and the degree of antagonism that should be obtained (Arunlakshana and Schild, 1959) and provides a basis for calculating antagonist potency (affinity). Other types of antagonism may be defined theoretically as non-competitive and uncompetitive but, in practice, rarely conform to all the assumptions underlying the models. The cardinal features of non-competitive antagonism are that antagonist binding and agonist binding to their separate recognition sites are independent of each other and that antagonist binding prevents the isomerization of the agonist–receptor combination that is necessary to initiate a response. An uncompetitive antagonist, on the other hand, will bind to its recognition site and exert its antagonistic effect only when the agonist is already bound to its receptor. Such a situation occurs with antagonists that are open-channel blockers where not only is agonist required to bind to the receptor but also the agonist–receptor complex is required to isomerize to open the ion channel before the antagonist can access its binding site within the channel. The consequent curtailment of the response to the agonist is often described as ‘use-dependent’. With regard to quantitative studies on slices of CNS tissue, it is of interest to see how these different types of antagonism can be identified and characterized.

III.1 Competitive Antagonism

This requires that agonist log dose–response curves be shifted to the right in a parallel fashion with no reduction in the maximal response. However, these criteria alone are not reliably predictive of competitive antagonism. I pointed out earlier some of the factors that could influence the observed maximal response to agonist. Interaction of antagonists with one or more of these factors may explain certain unexpected findings. For example, the $GABA_A$ antagonist bicuculline and the glycine antagonist strychnine each can cause *elevations* in the maximal responses to GABA and glycine, respectively (Simmonds, 1978, 1983a), at the same time as causing antagonism seen as a shift of the agonist log dose–response curve to the right (Figure 4). The explanation offered for these effects was a possibly reduced degree of desensitization to agonist in the presence of a competitive antagonist. (It will be seen later that certain potentiators of muscimol do the reverse and reduce the maximal response whilst shifting the log concentration–response curve to the left.)

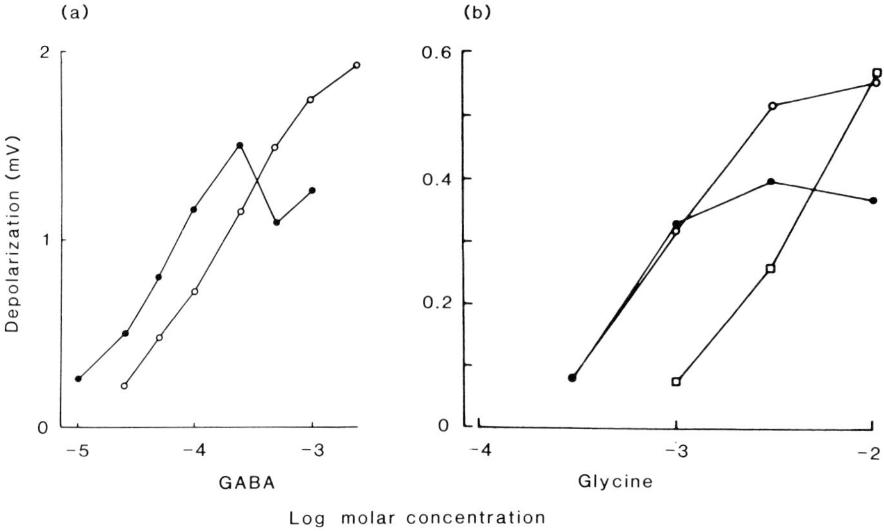

Figure 4 (a) Effect of (+)-bicuculline on the log dose–response curve to GABA in a slice of rat cuneate nucleus. Each point represents a single response. Compared with the control data (●), the curve in the presence of 10^{-6} M bicuculline (○) is shifted to the right, indicating antagonism, but the maximum is increased. (Taken from Simmonds, 1978.) (b) Effect of strychnine on the log dose–response curve to glycine in the rat isolated optic nerve. Each point represents a single response. Compared with the control data (●), the curve in the presence of 10^{-7} M strychnine (○) was not shifted laterally but the maximum was increased. In the presence of 10^{-6} M strychnine (□), the curve was shifted to the right and the increased maximum was sustained. (Taken from Simmonds, 1983a)

With respect to the parallel shift of the log dose–response curve, all that this signifies on its own is that the antagonism is surmountable by increasing the dose of agonist. In practice it is found that antagonists which are not competitive can, nevertheless, cause a parallel shift. This presumably results from the situation described earlier whereby an increase in dose of the agonist allows the agonist to diffuse further into the tissue and achieve effective concentrations at an increased population of receptors. Thus, without displacing antagonist, the response may be restored. It might be anticipated, however, that the quantitative relationship between the concentration of a non-competitive antagonist and the degree of the parallel shift to the right of the agonist log dose–response curve would not conform with the predictions of occupancy theory for a competitive antagonist. Such a distinction should become apparent if Schild plots are constructed.

The equation for the Schild plot derived from occupancy theory is

$$\log(\mathrm{dr}^n - 1) = \log[\text{antagonist}] - \log K_A$$

where dr is the agonist dose ratio for equal responses in the presence and absence of antagonist, n is the stoichiometry of the agonist–receptor interaction for activation of the receptor complex and K_A is the dissociation constant of the antagonist. Thus, the regression of $\log(\mathrm{dr}^n - 1)$ against log [antagonist] should yield a slope of 1 with K_A given by the concentration of antagonist that yields a value of $\mathrm{dr}^n = 2$. This assumes a stoichiometry of 1 for the competitive antagonist–receptor interaction necessary to antagonize the response to agonist, whatever may be the value of n for the agonist stoichiometry. In other words, only one of the n molecules of agonist needs to be displaced by antagonist to stop receptor activation. Although this seems entirely logical, provision can be made in the Gaddum–Schild equation for an antagonist receptor stoichiometry of more than 1, the value being given by the slope of the Schild plot (see Williams *et al*., 1988).

In my studies of antagonists of the $\mathrm{GABA_A}$ receptor complex in slices of rat cuneate nucleus (Simmonds, 1980a, 1982), the stoichiometry of the agonist–receptor interaction was assumed to be 1. Experiments with bicuculline yielded a linear Schild plot that had a slope close to 1 (1.07 ± 0.032) provided that muscimol was used as the agonist. When GABA was used as the agonist, the Schild plot had a slope of less than 1 over most of its range and only at low levels of antagonism did the slope approximate to 1. The explanation given for this was that uptake processes for GABA were distorting the assumed constant relationship between agonist concentration in the superfusing medium and the concentration in the vicinity of the receptors. Since the value of n for GABA is no greater than 1, as shown in Figure 2b, this still seems to be the most plausible explanation. With respect to the data based on responses to muscimol, however, the question does now arise as to whether the assumption of $n = 1$ for muscimol was justified since the Hill slope shown in Figure 2b indicates a value of n approaching 2. I have, therefore, re-calculated the Schild plot for bicuculline antagonism of muscimol with a value of $n = 2$ inserted into the equation. As might be anticipated, this increased the slope of the Schild plot and, in fact, yielded a reasonably linear relationship with a slope of 1.77. The value of the apparent dissociation constant (K_A) for bicuculline was also substantially changed. In the original experiments, the pA_2 value obtained was 5.98, which translates into an apparent K_A of 1.05×10^{-6} M. As a result of the re-calculation setting $n = 2$ for muscimol and taking into account the slope of the Schild plot, the apparent K_A for bicuculline becomes 6.89×10^{-12} M. This seems unrealistic when compared with data from binding studies indicating a dissociation constant for bicuculline in the nanomolar range, e.g. 6.8×10^{-8} M for bicuculline displacement of [^{3}H]bicuculline methiodide binding to $GABA_A$ receptors (Möhler and Okada, 1977). Likewise, other antagonists that share a common site of action with bicuculline and yielded Schild slopes close to 1 when the value of n for muscimol was assumed to be 1 also have their apparent affinities substantially increased when the data are re-calculated with $n = 2$ for muscimol. Indeed, for strychnine, which has reasonable selectivity as a glycine antagonist (Krnjević, 1974), this yields an apparent affinity for the $GABA_A$ receptor which is considerably greater than the binding affinity of [^{3}H]strychnine for the glycine receptor. Such a nonsensical outcome, taken together with the difficulty of envisaging why an antagonist stoichiometry of greater than 1 should be necessary when displacement of just one of the two molecules of agonist should be sufficient for antagonism, strongly suggests that something is wrong. Since the problem appeared when a muscimol stoichiometry of 2 was used, it looks as if it is this value which is incorrect. A similar critique could be made of the analysis by Williams *et al.* (1988) of the antagonism of NMDA responses by kynurenate and 2-amino-5-phosphonopentanoic acid (AP5) in slices of mouse neocortex. When a stoichiometry of 2 was accepted for NMDA, Schild slopes of 2.24 and 1.83 were obtained for kynurenate and AP5, respectively. This again raises doubts as to the correctness of the apparent stoichiometry of the agonist, NMDA in this case.

In contrast, it is much more tempting to accept the Hill slope of about 2 calculated earlier for strychnine-sensitive glycine receptors on rat optic nerve. This is because the original assumption of $n = 1$ for glycine yielded a Schild plot for strychnine antagonism of glycine that had a slope well below 1 at glycine dose ratios in excess of 4 (Simmonds, 1983a). When these data are re-calculated with a value of $n = 2$ for glycine, the Schild slope becomes linear over its entire range with a slope of 1.04 (Figure 5). This makes strychnine appear to be more convincingly competitive with a strychnine–receptor stoichiometry of 1 and an apparent K_A of 8.3×10^{-8} M rather than the value of 2.6×10^{-7} M obtained when $n = 1$ was used for glycine. The new K_A value is slightly closer to the dissociation constant of 7.0×10^{-9} M for [^{3}H]strychnine binding to glycine receptors (Braestrup *et al.*, 1986).

Since uncertainty over the true stoichiometry of the agonist–receptor interaction, in turn, creates uncertainty over the values derived from the Schild plots for the apparent dissociation constants of competitive antagonists, there seems to be little justification in going beyond the general practice of assuming an agonist stoichiometry of 1. Consequently, it must be accepted that the pA_2 values obtained may not be accurate estimates of the antagonists' dissociation constants although they remain useful for comparing antagonist potencies. It is, however, reassuring to note the consistency of the pA_2 values obtained for bicuculline over a range of different types of preparation; e.g. 5.98 for antagonism of muscimol depolarizations of the dorsal funiculus in slices of rat cuneate nucleus (Simmonds, 1980a), 5.93 for antagonism of GABA-induced increases in membrane conductance in *Xenopus* oocytes injected with rat brain mRNA (Smart *et al.*, 1986), 6.29 for antagonism of muscimol-induced potentiation of [^{3}H]diazepam binding to rat

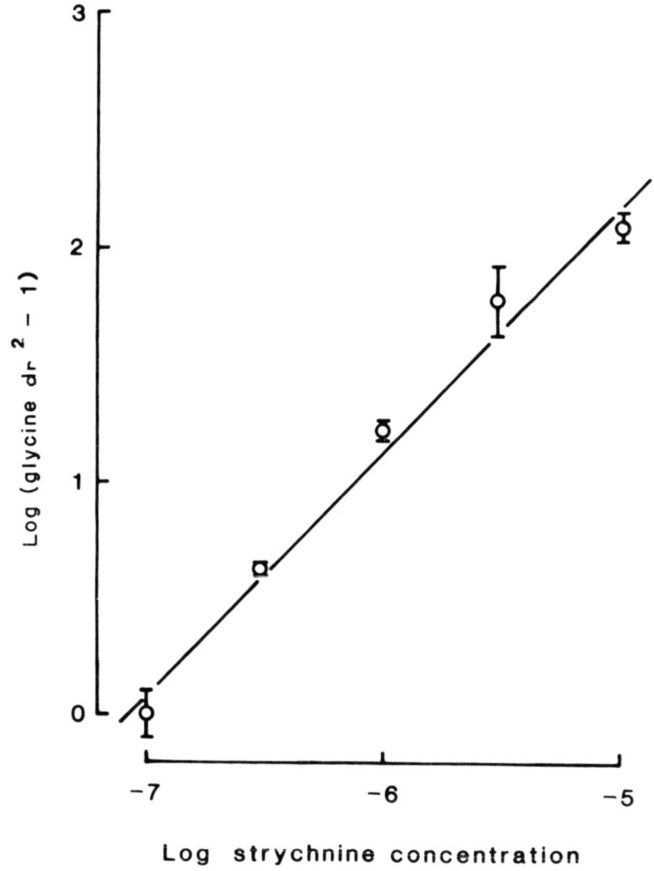

Figure 5 Schild plot for antagonism of glycine by strychnine on rat isolated optic nerve. A value of $n = 2$ for the stoichiometry of the glycine–receptor interaction has been incorporated into the ordinate. Each point shows the mean ± SEM of 4–9 values and these are derived from the same data as presented in an earlier paper (Simmonds, 1983a). The slope of the line calculated by least squares regression is 1.04 and the value of pK_A derived from it is 7.08

forebrain membranes (Braestrup and Nielsen, 1980). Likewise, there has been consistency in the pA_2 values obtained for AP5 antagonism of NMDA; e.g. 5.2 for (−)AP5 on rat cerebral cortical slices (Harrison and Simmonds, 1985), 5.0 for (±)AP5 on rat spinal cord (Wheatley and Collins, 1986), 4.9 for (±)AP5 on mouse neocortical slices (Burton *et al*., 1988).

III.2 Non-competitive Antagonism

A different range of interpretational problems arises with non-competitive antagonists in quantitative studies on slices of CNS tissue. The first of these is that the generally found parallel shift of the agonist log dose–response curve (see Simmonds, 1982) is not predicted by theory. An explanation for this was given above where it was also pointed out that the observed relationship between the shift and the concentration of non-competitive antagonist would not be expected to conform with the Gaddum–Schild equation. This is what has been found with the picrotoxin-like $GABA_A$ antagonists in slices of rat cuneate nucleus (Simmonds, 1982). However, there are other examples of an antagonist yielding a linear Schild plot of slope close to 1 and therefore appearing to be competitive when, on other grounds, there is reason to believe that it is not competitive. An example of this is the antagonism of quisqualate by barbiturates on the rat cortical slice preparation (Simmonds and Horne, 1987). This is most likely to arise when the range of antagonist concentrations employed is limited.

The question arises, therefore, how best to distinguish between truly competitive and non-competitive antagonists in this situation. A simple method is to estimate the potency of the antagonist in question, firstly alone and then in combination with a known competitive antagonist. This was used to distinguish between bicuculline as a competitive $GABA_A$ antagonist and picrotoxin as a non-competitive antagonist (Simmonds, 1980a). Furthermore, the method clearly established that antagonism of $GABA_A$ responses by strychnine, (+)tubocurarine and the steroid RU5135 is bicuculline-like while antagonism by isopropylbicyclophosphate and pentylenetetrazole (leptazol) is picrotoxin-like (Simmonds, 1982; Simmonds and Turner, 1985). Also, with excitatory amino acid antagonists, antagonism of NMDA on the rat cortical slice by AP5 and by ketamine each gave a Schild slope of about 1. Yet the combination of the two antagonists gave NMDA dose ratios that clearly indicated separate sites of action of these two antagonists (Harrison and Simmonds, 1985).

The theoretical basis of the procedure has been set out by Abramson *et al.* (1969) and the procedure itself may best be described here by an example. For each of the two antagonists to be compared, a concentration is found that gives an agonist dose ratio of, say, 5. If both antagonists share a common site of action, the effect of a combination of the two at these concentrations should be equal to the effect of twice the concentration of either one of them. With a Schild slope of 1, this combination effect will be an agonist dose ratio of 9, i.e. $(dr_1 + dr_2 - 1)$. On the other hand, if the two antagonists operate through independent recognition sites, the shifted agonist log dose–response curve in the presence of one of the antagonists can be regarded as the control curve for the second antagonist. This will yield a combined dose ratio of 25 for the agonist in this example, i.e. $(dr_1 \times dr_2)$. In practice, I have found that the agonist dose ratio for the combination of antagonists comes close to one or other of the predicted values, thus allowing unambiguous interpretation. However, the underlying theory does require that one of the antagonists be established as competitive since, as pointed out by Johnson and Ascher (1988), this method cannot satisfactorily distinguish between common or independent sites of action for two antagonists neither of which is competitive.

Having established that an antagonist is not competitive, some would say that the construction of a Schild plot for such an antagonist is not justified. Nevertheless, a Schild plot remains a useful way to express the concentration–effect profile of the antagonist. The slope of the plot and whether it is linear or curved have no interpretational significance but the pA_2 value that can be determined provides a useful index of potency for comparison within a group of non-competitive antagonists. However, values obtained on one type of slice preparation cannot be expected to be the same in another type of preparation.

III.3 Other Types of Antagonism

Antagonists which operate by blocking the ion channel opened by the agonist may show two characteristics that distinguish them from non-competitive antagonists in population response studies on slices of CNS tissue. The first is that the degree of antagonism at any given concentration of antagonist should increase with the dose of agonist, as more channels are opened to admit the antagonist. This should be manifest as a depressed log dose–response curve that is rather flattened in appearance and may even begin to fall at the higher doses of agonist. Whilst this will, in part, be obscured by the same factors that affect the manifestation of non-competitive antagonism described above, the depression of the log dose–response curve may be sufficiently marked still to be apparent. This is certainly the case with the antagonism of $GABA_A$ responses by penicillin on the rat cuneate nucleus slice (Pickles and Simmonds, 1980). The second characteristic is that the antagonism should be use-dependent, i.e. the antagonism of a given dose of agonist progressively increases to a stable level with repeated application of the agonist, but not in the absence of agonist over the same period of time. With population responses to agonist, this is likely to be seen best with low doses of agonist and an antagonist that binds with high affinity once it has entered the channel. Perhaps the clearest example of this to date is found with the antagonism of NMDA by MK-801 on slices of rat cerebral cortex (Davies *et al*., 1988).

Distinct from channel blockade is a form of antagonism that may be described as 'functional' in that the antagonist does not directly bind to or affect the protein complex that contains the agonist recognition site. This form of antagonism can operate only on those responses that depend on a sequence of events subsequent to the primary receptor-controlled event. For example, measurement of population field potentials, as in the grease-gap apparatus, in response to an agonist-induced opening of ion channels, depends on ions flowing through the opened channels. Drugs that reduce the electrochemical gradient of the ion concerned will reduce the field potential response for a given degree of receptor activation and associated conductance increase. This may be the mechanism by which frusemide, an inhibitor of Cl^- transport, antagonizes depolarization responses to $GABA_A$ receptor agonists (Simmonds, 1982). In so far as the conductance increase to GABA should not be affected, frusemide may have little effect on the inhibitory action of GABA.

IV STUDIES WITH POTENTIATORS

Potentiations of responses to agonists superfused over slices of CNS tissue are of two main types: (1) potentiation resulting from increased concentrations of free agonist within the tissue due to inhibition of uptake or some other removal process; (2) increased responsiveness of the receptor–effector protein to the agonist.

The former process is not always as readily demonstrable as might be expected. For example, it is difficult to achieve a sufficient concentration of the GABA uptake inhibitor nipecotic acid to potentiate superfused GABA in the rat cuneate nucleus slice without the nipecotic acid having an apparent agonist action of its own (unpublished observations). Such an effect of nipecotic acid is probably due to block of re-uptake of spontaneously released endogenous GABA and it can result in occlusion of the responses to superfused GABA rather than their potentiation. In contrast, potentiations due to the actions of drugs on the $GABA_A$ receptor protein are generally much more readily revealed *in vitro*. These include potentiations by benzodiazepines and barbiturates (Simmonds, 1981; Harrison and Simmonds, 1983) and by certain steroids (Harrison and Simmonds, 1984; Simmonds and Turner, 1987; Turner and Simmonds, 1989).

In general terms, a potentiator acting on the agonist receptor protein must act through a separate recognition site from that for the agonist itself and several possible mechanisms of potentiation can be envisaged. In terms of the simplified state equation for agonist–receptor interaction

$$D + R \underset{k_{-1}}{\overset{k_{+1}}{\rightleftharpoons}} DR \underset{\alpha}{\overset{\beta}{\rightleftharpoons}} DR^*$$

it is not easy to distinguish in practice whether the bound potentiator causes an increase in affinity of the receptor for agonist, i.e. a decrease in k_{-1}, or an increased rate of isomerization (β) of the agonist–receptor complex (DR) to the active form (DR^*). Much more readily discernible electrophysiologically is an increase in the average open time of the agonist-operated channel in single-channel recordings and in the power spectrum of the membrane noise in whole-cell recordings, i.e. a decrease in α. The latter mechanism has been clearly established for barbiturate and steroid potentiations of $GABA_A$ responses (Study and Barker, 1981; Mathers, 1985, 1987; Barker *et al*., 1987; Callachan *et al*., 1987) whilst, largely by default, the former mechanism(s) may be supposed to underlie benzodiazepine potentiations of GABA.

No matter which rate constant a potentiator might affect, it is theoretically possible that the same rate constant could also be modulated in the opposite direction. Drugs that did this would, of course, be antagonists and they could act either at the same recognition site as the potentiator or at a separate site. Such a situation has been postulated for the spectrum of drugs acting at the benzodiazepine receptor in the $GABA_A$ complex, i.e. drugs extending from the benzodiazepine potentiators of $GABA_A$ responses through to the β-carboline antagonists of $GABA_A$ responses (Braestrup *et al*., 1984). The terminology 'benzodiazepine agonist' and 'inverse agonist' has been widely adopted for these two ends of the spectrum with 'benzodiazepine antagonist' in the middle and 'partial agonist' and 'partial inverse agonist' in between. In the context of the benzodiazepine–β-carboline interaction with the $GABA_A$ system, it seems a pity that they were not designated 'modulators', with either potentiating or antagonistic activity at the ends of the spectrum and neutral (stabilizing) activity in the middle.

In grease-gap recordings from the rat cuneate nucleus slice, potentiations of $GABA_A$ receptor-mediated responses to muscimol usually appear as approximately parallel leftward shifts of the muscimol log dose–response curve. The maximum shift obtainable with benzodiazepines is rather small, a factor of about 2, compared with the shift that can be induced by barbiturates, up to a factor of 8 (Simmonds, 1980b, 1981). The antagonistic rightward shift obtainable with the β-carbolines is likewise small (see Simmonds and Turner, 1989). Whether these differences are a necessary concomitant of the different mechanisms of action is not clear. Steroids, which have a similar mechanism of action to the barbiturates (Barker *et al*., 1987; Callachan *et al*., 1987), produce only rather small maximal potentiations (Turner and Simmonds, 1989) but there is some evidence (see below) that enhanced desensitization may be a limiting factor in this case.

The parallel shifts make it easy to assign a single numerical value to the potentiation induced by a given concentration of potentiator. This has enabled structure–activity profiles to be constructed with a series of barbiturates and also with a series of steroids. As far as the barbiturates are concerned, the profile for $GABA_A$ potentiation corresponds closely to that for the hypnotic/anaesthetic potencies of these compounds (Harrison and Simmonds, 1983). The situation is less clear with the steroids except to say that all the active compounds are known to be hypnotic/anaesthetic (Atkinson *et al*., 1965). One of the difficulties with the steroids is that they appear to cause an enhanced degree of desensitization of the responses to muscimol. This is illustrated in Figure 6 where the pregnanolone 3α-hydroxy,5β-pregnane-20-one shifts the muscimol log dose–response curve to the left but at the same time reduces the maximum of the curve. Whether or not it is desensitization that reduces the maximum, the same influence is likely also to limit the leftward shift of the curve. Indeed, with increasing concentrations of the steroid anaesthetic alphaxalone, the leftward shift increases to a maximum at 1 μM alphaxalone and the potentiation then tends to decline with further increases in alphaxalone concentration (Turner and Simmonds, 1989). An even more marked limitation of maximal potentiating action was shown by the 3β-hydroxy isomers of pregnanolone. This sort of problem was not experienced with the barbiturates and it is interesting to note that although both groups of drugs potentiate GABA by prolonging the average open time of the channel, they appear to act through distinct recognition sites (Cottrell *et al*., 1987; Peters *et al*., 1988).

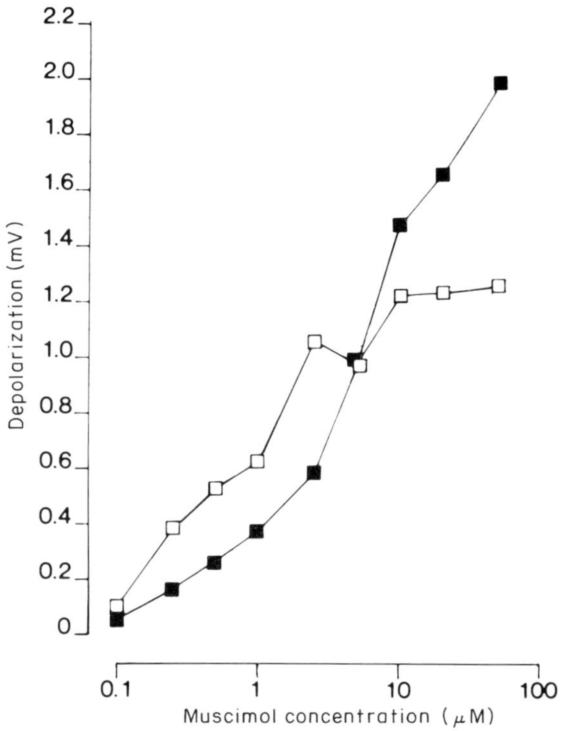

Figure 6 Effect of the pregnanolone 3α-hydroxy,5β-pregnane-20-one on the log dose–response curve to muscimol in a slice of rat cuneate nucleus. Each point represents a single response. Compared with the control curve (■), the curve in the presence of 1 μM pregnanolone (□) is shifted to the left but the maximum is reduced. (Taken from Turner and Simmonds, 1989)

At high concentrations relative to those required to potentiate superfused GABA or muscimol, both barbiturates and steroids can evoke GABA-mimetic responses. In slices of CNS tissue, this is quite likely to be, in part, ascribable to the potentiation of subthreshold effects of endogenous GABA so that they become suprathreshold, although it has been claimed that GABA is not necessary for the GABA-mimetic effects of barbiturates (Mathers and Barker, 1980). If the GABA-mimetic effects are too large, responses to superfused GABA or muscimol become occluded (Simmonds, 1981), as might be expected. Thus, it seems quite possible that, in slices, a common mechanism of action underlies both the potentiation of superfused GABA and the direct GABA-mimetic effects evoked by barbiturates and steroids.

Rather different circumstances surround the phenomenon of glycine potentiation of responses to the excitatory amino acid NMDA. The phenomenon was first observed on isolated neurons and required less than 1 μM glycine (Johnson and

Ascher, 1987, 1988). When glycine was absent, the responses to NMDA were very small. In slices of rat cerebral cortex, addition of glycine failed to enhance responses to NMDA. However, antagonism of NMDA by 3-amino-1-hydroxy-2-pyrrolidone (HA966), kynurenate and 7-chlorokynurenate could be reversed by glycine (Fletcher *et al*., 1988; Kemp *et al*., 1988), whereas antagonism of NMDA by the competitive antagonist AP5 was not reversed by glycine (Fletcher and Lodge, 1988). This suggests that HA966 and kynurenate are glycine antagonists and that under the experimental conditions the glycine site may have been fully activated by endogenous glycine. Interestingly, HA966 was first described as an excitatory amino acid L-antagonist by Davies and Watkins (1972) and only recently was this mode of action proposed.

This example raises the intriguing question of whether non-competitive antagonists on other systems might also be reversing an endogenous potentiating influence. The antagonism of $GABA_A$ responses by picrotoxin is a particularly interesting example since this antagonism can be reversed by certain barbiturates and steroids without competitive antagonism by bicuculline being affected (Simmonds, 1981; Turner and Simmonds, 1989). However, the substantial antagonism that picrotoxin can cause would require that the potentiating influence, it was postulated to reverse, should be an essential factor permitting responses to GABA. There is no evidence from studies on isolated neurons bathed in artificial medium that such a factor is required. It appears unlikely, therefore, that the action of picrotoxin depends on the reversal of an endogenous potentiating influence. Nevertheless, the reversal of picrotoxin antagonism of $GABA_A$ responses by some steroids does suggest the possibility that similar endogenous steroids could affect the action of picrotoxin *in vivo*.

The structure–activity profiles of both barbiturates and steroids for their reversal of picrotoxin are different from the profiles for their potentiation of muscimol responses (Harrison and Simmonds, 1983; Turner and Simmonds, 1989). The way in which drug interactions with picrotoxin potency were assessed quantitatively made use of the Schild plot for picrotoxin as a convenient log concentration–effect relationship, the effect of picrotoxin being represented as log (muscimol dose ratio − 1). The barbiturates and steroids that reduced picrotoxin potency shifted the Schild plot to the right in an approximately parallel fashion and the extent of the shift was measured. At the same time, however, responses to muscimol had already been potentiated by the barbiturates and steroids and there was a possibility that picrotoxin might affect the potentiation which would, in turn, affect the apparent potency of picrotoxin. To standardize the influence of this potential complication, comparisons of the effects of barbiturates and steroids on picrotoxin potency were made with concentrations that were equi-effective for potentiation of muscimol. In the barbiturate series, (±)-5-(1,3-dimethylbutyl)-5-ethylbarbituric acid (DMBB), phenobarbitone and (−)methylphenobarbitone were more effective at reducing the potency of picrotoxin than were the other barbiturates tested, i.e. pentobarbitone, quinalbarbitone, butobarbitone and barbitone (Harrison and Simmonds, 1983). Similarly, in a series of steroids, the 3α-hydroxypregnanolones, the 5β-isomer of alphaxalone and androsterone were more effective at reducing picrotoxin potency than the 5β-hydroxypregnanolones or alphaxalone (Turner and Simmonds, 1989).

These data were taken to suggest that there may be two distinct modes of interaction of barbiturates and steroids with the $GABA_A$ receptor complex. On further reflection, another possibility comes to mind. This is that the barbiturates and steroids together with picrotoxin and other $GABA_A$ antagonists that share the same site of action as picrotoxin, i.e. bicyclophosphates, the bicyclophosphorothionate TBPS and pentylenetetrazole (Simmonds, 1982; Squires *et al*., 1983; Trifiletti *et al*., 1985), form a spectrum of modulators of the $GABA_A$ receptor complex, all operating on a common mechanism. By analogy with the benzodiazepine and β-carboline 'agonist' and 'inverse agonist' concept described earlier, potentiation of responses to GABA or muscimol might be regarded as an 'agonist' modulation and antagonism as an 'inverse agonist' modulation. Those substances that caused both potentiation of GABA and a reduction in picrotoxin potency might be classified as 'partial agonist' modulators. If this interpretation is correct, it would be expected that potentiation of GABA by barbiturates and steroids and antagonism by picrotoxin should result in opposite changes in the response characteristics to GABA. In this respect, the available data are not very clear since the prolongation of the average open time of the $GABA_A$ channel by barbiturates and steroids is not matched by a shortening of the open time by picrotoxin although pentylenetetrazole does reduce the open time (Barker *et al*., 1981). Further studies are required to clarify the situation and to test other predictions arising from this interpretation.

V PROBLEMS WITH MIXED RECEPTOR POPULATIONS

It is now becoming common to find multiple receptor types for a given neurotransmitter, even on the same cell. If more than one of the receptor types activated by a single agonist contribute to the response recorded from a slice in a grease-gap experiment, it is not always a simple matter to distinguish the component contributions. Incubation with selective antagonists in attempts to eliminate all but one of the components may not be entirely robust when high doses of agonist are used to define the antagonist susceptibility of the component being studied. The alternative of using selective agonists may also not be entirely satisfactory in similar circumstances if the selectivity is insufficient. An additional problem with agonists is that their differing susceptibilities to removal by uptake can distort quantitative profiles.

One example of this concerns responses to the excitatory amino acid L-glutamate and the question of what proportions of the overall response are contributed by NMDA receptors, quisqualate receptors, kainate receptors and any other receptors.

On the rat cortical slice preparation, responses to L-glutamate and L-aspartate were only partially blocked by concentrations of AP5 and ketamine that almost completely blocked responses to NMDA without affecting responses to quisqualate (Harrison and Simmonds, 1985). The responses to L-aspartate were reduced to a greater extent than those to L-glutamate. Likewise, on the rat olfactory cortex slice, it was found that 36% of the response to L-aspartate and 21% of the response to L-glutamate could be blocked by AP5 and, therefore, attributed to an action at NMDA receptors (Surtees and Collins, 1985). Attempts to block, additionally, quisqualate and kainate receptors by adding (±)*cis*-2,3-piperidine dicarboxylate and γ-glutamylglycine to the AP5 resulted in an 80% reduction in responses to quisqualate and kainate as well as a complete block of NMDA responses but only a 50% reduction in responses to L-glutamate and L-aspartate. Even the use of receptor desensitization prior to addition of antagonists failed to cause much more of an antagonism of L-glutamate and L-aspartate (Collins and Brown, 1986). It is of interest in the latter paper that, of six amino acids tested, the four that were incompletely blocked are all substrates for a common uptake system. This raises the question of whether the residual responses were due to electrogenic uptake of the amino acids concerned.

In the field of inhibitory amino acids, a particularly difficult problem is to determine whether responses to taurine can be adequately accounted for by activation of $GABA_A$ receptors and glycine receptors or whether an additional 'taurine receptor' is also involved. Pharmacological evidence from a variety of studies, both *in vivo* and *in vitro*, indicates that responses to taurine are modulated by $GABA_A$ potentiators and antagonists and by the glycine antagonist strychnine (see Simmonds, 1986). On the rat cuneate nucleus slice, the maximum response to taurine was substantially less than that to muscimol and about the same as that to glycine. On the optic nerve preparation, however, the maximum response to taurine was substantially less than that to glycine and about the same as that to muscimol. One interpretation of these observations is that taurine acts as a partial agonist at both $GABA_A$ receptors and strychnine-sensitive glycine receptors. Since taurine is potently antagonized by strychnine on both preparations and is little affected by bicuculline, it may be that the major component of its effect is on the glycine receptor. However, in the presence of strychnine, the partially antagonized responses to higher doses of taurine are more sensitive to bicuculline. This suggests that $GABA_A$ receptors are activated at these higher doses. An alternative means of selectively enhancing a $GABA_A$ receptor-mediated component of taurine's action is to incubate the tissues with pentobarbitone. The potentiated responses to taurine are now much less sensitive to strychnine (Figure 7) whereas the potency of strychnine as a glycine antagonist is unaffected. Thus, it appears that potentiation of taurine by pentobarbitone shifts its normally predominant action on glycine receptors to a predominant action on $GABA_A$ receptors.

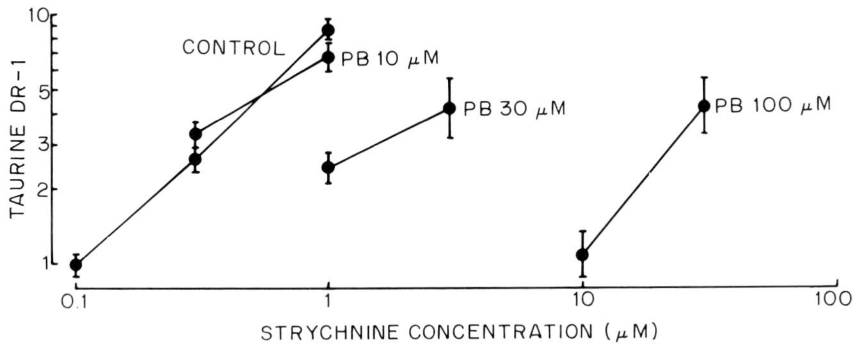

Figure 7 Schild plots for antagonism of taurine-induced depolarizations of rat optic nerve by strychnine. Note the substantial reduction in strychnine potency in the presence of pentobarbitone (PB)

VI REFLECTION OF CHRONIC PRETREATMENTS IN THE RESPONSIVENESS OF SLICES

In both the cuneate nucleus and the cortical wedge slice preparations, the quantitative pharmacology of the amino acid receptor systems has proved sufficiently reproducible over a period of time that it should be possible to detect changes in receptor properties induced by chronic pretreatment of the donor animals. Two situations have been examined with regard to the cuneate nucleus.

We were concerned that a variable rise in circulating corticosteroids immediately prior to killing the rat might result in variable levels of steroids dissolved in the lipids in the cuneate nucleus slice and thereby influence the sensitivity of the $GABA_A$ receptor complex to exogenous steroids. Male rats were, therefore, bilaterally adrenalectomized or sham operated and were provided with both 0.9% NaCl solution and tap water to drink. After 14 days during which both groups of rats thrived equally well, the rats were killed and cuneate nucleus slices prepared. We could find no influence of the pretreatment on the responsiveness of the slices to muscimol or in the potency of alphaxalone as a potentiator of muscimol, despite a good level of reproducibility in each group (Andres-Trelles and Simmonds, unpublished observations). This could mean that circulating corticosteroids do not modulate the $GABA_A$ receptor complex in the CNS or that the 2–3-h preincubation of the slices prior to testing washed the endogenous steroids away from their site of action. The first of these possibilities is strengthened by the absence of any adaptive change in sensitivity to alphaxalone after the 14 days of adrenalectomy.

In contrast, chronic elevation of endogenous GABA in the brains of rats by treatment with the GABA-transaminase inhibitor ethanolamine-O-sulphate (EOS) in the drinking water (Fletcher and Fowler, 1980) caused a distinct pattern of changes in the sensitivity of the $GABA_A$ receptor complex to various modulators (Lindgren and Simmonds, 1987). Slices of cuneate nucleus taken from rats after 15–30 days of pretreatment with EOS showed a reduced responsiveness to muscimol, enhanced potentiations of muscimol by flurazepam and by pentobarbitone and a reduced antagonism of muscimol by picrotoxin. None of these changes were found in slices from rats in which endogenous GABA had been increased only for 1 h by injection of γ-acetylenic GABA (GAG). Thus, it appears that these changes in drug sensitivity were not due to raised concentrations of endogenous GABA within the slice but more likely to adaptation in the sensitivity of the $GABA_A$ receptor complex. It might be expected that adaptation would be in the direction of ameliorating the effect of the chronically elevated GABA, i.e. reduced responsiveness to GABA agonists, reduced action of endogenous facilitatory influences on GABA responses and increased action of endogenous antagonistic influences. The first of these was, in fact, seen. The data with flurazepam, pentobarbitone and picrotoxin would be compatible with the latter two types of adaptation if they occurred largely by changes in the release of endogenous modulatory substances, since it is generally found that a reduction in endogenous release of a mediator results in increased receptor sensitivity to exogenously applied mediator. The fact that these changes occurred at all suggests that each of the modulatory sites at which flurazepam, pentobarbitone and picrotoxin have their effects are under physiological control. The only response change that was most probably due to the elevated GABA in the slice was an increase in response to the uptake inhibitor nipecotic acid, seen following both chronic elevation of GABA by EOS and acute elevation by GAG.

Observations such as those just described suggest that postmortem assessment of the quantitative pharmacological profile of a receptor complex can reflect antemortem adaptations of the receptor complex. Whilst this may be exploited for specific purposes, it is also possible that unknown or uncontrolled influences *in vivo* may contribute to the variability of drug responses obtained *in vitro*. During the last ten years of work on the $GABA_A$ system, the drug effects that have been most troublesome in this respect are the potentiations of muscimol by flurazepam (Simmonds, 1983b) and by alphaxalone (Simmonds and Turner, 1987).

VII CONCLUSION

I have not attempted to make a comprehensive survey of the published literature on quantitative pharmacological studies in slice preparations. My intention has been to explore how far the data obtained in this type of study can be interpreted in terms of receptor theory and models. Instances of uncertainty and examples of possible misinterpretation have been discussed. Provided that a realistic view is taken of these, it is possible to distinguish reliably between competitive and non-competitive antagonists and to make quantitative comparisons of the potencies within each group of antagonists and within groups of potentiating drugs. It is doubtful, however, whether agonist dose–response relationships are robust enough to indicate reliably the stoichiometry of the agonist–receptor interaction. The reproducibility of many of the drug effects described is quite good, sufficient to detect in the slices changes in receptor sensitivity induced by antemortem manipulations.

ACKNOWLEDGEMENT

I am grateful to the Medical Research Council and the Wellcome Trust for financial support.

REFERENCES

Abramson, F.B., Barlow, R.B., Mustafa, M.G., and Stephenson, R.P. (1969). 'Relationships between chemical structure and affinity for acetylcholine receptors', *Br. J. Pharmacol.*, **37**, 207–233.

Akaike, N., Inoue, M., and Krishtal, O.A. (1986). '"Concentration-clamp" study of γ-aminobutyric-acid-induced chloride current kinetics in frog sensory neurones', *J. Physiol. (Lond.).*, **379**, 171–185.

Arunlakshana, O., and Schild, H.O. (1959). 'Some quantitative uses of drug antagonists', *Br. J. Pharmacol.*, **14**, 48–58.

Atkinson, R.M., Davis, B., Pratt, M.A., Sharpe, H.M., and Tomich, E.G. (1965). 'Action of some steroids on the central nervous system of the mouse. II. Pharmacology', *J. Med. Chem.*, **8**, 426–432.

Barker, J.L., MacDonald, J.F., Mathers, D.A., McBurney, R.N., and Oertel, W. (1981). 'GABA receptor functions in cultured mouse spinal neurons', in *Amino Acid Neurotransmitters* (Eds F.V. De Feudis and P. Mandel), pp. 281–293, Raven Press, New York.

Barker, J.L., Harrison, N.L., Lange, G.D., and Owen, D.G. (1987). 'Potentiation of γ-aminobutyric-acid-activated chloride conductance by a steroid anaesthetic in cultured rat spinal cord neurones', *J. Physiol. (Lond.).*, **386**, 485–501.

Barnard, E.A., Darlison, M.G., and Seeburg, P. (1987). 'Molecular biology of the $GABA_A$ receptor: the receptor/channel superfamily', *Trends Neurosci.*, **10**, 502–509.

Blake, J.F., Brown, M.W., and Collingridge, G.L. (1988). 'A quantitative study of the actions of excitatory amino acids and antagonists in rat hippocampal slices', *Br. J. Pharmacol.*, **95**, 291–299.

Bormann, J. (1988). 'Electrophysiology of $GABA_A$ and $GABA_B$ receptor subtypes', *Trends Neurosci.*, **11**, 112–116.

Braestrup, C., and Nielsen, M. (1980). 'Strychnine as a potent inhibitor of the brain GABA/benzodiazepine receptor complex', *Brain Res. Bull.*, **5** (Suppl. 2), 681–684.

Braestrup, C., Honoré, T., Nielsen, M., Petersen, E.N., and Jensen, L.H. (1984). 'Ligands for benzodiazepine receptors with positive and negative efficacy', *Biochem. Pharmacol.*, **33**, 859–862.

Braestrup, C., Nielsen, M., and Krogsgaard-Larsen, P. (1986). 'Glycine antagonists structurally related to 4,5,6,7-tetrahydroisoxazolo[5,4-c]pyridin-3-ol inhibit binding of $[^3H]$strychnine to rat brain membranes', *J. Neurochem.*, **47**, 691–696.

Burton, N.R., Smith, D.A.S., and Stone, T.W. (1988). 'A quantitative pharmacological analysis of some excitatory amino acid receptors in the mouse neocortex in vitro', *Br. J. Pharmacol.*, **93**, 693–701.

Callachan, H., Cottrell, G.A., Hather, N.Y., Lambert, J.J., Nooney, J.M., and Peters, J.A. (1987). 'Modulation of the $GABA_A$ receptor by progesterone metabolites', *Proc. R. Soc. Lond. B*, **231**, 359–369.

Cohen, M.W. (1970). 'The contribution by glial cells to surface recordings from the optic nerve of an amphibian', *J. Physiol. (Lond.)*., **210**, 565–580.

Collins, G.G.S., and Brown, G. (1986). 'A quantitative estimate of the contribution made by various receptor categories to the depolarizations evoked by some excitatory amino acids in the olfactory cortex', *Brain Res*., **371**, 9–16.

Colquhoun, D. (1973). 'The relation between classical and co-operative models for drug action', in *Drug Receptors* (Ed. H.P. Rang), pp. 149–182, Macmillan, London.

Cottrell, G.A., Lambert, J.J., and Peters, J.A. (1987). 'Modulation of $GABA_A$ receptor activity by alphaxalone', *Br. J. Pharmacol*., **90**, 491–500.

Davies, J., and Watkins, J.C. (1972). 'Is 1-hydroxy-3-aminopyrrol-idone-2 (HA966) a selective excitatory amino-acid antagonist?', *Nature New Biol*., **238**, 61–63.

Davies, S.N., Martin, D., Millar, J.D., Aram, J.A., Church, J., and Lodge, D. (1988). 'Differences in results from in vivo and in vitro studies on the use-dependency of N-methylaspartate antagonism by MK-801 and other phencyclidine receptor ligands', *Eur. J. Pharmacol*., **145**, 141–151.

Evans, R.H. (1978). 'The effects of amino acids and antagonists on the isolated hemisected spinal cord of the immature rat', *Br. J. Pharmacol*., **62**, 171–176.

Fletcher, A., and Fowler, L.J. (1980). 'γ-Aminobutyric acid metabolism in rat brain following chronic oral administration of ethanolamine-O-sulphate', *Biochem. Pharmacol*., **29**, 1451–1454.

Fletcher, E.J., and Lodge, D. (1988). 'Glycine modulation of NMDA antagonist action in rat neocortical slices', *J. Physiol. (Lond.)*., **406**, 13P.

Fletcher, E.J., Jones, M.G., and Lodge, D. (1988). 'Strychnine-resistant effects of glycine on central neurones', *Br. J. Pharmacol*., **95**, 756P.

Garthwaite, J. (1985). 'Cellular uptake disguises action of L-glutamate on N-methyl-D-aspartate receptors', *Br. J. Pharmacol*., **85**, 297–307.

Garthwaite, J., Garthwaite, G., and Hajós, F. (1986). 'Amino acid neurotoxicity: relationship to neuronal depolarization in rat cerebellar slices', *Neuroscience*, **18**, 449–460.

Harrison, N.L., and Simmonds, M.A. (1983). 'Two distinct interactions of barbiturates and chlormethiazole with the $GABA_A$ receptor complex in rat cuneate nucleus in vitro', *Br. J. Pharmacol*., **80**, 387–394.

Harrison, N.L., and Simmonds, M.A. (1984). 'Modulation of the GABA receptor complex by a steroid anaesthetic', *Brain Res*., **323**, 287–292.

Harrison, N.L., and Simmonds, M.A. (1985). 'Quantitative studies on some antagonists of N-methyl D-aspartate in slices of rat cerebral cortex', *Br. J. Pharmacol*., **84**, 381–391.

Horne, A.L., and Simmonds, M.A. (1989). 'The pharmacology of quisqualate and AMPA in the cerebral cortex of the rat *in vitro*', *Neuropharmacology*, **28**, 1113–1118.

Johnson, J.W., and Ascher, P. (1987). 'Glycine potentiates the NMDA response in cultured mouse brain neurons', *Nature*, **325**, 529–531.

Johnson, J.W., and Ascher, P. (1988). 'The NMDA receptor and its channel. Modulation by magnesium and by glycine', in *Excitatory Amino Acids in Health and Disease* (Ed. D. Lodge), pp. 143–164, Wiley, Chichester.

Kemp, J.A., Priestley, T., and Woodruff, G.N. (1988). 'Differences in the N-methyl-D-aspartate antagonist profiles of two compounds acting at the glycine modulatory site', *Br. J. Pharmacol*., **95**, 759P.

Krnjević, K. (1974). 'Chemical nature of synaptic transmission in vertebrates', *Physiol. Rev*., **54**, 418–540.

Levitan, E.S., Schofield, P.R., Burt, D.R., Rhee, L.M., Wisden, W., Köhler, M., Fujita, N., Rodriguez, H.F., Stephenson, A., Darlison, M.G., Barnard, E.A., and Seeburg, P.H. (1988). 'Structural and functional basis for $GABA_A$ receptor heterogeneity', *Nature*, **335**, 76–79.

Lindgren, S., and Simmonds, M.A. (1987). 'Adaptation of the $GABA_A$–receptor complex in rat brain during chronic elevation of GABA by ethanolamine O-sulphate', *Br. J. Pharmacol.*, **91**, 617–625.
Lodge, D., Curtis, D.R., Johnston, G.A.R., and Bornstein, J.C. (1980). 'In vivo inactivation of quisqualate: studies in the cat spinal cord', *Brain Res.*, **182**, 491–495.
Mathers, D.A. (1985). 'Pentobarbital promotes bursts of γ-aminobutyric acid-activated single channel currents in cultured mouse central neurons', *Neurosci. Lett.*, **60**, 121–126.
Mathers, D.A. (1987). 'The $GABA_A$ receptor: new insights from single-channel recording', *Synapse*, **1**, 96–101.
Mathers, D.A., and Barker, J.L. (1980). '(−)Pentobarbital opens ion channels of long duration in cultured mouse spinal neurons', *Science*, **209**, 507–509.
Möhler, H., and Okada, T. (1977). 'GABA receptor binding with ^{3}H(+)bicuculline-methiodide in rat CNS', *Nature*, **267**, 65–67.
Peters, J.A., Kirkness, E.F., Callachan, H., Lambert, J.J., and Turner, A.J. (1988). 'Modulation of the $GABA_A$ receptor by depressant barbiturates and pregnane steroids', *Br. J. Pharmacol.*, **94**, 1257–1269.
Pickles, H.G., and Simmonds, M.A. (1980). 'Antagonism by penicillin of γ-aminobutyric acid depolarizations at presynaptic sites in rat olfactory cortex and cuneate nucleus in vitro', *Neuropharmacology*, **19**, 35–38.
Podleski, T.R. (1973). 'Cooperativity of the electroplax membrane', in *Drug Receptors* (Ed. H.P. Rang), pp. 135–148, Macmillan, London.
Simmonds, M.A. (1978). 'Presynaptic actions of γ-aminobutyric acid and some antagonists in a slice preparation of cuneate nucleus', *Br. J. Pharmacol.*, **63**, 495–502.
Simmonds, M.A. (1980a). 'Evidence that bicuculline and picrotoxin act at separate sites to antagonize γ-aminobutyric acid in rat cuneate nucleus', *Neuropharmacology*, **19**, 39–45.
Simmonds, M.A. (1980b). 'A site for the potentiation of GABA-mediated responses by benzodiazepines', *Nature*, **284**, 558–560.
Simmonds, M.A. (1981). 'Distinction between the effects of barbiturates, benzodiazepines and phenytoin on responses to γ-aminobutyric acid receptor activation and antagonism by bicuculline and picrotoxin', *Br. J. Pharmacol.*, **73**, 739–747.
Simmonds, M.A. (1982). 'Classification of some GABA antagonists with regard to site of action and potency in slices of rat cuneate nucleus', *Eur. J. Pharmacol.*, **80**, 347–358.
Simmonds, M.A. (1983a). 'Depolarizing responses to glycine, β-alanine and muscimol in isolated optic nerve and cuneate nucleus', *Br. J. Pharmacol.*, **79**, 799–806.
Simmonds, M.A. (1983b). 'Variations in response of the GABA-picrotoxin-benzodiazepine receptor complex to flurazepam', *Pharmacol. Biochem. Behav.*, **18**, 299–301.
Simmonds, M.A. (1986). 'Classification of inhibitory amino acid receptors in the mammalian nervous system', *Med. Biol.*, **64**, 301–311.
Simmonds, M.A., and Horne, A.L. (1987). 'Antagonism of excitatory amino acids by barbiturates', in *Excitatory Amino Acid Transmission* (Eds T.P. Hicks, D. Lodge and H. McLennan), pp. 99–105. Liss, New York.
Simmonds, M.A., and Turner, J.P. (1985). 'Antagonism of inhibitory amino acids by the steroid derivative RU5135', *Br. J. Pharmacol.*, **84**, 631–635.
Simmonds, M.A., and Turner, J.P. (1987). 'Potentiators of responses to activation of γ-aminobutyric acid ($GABA_A$) receptors', *Neuropharmacology*, **26**, 923–930.
Smart, T.G., Constanti, A., Houamed, K., Bilbe, G., Brown, D.A., Barnard, E.A., and Van Renterghem, C. (1986). 'Expression of vertebrate amino acid receptors in Xenopus oocytes', in *Excitatory Amino Acids and Epilepsy* (Eds R. Schwarcz and Y. Ari), pp. 525–537, Plenum, New York.
Squires, R.F., Casida, J.E., Richardson, M., and Saederup, E. (1983). '[^{35}S]t-Butylbicyclophosphorothionate binds with high affinity to brain-specific sites coupled to

γ-aminobutyric acid-A and ion recognition sites', *Mol. Pharmacol.*, **23**, 326–336.

Study, R.E and Barker, J.L. (1981). 'Diazepam and (−)pentobarbital: fluctuation analysis reveals different mechanisms for potentiation of γ-aminobutyric acid responses in cultured central neurons', *Proc. Natl. Acad. Sci. USA*, **78**, 7180–7184.

Surtees, L., and Collins, G.G.S. (1985). 'Receptor types mediating the excitatory actions of exogenous L-aspartate and L-glutamate in rat olfactory cortex', *Brain Res.*, **334**, 287–295.

Tokutomi, N., Kaneda, M., and Akaike, N. (1989). 'What confers specificity on glycine for its receptor site?', *Br. J. Pharmacol.*, **97**, 353–360.

Trifiletti, R.R., Snowman, A.M., and Snyder, S.H. (1985). 'Barbiturate recognition site on the GABA/benzodiazepine receptor complex is distinct from the picrotoxin/TBPS recognition site', *Eur. J. Pharmacol.*, **106**, 441–447.

Turner, J.P., and Simmonds, M.A. (1989). 'Modulation of the $GABA_A$ receptor complex by steroids in slices of rat cuneate nucleus', *Br. J. Pharmacol.*, **96**, 409–417.

Wheatley, P.L., and Collins, K.J. (1986). 'Quantitative studies of N-methyl-D-aspartate, 2-amino-5-phosphonovalerate and cis-2,3-piperidine dicarboxylate interactions on the neonatal rat spinal cord in vitro', *Eur. J. Pharmacol.*, **121**, 257–263.

Williams, T.L., Smith, D.A.S., Burton, N.R., and Stone, T.W. (1988). 'Amino acid pharmacology in neocortical slices: evidence for bimolecular actions from an extension of the Hill and Gaddum–Schild equations', *Br. J. Pharmacol.*, **95**, 805–810.

Preparations of Vertebrate Central Nervous System *In Vitro*
Edited by H. Jahnsen
Published by John Wiley & Sons Ltd

CHAPTER 4

Cultures of Brain Slices

BEAT H. GÄHWILER and THOMAS KNÖPFEL
Brain Research Institute, University of Zurich, August-Forel-Strasse 1, CH-8029 Zurich, Switzerland

I INTRODUCTION

Slice cultures of nervous tissue are an *in vitro* preparation which has been developed over a period of several years in an attempt to obtain a preparation which is thin enough to allow experimental manipulations with and visualization of individual cells, but on the other hand keeps as much as possible of the original tissue organization. This chapter summarizes the details of the roller-tube culturing technique and illustrates a number of typical applications of the slice culture method which have made use of several morphological, physiological and pharmacological techniques.

II CULTURING TECHNIQUES

Tissue is collected from infant rats. The best age for culturing appears to vary between regions and must be determined empirically. Survival of neurons is definitely superior with tissue derived from young postnatal or even fetal animals, but the requirements for an organotypic tissue organization would favor the use of older animals. Early postnatal age (1–6 days) is a good compromise, since most of the cytoarchitecture is already present and the peak of neuronal migration (which will invariably lead to changes in tissue morphology) has already passed.

The tissue to be cultured is dissected out of the brain of young rats, washed in a balanced salt solution (BSS) and quickly cut into 400-μm thick slices by means of a McIlwain tissue chopper (Figure 1). The slices are kept at room temperature in a plastic Petri dish filled with Gey's BSS which is enriched with glucose to a concentration of 6.5 mg/ml. This recovery period of about 1–2 h is helpful

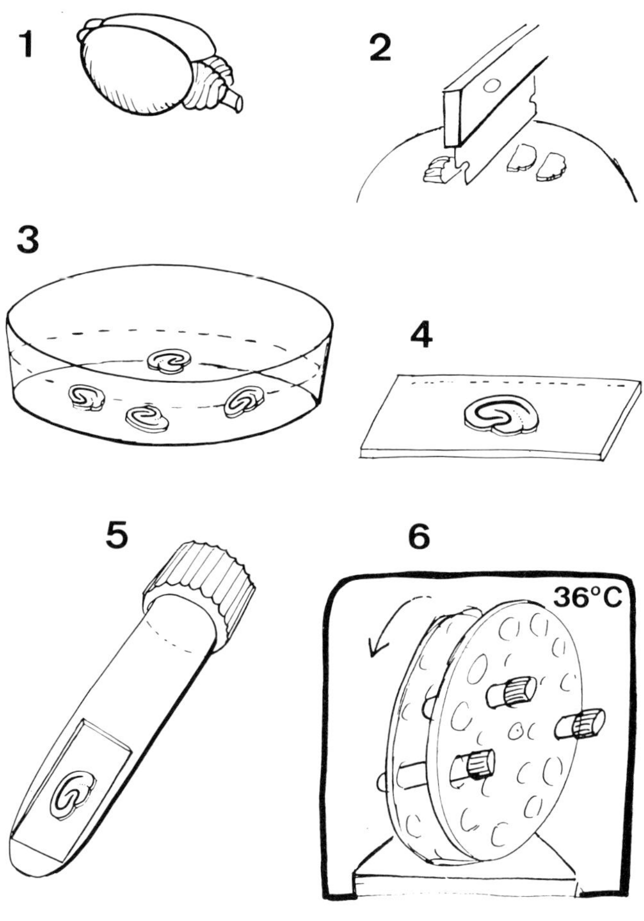

Figure 1 Cultivation procedure. The tissue to be cultured is dissected out of the brain of young rats (1), quickly cut into 400-μm thick slices by means of a McIlwain tissue chopper (2) and washed in a balanced salt solution (3). The slices are embedded on glass coverslips in a plasma clot (4). The coverslips bearing the cultures are inserted into plastic test tubes (5) which are placed in a roller drum (6) and rotated at 10 revolutions/h in an oven in dry air at 36°C. The slow rotation results in a periodic washing and feeding of the cultures

because potentially toxic substances such as excitatory amino acids may be washed from damaged neurons into the BSS (Gähwiler, 1981, 1984a, 1988).

The slices are placed on glass coverslips (12 × 24 mm) in a drop (20 μl) of chicken plasma which had been spread over the entire surface to coat the slide. To keep the explants in place, the plasma is coagulated by adding 20 μl of thrombin (Hoffmann-La-Roche, 20 NIH U/ml) resulting within minutes in the formation of a firm plasma clot. The coverslips bearing the cultures are inserted into plastic test tubes (Nunc 1-56758) to which 0.5 ml of medium has been added, placed in a roller drum and rotated at 10 revolutions/h in an oven in dry air at 36°C. The slow rotation results in a periodic washing and feeding of the cultures.

The medium consists of heat-inactivated horse serum (25%), basal medium (Eagle, 50%), and Hanks' or Earle's BSS (25%), supplemented with glucose to a final concentration of 6.5 mg/ml. To facilitate control of medium pH, phenol red is added to all solutions in a concentration of 10 μg/ml. No antibiotics are added to the culture medium, and no additional CO_2 is required in the incubator. Due to the fact that the plastic test tubes are semipermeable for CO_2 (produced by the metabolism of the tissue), stable pH values of 7.2–7.4 can be achieved by proper choice of the medium concentration of bicarbonate. To reduce the number of non-neuronal cells, antimitotic drugs (ARA-C or FUDR, 10^{-6} M) are usually added for 24 h about 4–5 days after the cultures were started. Cultures used for patch-clamp recordings (Llano *et al.*, 1988) are exposed to ionizing radiation (X-rays, about 800 rad) at the day of explantation immediately before embedding in the plasma clot on the glass coverslips. The cultures are fed once or twice weekly and survive under these culture conditions for several weeks up to 3 months *in vitro*.

III MORPHOLOGICAL DIFFERENTIATION

During the first 2 weeks in culture, a dramatic flattening and spreading of the cultures takes place. Several factors contribute to this development. First, all the afferent and efferent fibers and many cells that were damaged during the slicing procedure degenerate. Second, due to cellular migration, the overall surface area covered by the slices greatly enlarges. And third, neurons tend to move away from the compact layers such as the pyramidal or Purkinje cell layer, presumably because oxygen supply is limited within the deeper layers.

After 2–4 weeks *in vitro*, the cultures have flattened from the original 400 μm into a quasi-monolayer so that individual nerve cells can now be observed in the living state with either phase or differential interference contrast microscopy. Figure 2 illustrates groups of cerebellar Purkinje cells which were grown for several weeks *in vitro* and which display well-characterized nuclei and one or two prominent nucleoli. Purkinje cell axons become myelinated during the first 2 weeks in culture.

Although cerebellar cultures derived from newborn animals show particularly dramatic spreading (the surface area enlarges several-fold), the overall tissue organization with cortical structures and deep cerebellar nuclei can still be easily

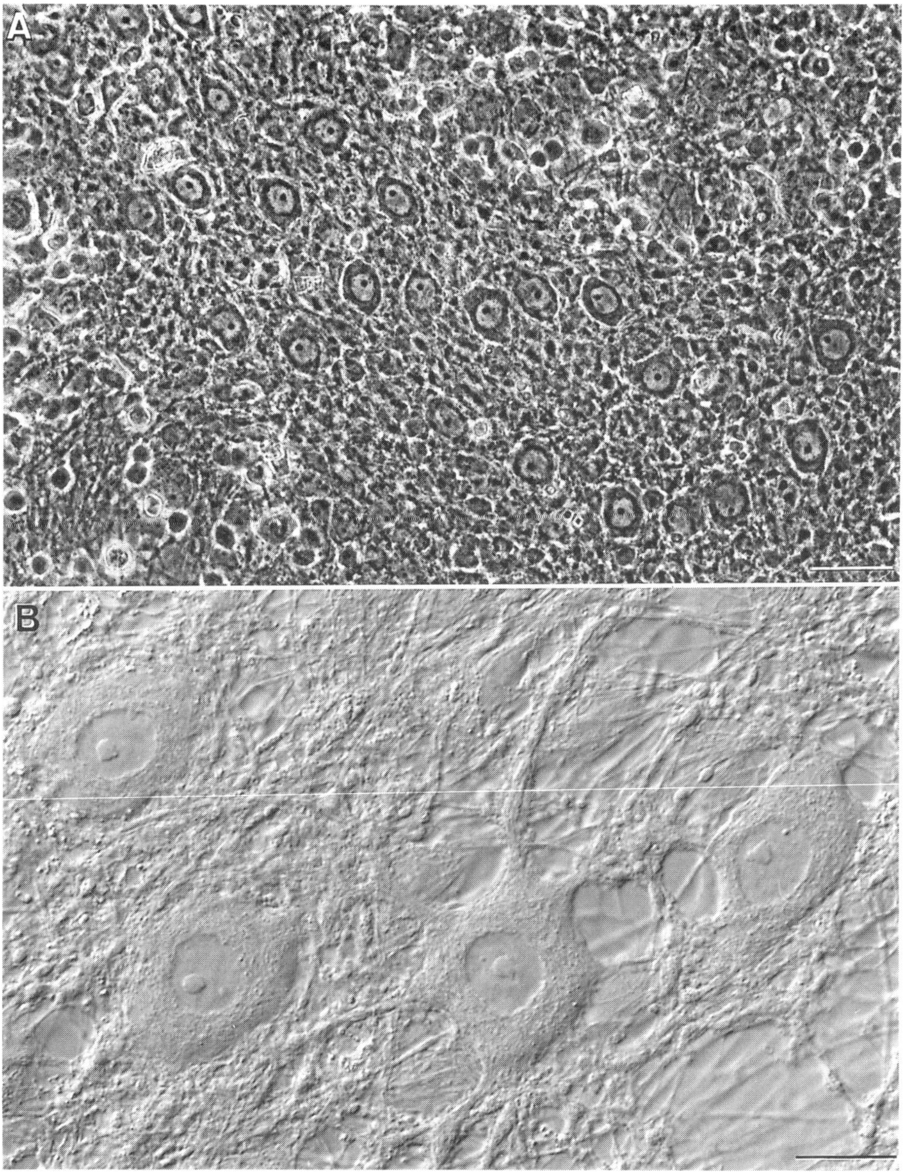

Figure 2 Living cerebellar Purkinje cells. (A) Phase contrast microscopy, after 20 days *in vitro*. (B) Differential interference contrast optics, 18 days *in vitro*. Scale bar is 50 μm in (A) and 25 μm in (B)

recognized. In cerebellar slice cultures, Purkinje cells no longer align in a single row, but rather cluster to form folia-like structures (Figure 3). Immunocytochemical analysis with antibodies raised against GABA (cf. Streit *et al.*, 1989) revealed that the somata of Purkinje cells remain unlabeled, whereas their axons are seen to grow toward the deep cerebellar nuclei located in the center of these cultures (Figure 4). Purkinje cells including somata, axonal as well as dendritic processes, can be stained with antibodies to the 28-kDa Ca^{2+}-binding protein (Figure 3). Dendritic trees of cerebellar Purkinje cells in culture never attain the full arborization of adult Purkinje cells (Gähwiler, 1981, 1984a) (Figure 3).

Tissue organization and cellular differentiation has been shown to occur in all brain areas cultured by means of the roller-tube technique. Hippocampal cultures display an organotypic organization reminiscent of their *in situ* counterparts. Although the hippocampal nerve cells are tightly packed within cell layers, the cell bodies of pyramidal cells located in areas CA1 to CA4 can be viewed in the living state using various optical techniques (Figure 5). Studies using dye-injection techniques have shown that in contrast to cerebellar Purkinje cells, pyramidal cells grow dendritic arborizations which, as far as shape, complexity of arbors, and spine investment are concerned, closely resemble those observed in pyramidal cells *in situ*. Their dendrites show bipolar orientation with basal as well as apical arborizations (Figure 6). With respect to their dendritic morphology, cultured granule cells also show normal dendritic trees with monopolar dendrites covered by spinous processes (Gähwiler, 1984b; Zimmer *et al.*, 1985).

The connective organization of slice cultures has been extensively studied in hippocampal tissue, where possible synaptic rearrangements of one of the intrinsic pathways, the mossy fiber system, was analyzed by means of electron microscopy (Frotscher and Gähwiler, 1988) and by the Timm's sulfide silver stain (Zimmer and Gähwiler, 1987; Zimmer *et al.*, 1985). In the ultrastructural analysis, mossy fiber endings were identified by their large size and their numerous clear synaptic vesicles, which were intermingled with some dense-core vesicles (Frotscher and Gähwiler, 1988). The histological studies using the heavy metal stains demonstrated that in culture mossy fibers essentially project to their normal terminal fields. In addition, new aberrant connections formed, in particular some supragranular mossy fiber collaterals, suggesting that synaptic reorganization takes place in response to deafferentation. Interestingly, these aberrant connections developed according to the rules known from lesion and transplant studies (Zimmer and Gähwiler, 1987; Zimmer *et al.*, 1985).

IV PHYSIOLOGICAL PROPERTIES

IV.1 Microelectrode Experiments

For physiological studies, the cultures are transferred to a special microchamber which is located between the optics of an inverted microscope. The microscope is

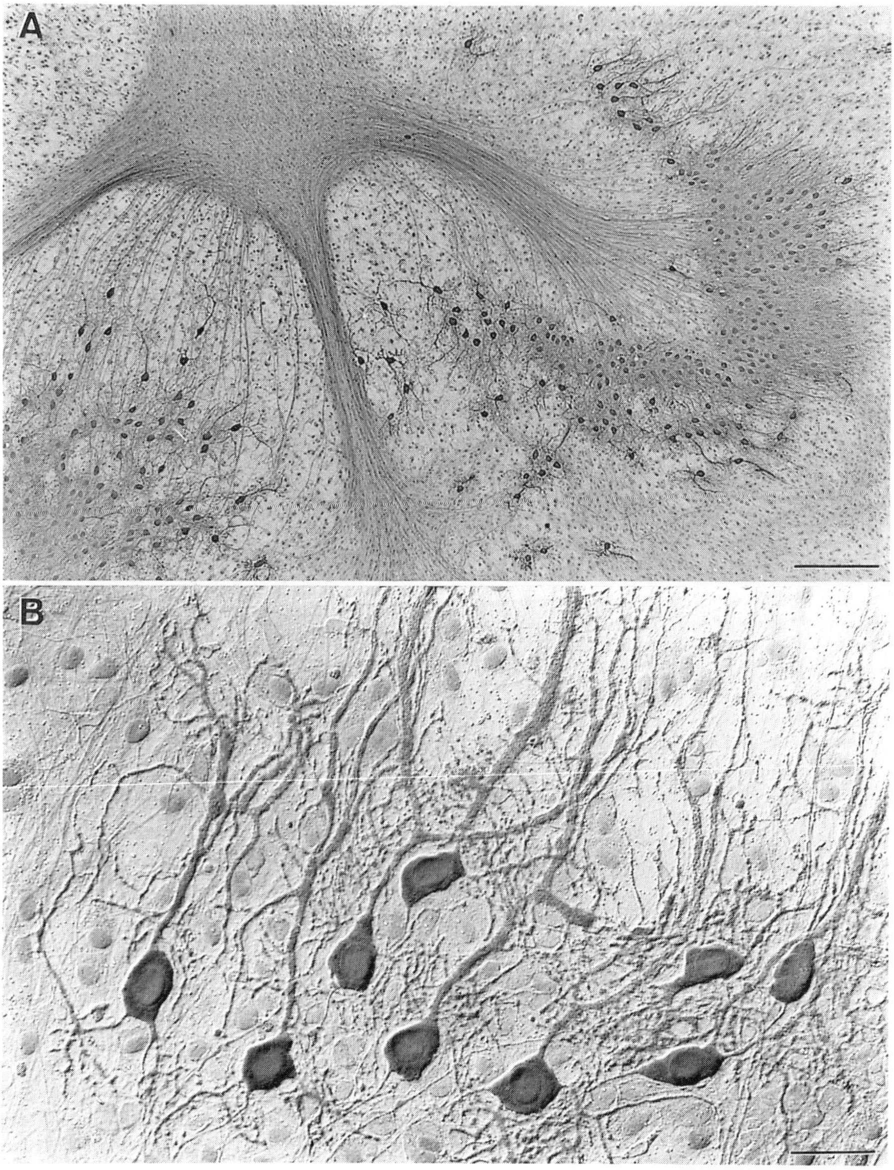

Figure 3 Cerebellar slice cultures. (A) 33-day-old culture stained with antibodies raised against the 28-kDa Ca^{2+}-binding protein. This antibody labels somata, dendritic as well as axonal, of Purkinje cells. (B) Detail of group of Purkinje cells immunocytochemically stained with antibodies to the 28-kDa Ca^{2+}-binding protein. Scale bar is 0.5 mm in (A) and 50 μm in (B)

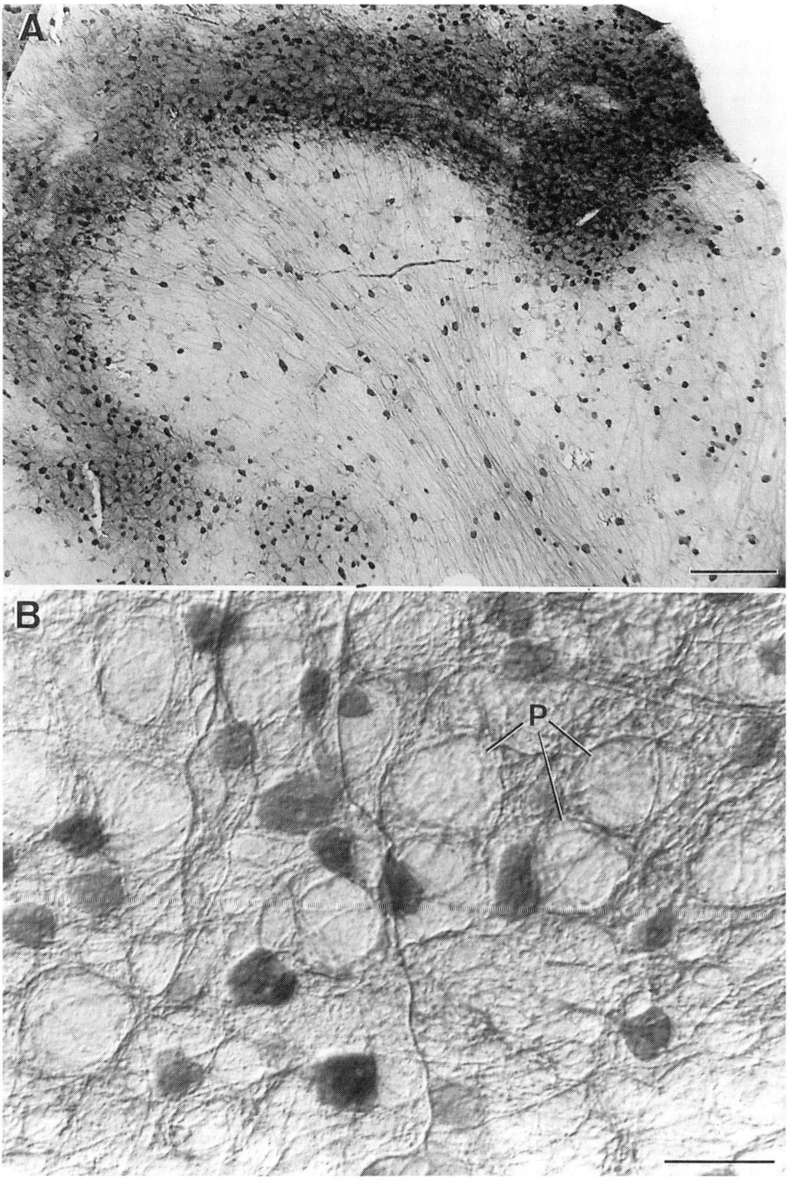

Figure 4 Cerebellar culture stained with anti-GABA antibodies. (A) A 19-day-old culture stained with anti-GABA antibodies (mAb 3A12; Matute and Streit, 1986). The darkly stained cell bodies presumably represent cerebellar interneurons. The somata of Purkinje cells (P) are not stained, but their labeled axons are seen to grow in the direction toward the deep cerebellar nuclei. (B) Detail enlarged from (A). Scale bar is 0.2 mm in (A) and 25 μm in (B). (P. Streit and B.H. Gähwiler, unpublished observations)

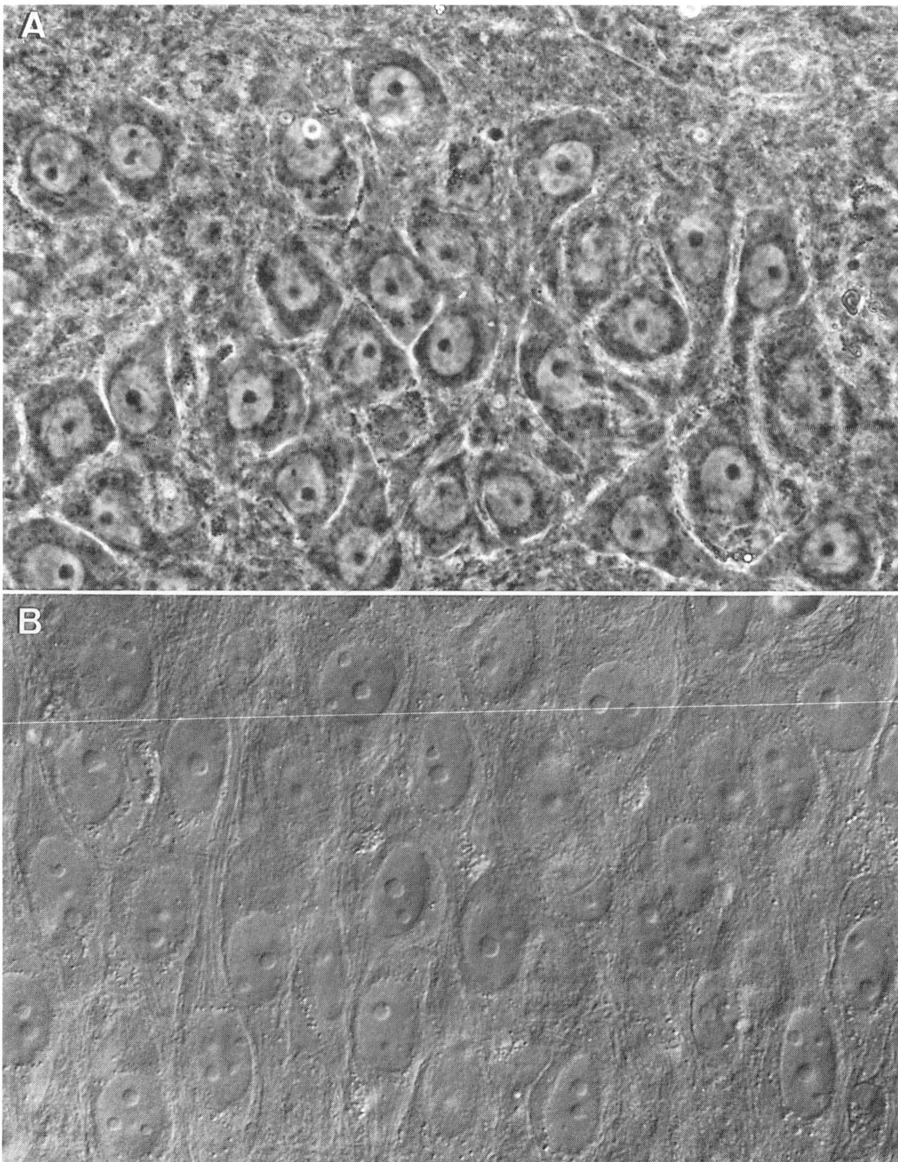

Figure 5 Living hippocampal pyramidal cells in 20-day-old culture viewed with phase contrast (A) or differential interference (B) optics. Scale bar = 25 μm

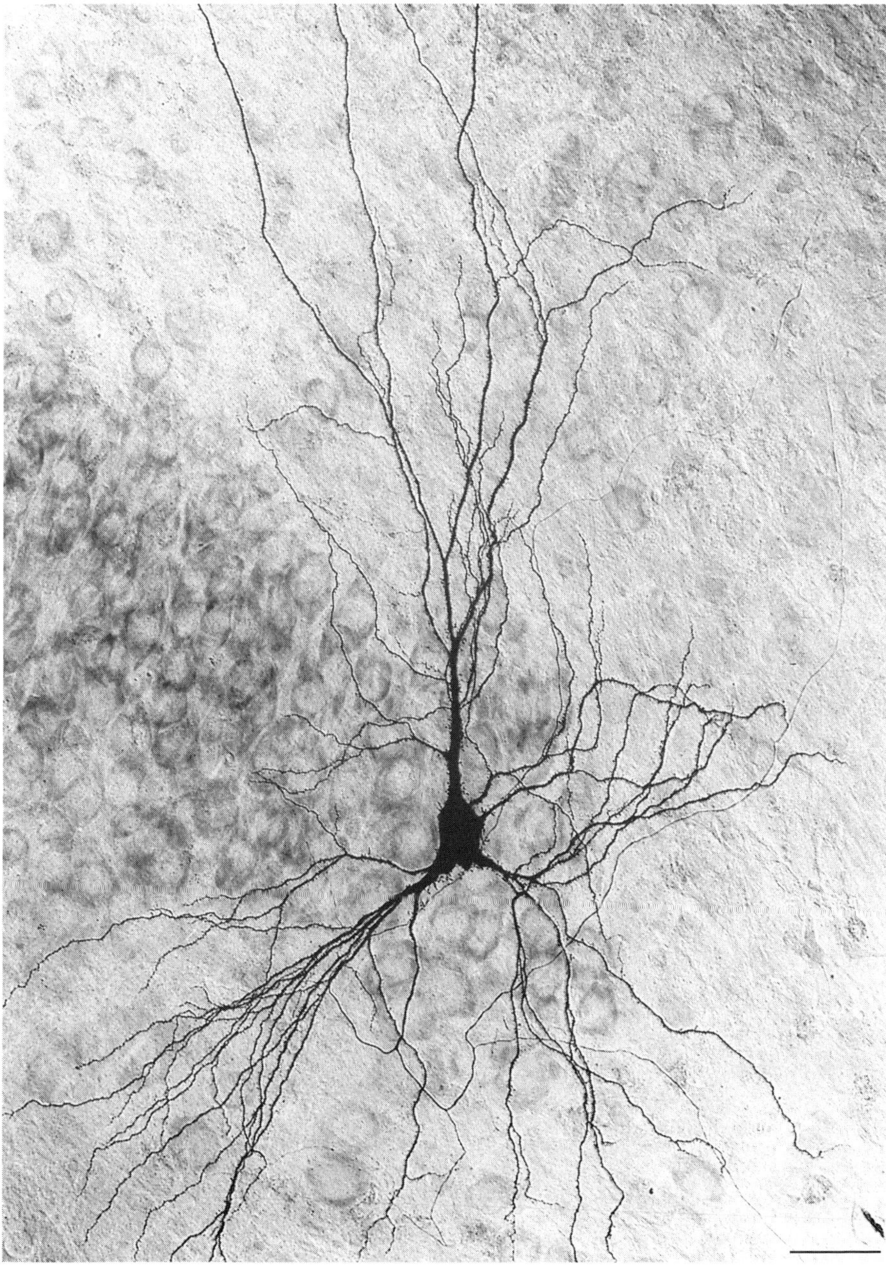

Figure 6 Cultured pyramidal cell from area CA3 of hippocampal culture after 32 days *in vitro*. The cell was labeled by intracellular injection of horseradish peroxidase and counterstained with toluidine blue. Scale bar = 50 μm

mounted on specially designed horizontally movable bearings so that the whole culture can be scanned at high power even if a microelectrode has already been positioned (Figure 7). This movable system, therefore, allows for accurate placement of several microelectrodes under visual control (Gähwiler, 1981).

The microchamber is made out of stainless steel (Figure 8). Insulated thermocoax wires (Philips) are inserted into the metal frame and the temperature is sensed by means of a thermistor and fed back into a bridge circuit. Temperature measurements

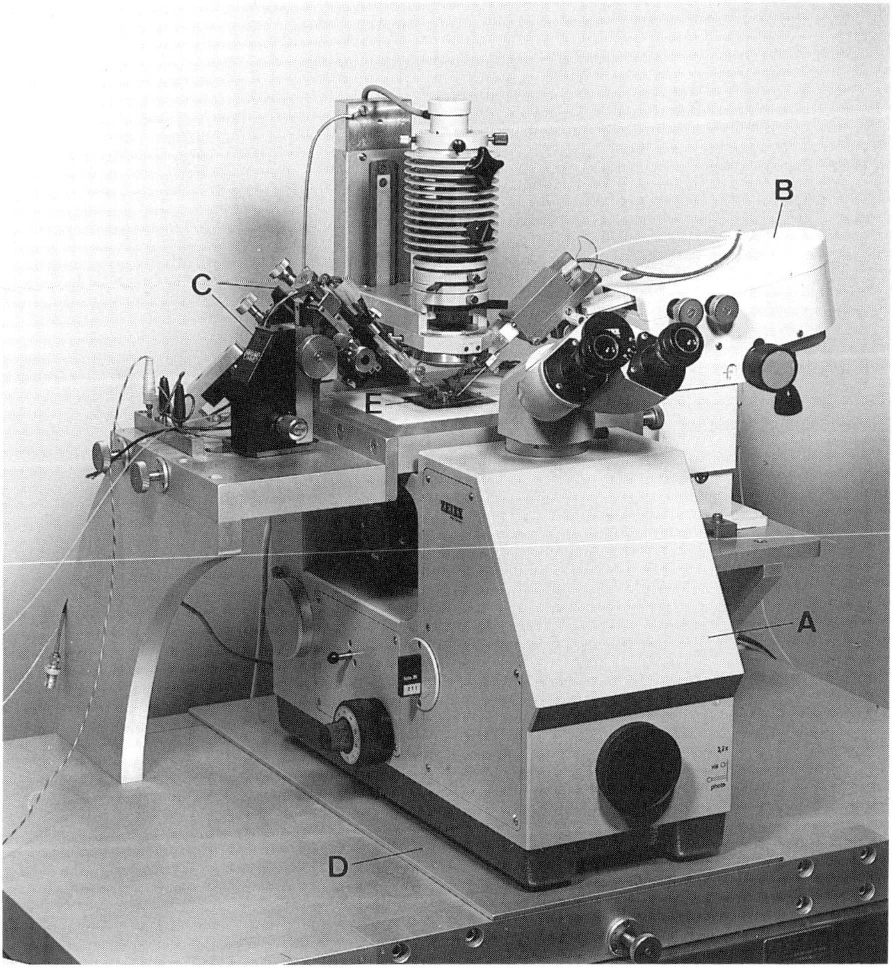

Figure 7 Set-up for electrophysiological experiments. (A) Inverted microscope. (B) Mechanical micromanipulator with additional hydraulic microdrive. (C) Additional micromanipulators for stimulation and drug delivery. (D) Sliding device to move microscope. (E) Microscopic stage

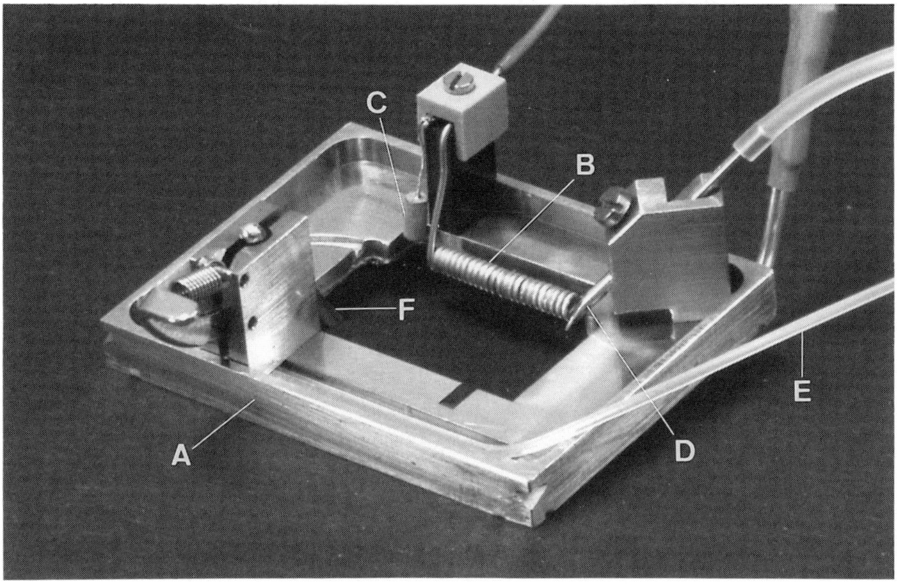

Figure 8 Microchamber for electrophysiological recordings. (A) Stainless steel frame with inserted thermocoax wire. (B) Platinum coil for holding thermistor; perfusion outflow. (C) Reference electrode (Ag–AgCl pellet). (D) Needle for setting fluid level. (E) Perfusion inflow. (F) Pressure foot for stabilizing coverslip

at different points within the chamber indicated that the temperature stays uniform within 0.4°C. The chamber has a volume of 0.3 ml and is continuously perfused with BSS, usually at a rate of 0.5 ml/min. The fluid level in the chamber is kept constant by continuous aspiration of fluid and air into a cut surgical needle whose tip is placed at the desired fluid level.

Intracellular recordings are carried out from large nerve cells under optical control. Figure 9 shows representative records obtained from four different types of nerve cells derived from hippocampal and cerebellar cultures. Intracellular injection of small depolarizing current pulses results in the generation of action potentials in all neurons studied. The response of hippocampal pyramidal cells is characterized by a series of action potentials of high amplitude and relatively long duration, with a high initial discharge rate followed by rapid spike adaptation. In contrast, presumed hippocampal interneurons have very brief spikes, fast hyperpolarizing afterpotentials and do not show any spike adaptation in response to injection of a constant-current pulse. Action potentials of cerebellar Purkinje cells, on the other hand, appear to be superimposed on a small, slow-rising local response and are followed by prominent afterhyperpolarizations. Very high frequencies of repetitive discharges could be elicited in neurons derived from the deep cerebellar nuclei.

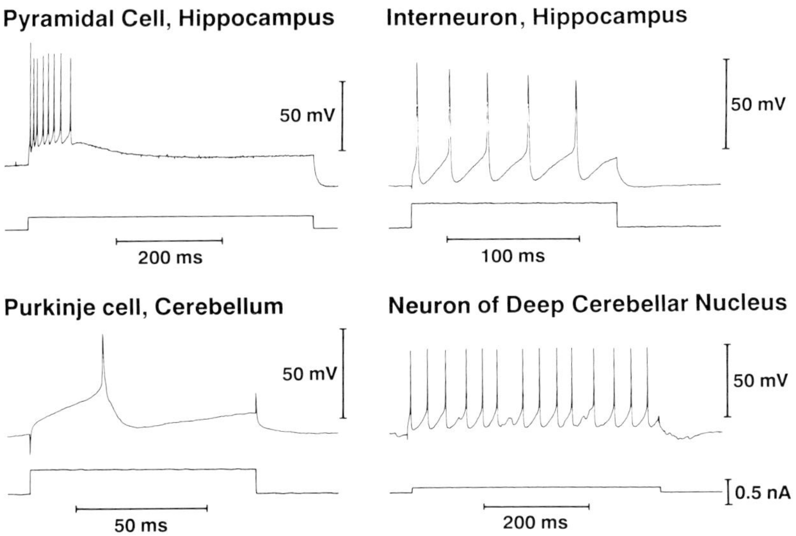

Figure 9 Intracellular recordings from different types of nerve cells in hippocampal and cerebellar cultures. Action potentials were evoked by intracellular injection of depolarizing current pulses

IV.2 Optical Recordings

Monolayered cultures grown on thin glass coverslips offer the unique opportunity to apply powerful optical recording techniques to organotypically organized neuronal networks. Methods for microfluorometric measurements of membrane potentials and intracellular Ca^{2+} have only recently been worked out in these cultures (Knöpfel *et al*., 1988) and will therefore be described in more detail. Optical recordings are performed with an experimental set-up similar to that described above for microelectrode techniques but which has been supplemented with additional equipment (Figure 10). For epifluorescence microscopy a mercury arc lamp (HBO 100, Osram, FRG), excitation and heat filters, a shutter (Prontor, FRG), a dichroic mirror and a barrier filter are used. The mercury arc lamp is driven by a stabilized d.c. power supply and carefully adjusted for minimal arc wandering noise. An 10×10 array of photodiodes is used to detect the optical signals with high temporal resolution (0.2 ms). Very low noise current-to-voltage converters for each photodiode output are positioned directly behind the array. Their outputs are fed into high-gain amplifiers having digital switches to select the gain and the low-pass cut-off frequency. High-spatial-resolution mapping of optical signals (256 $\times$ 256 picture elements) is obtained with a specially developed charge-coupled device (CCD) allowing a frame rate of 60 Hz with high dynamics. The combined implementation of these two detection systems in one experimental set-up allows for optical recordings with high spatio-temporal resolution. The experimental protocol including data acquisition, shutter opening, intracellular current injection and electrical stimulation is under computer control. For recordings of signals from less than sixteen photodiodes a personal computer suffices, but a more powerful workstation based on a Motorola 68020/30 CPU and the VME-bus is needed to acquire data from all photodiodes and the CCD.

Optical monitoring of membrane potentials takes advantage of the fact that the optical properties of certain dyes incorporated into the outer leaflet of the plasma membrane are a function of the membrane potential (Loew, 1988). For low-noise optical recordings it is critical that the neuronal plasma membrane is rather selectively stained with the voltage-dependent fluorescent dye (Knöpfel and Fromherz, 1987). The cultures are stained by bath application of the styryl dye RH 414 or the more lipophilic dye RH 421 (0.5–5 μm; Molecular Probes). The fluorescence of stained neurons is observed using an excitation filter centered around the intense 546-nm line of the mercury spectrum, a FT 510 (Zeiss)

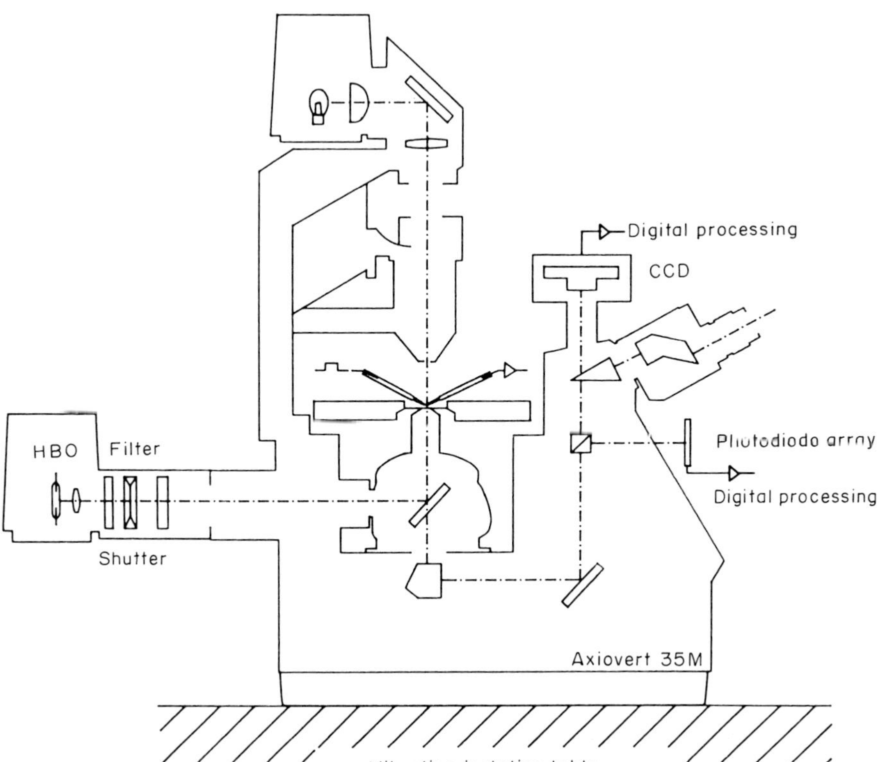

Figure 10 Schematic drawing of the experimental set-up used for optical recordings. An inverted microscope is mounted on a vibration isolation table and equipped for transmission and epifluorescence measurements

dichroic mirror and a RG 610 (Schott) cut-off filter. The dyes have been applied to hippocampal as well as cerebellar cultures. Although in cerebellar Purkinje cells the voltage-related fluorescence signals amount to only about 1%/100 mV of the total detected fluorescence (Figure 11), spontaneous action potentials and large postsynaptic potentials can be recorded from many cells simultaneously by means of low-noise amplification techniques (Knöpfel, Staub and Gähwiler, in preparation). Figure 11 shows an optical recording from a current-clamped Purkinje cell.

Changes in intracellular Ca^{2+} concentrations are measured with the fluorescent calcium indicator fura-2 (Grynkiewicz *et al*., 1985; Molecular Probes). Intracellular injection through the recording electrode has several advantages in comparison with exposing the cells to fura-2 AM esters. First, the optical signal is not obscured by signals arising from other cells. Second, difficulties arising from incomplete ester hydrolysis and loading of cellular compartments other than the cytosolic one are avoided. Most important, however, the Ca^{2+} measurements are combined with microelectrode recordings under current- and voltage-clamp conditions. For intracellular injection of fura-2, the very tip of the recording electrode is filled with a solution containing 1 mM fura-2 and 150 mM potassium acetate. The electrode is then backfilled with 2 M potassium acetate or potassium methylsulfate. The dye is injected by applying a constant hyperpolarizing current of −0.5 to −1 nA for 5–20 min. The electrode resistance drops from an initial

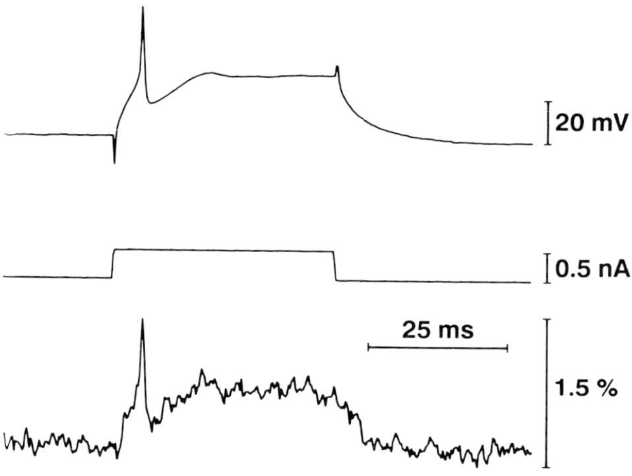

Figure 11 Simultaneous intracellular microelectrode and optical recording from a cerebellar Purkinje cell. An action potential was evoked by intracellular current injection. Upper trace, membrane potential; middle trace, injected current. The lower trace shows the simultaneously recorded fluorescence signal calibrated as a fraction of the overall fluorescence intensity. The cell was stained with RH 414 (5μM)

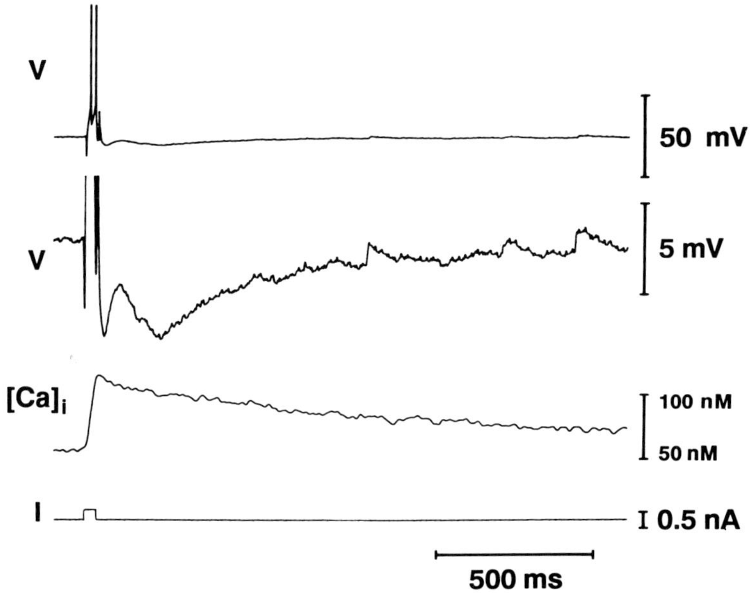

Figure 12 Intracellular recording from a hippocampal pyramidal cell combined with microfluorometric measurement of cytosolic free Ca^{2+}. Intracellular injection of a short current pulse resulted in the generation of two action potentials followed by a transient increase in intracellular Ca^{2+} and a slowly decaying afterhyperpolarization. Upper and second trace, membrane potential at high and low gain. Lower trace, injected current. Third trace, cytosolic free Ca^{2+} as monitored by fura-2 fluorescence

value of 150–200 MΩ to about 50 MΩ, thus allowing for switched current- and voltage-clamp recordings (Knöpfel, Brown and Gähwiler, in preparation). Figure 12 illustrates an example of recordings obtained from a hippocampal pyramidal cell. Intracellular injection of a short current pulse results in the generation of action potentials followed by a slowly decaying afterhyperpolarization which has been shown to be produced by a Ca^{2+}-dependent K^+ current. The depolarization-induced Ca^{2+} inflow is monitored by the fura-2 signal.

V PHARMACOLOGICAL PROPERTIES

The presence of receptors for putative neurotransmitters in slice cultures derived from various brain areas has been determined in a large number of studies by electrophysiological recording techniques whereby the drugs were applied either by iontophoresis, microperfusion or direct addition to the bathing solution. In particular, cultured cerebellar neurons are sensitive to putative amino acid transmitters (Gähwiler, 1975) and monoamines (Gähwiler, 1976), whereas hippocampal neurons show distinct responses to excitatory and inhibitory amino acids (Gähwiler and Brown, 1985a), monoamines (Gähwiler, 1980), cholinergic compounds (Gähwiler and Dreifuss, 1982) and opioid peptides (Gähwiler, 1980). Some of these effects have been analyzed with conventional electrophysiological techniques such as current- and voltage-clamp, and others by means of the patch-clamp approach (Llano *et al.*, 1988). Hypothalamic neurons (Wray *et al.*, 1988), on the other hand, additionally interact with various peptides such as angiotensin and opioid peptides, nicotinic agents, and amino acids (Gähwiler and Dreifuss, 1980).

As an example, the responses of cultured hippocampal neurons to muscarinic agents are summarized. Application of acetylcholine or muscarine to hippocampal cultures depolarizes and excites hippocampal pyramidal cells, an effect which is accompanied by a fall in input conductance and which is insensitive to anti-nicotinic drugs, but which is completely blocked by low concentrations of muscarinic antagonists such as atropine or pirenzepine (Gähwiler and Brown, 1985b). Single-electrode voltage-clamp experiments were used to determine the nature of the conductances which are affected by the muscarinic agonists. These studies clearly established that in cultured pyramidal cells, muscarine reduces at least three different ionic conductances. First, cholinergic agonists reduce a Ca^{2+}-independent K^+ current, I_M, which had previously been described in adult hippocampal pyramidal cells (Halliwell and Adams, 1982). Second, the Ca^{2+}-dependent K^+ current underlying the slowly decaying spike afterhyperpolarizations, I_{AHP} (Cole and Nicoll, 1984), is completely blocked by low concentrations of muscarine (Figure 13). Third, in more recent experiments it was found that in addition to its inhibiting effect on K^+ currents, muscarine can also inhibit Ca^{2+} currents in rat hippocampal pyramidal cells (Gähwiler and Brown, 1987a). The effects of muscarine on this latter current will be described in more detail.

In cells impaled with K^+-filled electrodes, muscarine produces a net inward current and reduces membrane input conductance. Under these conditions, inward Ca^{2+} currents (Gähwiler and Brown, 1987a) are transiently increased by muscarine (Figure 14 shows that the response to a depolarizing voltage command reverses from outward to net inward), but are later reduced in amplitude, even while the input conductance remains reduced. The initial increase in inward current is probably due to a suppression of outward current, since it is prevented or reduced by prior application of 1 mM Ba^{2+}, which replicates and occludes the effects of muscarine on outward currents (Gähwiler and Brown, 1987b).

To study the effect on muscarine on Ca^{2+} currents in more detail, the neurons were impaled with Cs^+-filled electrodes and K^+ currents were further reduced by extracellular application of TEA. Under these conditions, muscarine consistently reduces the amplitude of all components of inward current, as illustrated in Figure 15A and C. The action of muscarine was slow in onset and reversible. Norepinephrine, which has been reported to also block K^+ currents such as I_{AHP}, has no effect on inward Ca^{2+} currents (Figure 15B).

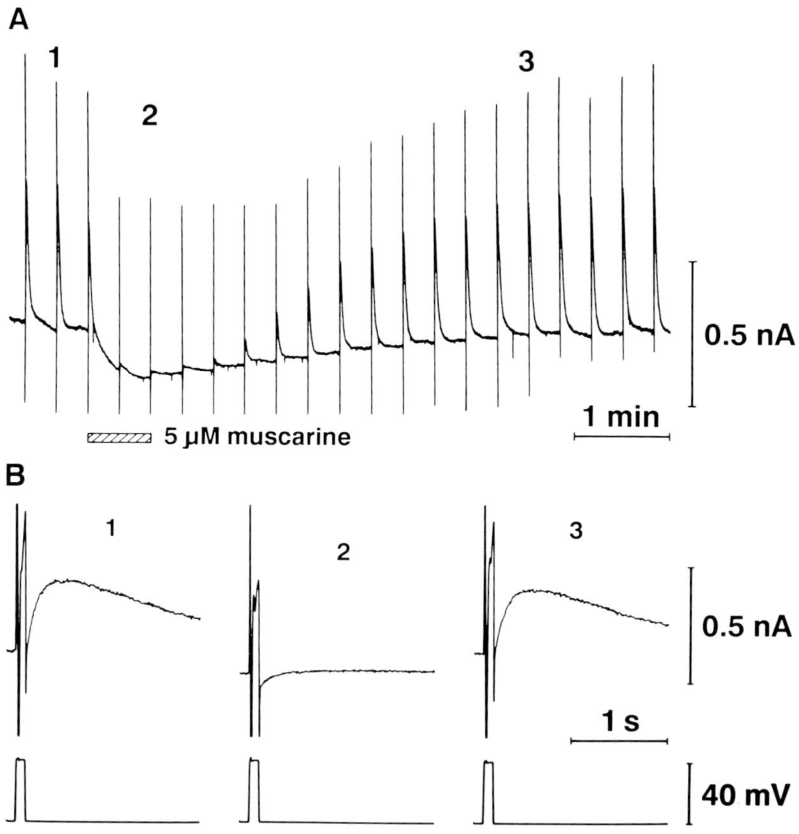

Figure 13 Effect of muscarine on the Ca^{2+}- activated K^+ conductance underlying the slowly decaying afterhyperpolarization in a hippocampal pyramidal cell. The cell was voltage-clamped at −57 mV in the presence of 1 μM TTX and 10 mM TEA. Voltage jumps of 100 ms duration to −17 mV induced outward tail currents which correspond to the voltage-clamped afterhyperpolarization. Bath-applied muscarine (5 μM) reversibly induced a slow inward current and abolished the outward tail current. (A) Continuous chart recording of clamp current. (B) Voltage jumps and corresponding clamp currents recorded at times indicated in (A)

IV CO-CULTURES

Several aspects of the development and function of a neuronal projection could be studied under well-controlled experimental conditions provided the neurons of the origin of a projection could be induced to innervate a co-cultured target tissue *in vitro*. In an attempt to establish the *in vitro* analog of a neuronal connection, we have co-cultured slices from anatomically remote brain areas which are interconnected *in situ*. A number of different tissues have so far been co-cultured, including the

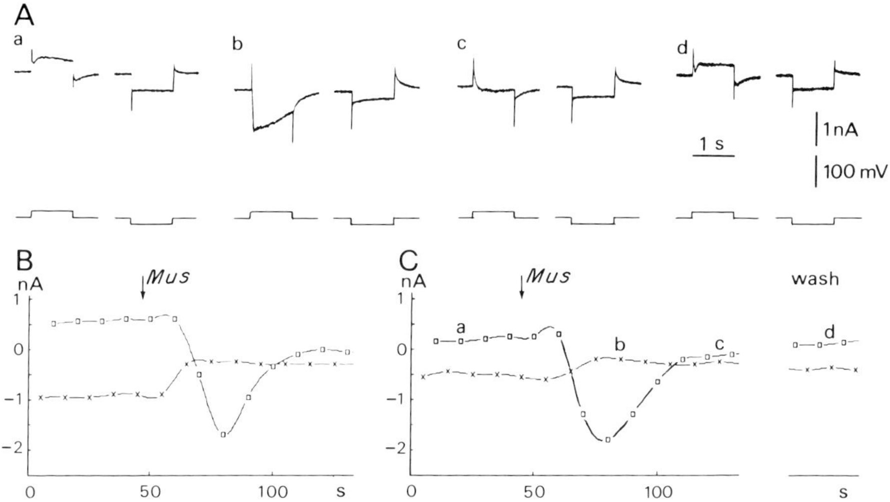

Figure 14 Effect of muscarinc on inward Ca^{2+} currents in a CA3 hippocampal pyramidal cell which was impaled with a KCl-filled microelectrode, clamped at a potential of −38 mV and subjected to 1 s, −20 mV and +20 mV voltage steps which alternated in polarity every 5 s. Application of 10 μM muscarine decreased the outward current as determined by application of hyperpolarizing commands and induced an inward current (+20 mV commands). Despite continuous perfusion with muscarine, this inward current rapidly declined. (A) Examples of currents recorded at the times indicated in (C). Upper traces, current; lower traces, voltage. (B) Time-course of conductance changes induced by muscarine. The graphs show the effect on 'leak' current (×) and peak inward current (o). (C) Muscarine applied a second time produced very similar effects. (B.H. Gähwiler and D.A. Brown, unpublished experiments; see Gähwiler and Brown, 1987b)

septohippocampal (Figure 16A), olivocerebellar, nigrostriatal, and hypothalamo-hypophysial pathways and projections from the dorsal root ganglia to the spinal cord and from the locus coeruleus to the hippocampus (Figure 16B).

The hippocampus receives a cholinergic innervation by fibers originating in the nuclei of the medial septum and the adjacent vertical limb of the diagonal band (Mesulam *et al.*, 1983; Milner *et al.*, 1983). This septohippocampal projection has traditionally been one of the favorite objects for studying mechanisms involved in neuronal development. We have, therefore, co-cultured slices of the septal region and of hippocampus and placed them side-by-side on glass coverslips (Gähwiler *et al.*, 1989; Gähwiler and Hefti, 1984). Histochemical stainings for the cholinergic enzymes acetylcholinesterase and choline acetyltransferase after several weeks *in vitro* show that cholinergic fibers originating in the septal explants have explored their new environment and grown into the adjacent hippocampal target tissue (Figure 16A). It is interesting that the trophic substance nerve growth factor (NGF) has a profound influence on the development of this cholinergic connection (Gähwiler *et al.*, 1987). Continuous application of anti-NGF antibodies during the

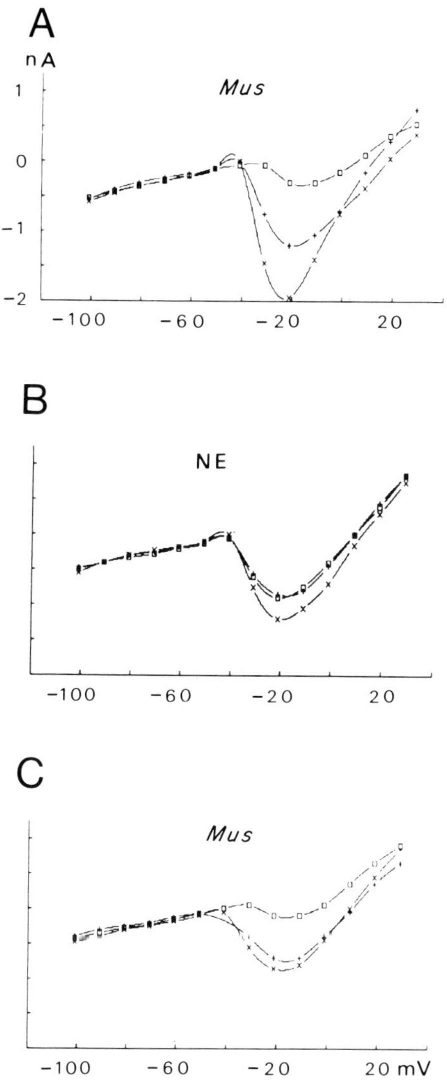

Figure 15 Effect of muscarine and norepinephrine on inward currents. The culture was perfused with BSS containing 1 μM TTX and 10 mM TEA. Holding potential, −40 mV. (A) Muscarine, applied at a concentration of 10 μM, reduced inward currents. The effect was partially reversible. (B) Application of norepinephrine (100 μM) had only a minor and irreversible effect on inward current. This decrease can probably be attributed to a gradual rundown of Ca^{2+} currents. (C) After washout of norepinephrine, the same concentration of muscarine again reversibly reduced inward currents. Graphs show current–voltage curves for peak inward current recorded before (×), during (□) and after (+) applications of muscarine (Mus) and noradrenaline (NE). (B.H. Gähwiler and D.A. Brown, unpublished experiments; see Gähwiler and Brown, 1987b)

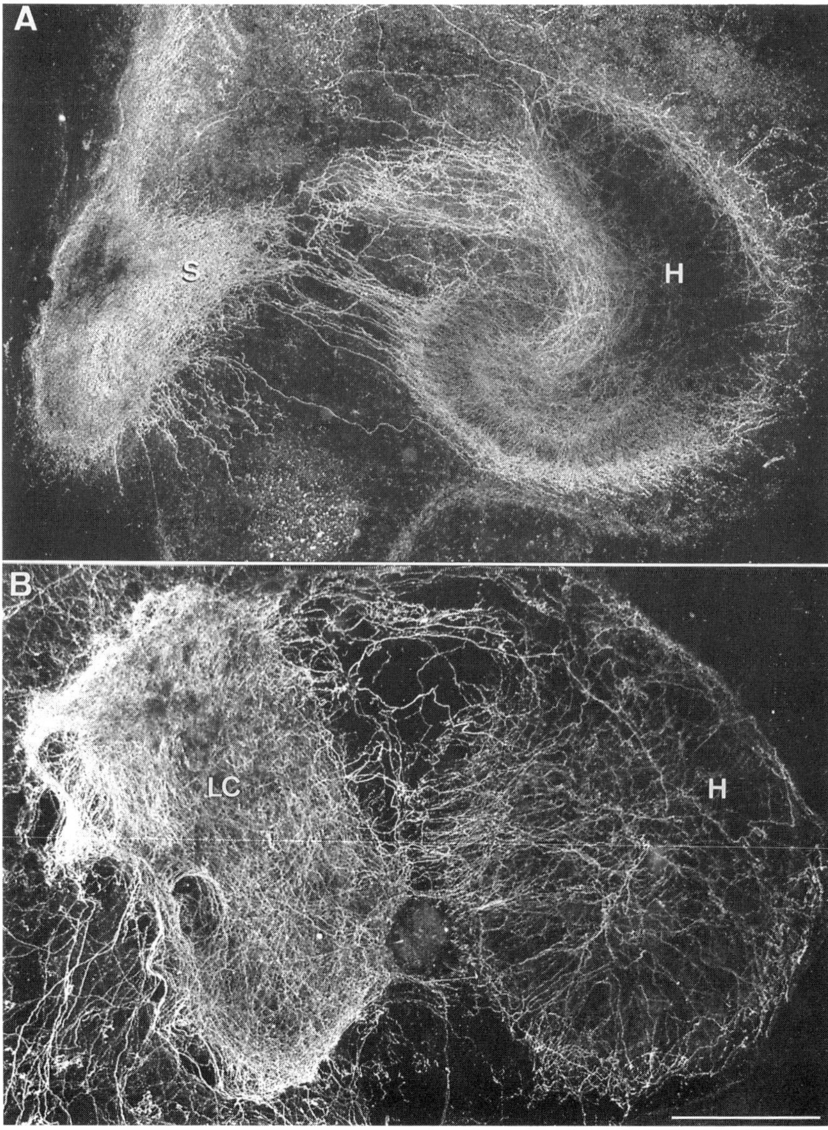

Figure 16 Co-cultures between different brain areas. (A) Co-culture of septum (S) and hippocampus (H) stained for acetylcholinesterase (Geneser-Jensen and Blackstad, 1972) and viewed with dark-field microscopy; 28 days in culture. (B) Co-culture derived from the locus coeruleus (LC) and the hippocampus (H) stained after 39 days *in vitro* with antibodies raised against tyrosine hydroxylase (TH). Dark-field illumination was again applied to illustrate hippocampal ingrowth by thin varicose TH-positive fibers that originate in presumed noradrenergic neurons of the locus coeruleus. Scale bar = 1 mm. (Antibody against TH kindly provided by M. Goldstein)

entire growth of the cultures reduces hippocampal ingrowth by cholinergic fibers, whereas a striking increase in the number of cholinergic fibers innervating the target is observed when NGF is added. Biochemical studies revealed that the activities of both cholinergic enzymes, acetylcholinesterase and choline acetyltransferase, are increased by a factor of 6 following treatment with NGF. These effects are completely abolished by simultaneous application of anti-NGF antibodies (Gähwiler *et al*., 1987, 1989).

The cholinergic fibers originating in the septal slices recognize the target neurons and establish functional muscarinic connections with hippocampal pyramidal cells. Electrophysiological studies have shown that septal stimulation induces in pyramidal cells a delayed, low-amplitude and very slow inward postsynaptic current which lasts up to a minute and which is enhanced by the anti-acetylcholinesterase drug neostigmine and blocked by the muscarinic antagonist atropine (Gähwiler and Brown, 1985b). Analogous to the effects of exogenous muscarine, synaptic stimulation reduces two distinct K^+ currents: the Ca^{2+}-independent current I_M and the Ca^{2+}-activated current which had been termed I_{AHP} (Figure 17).

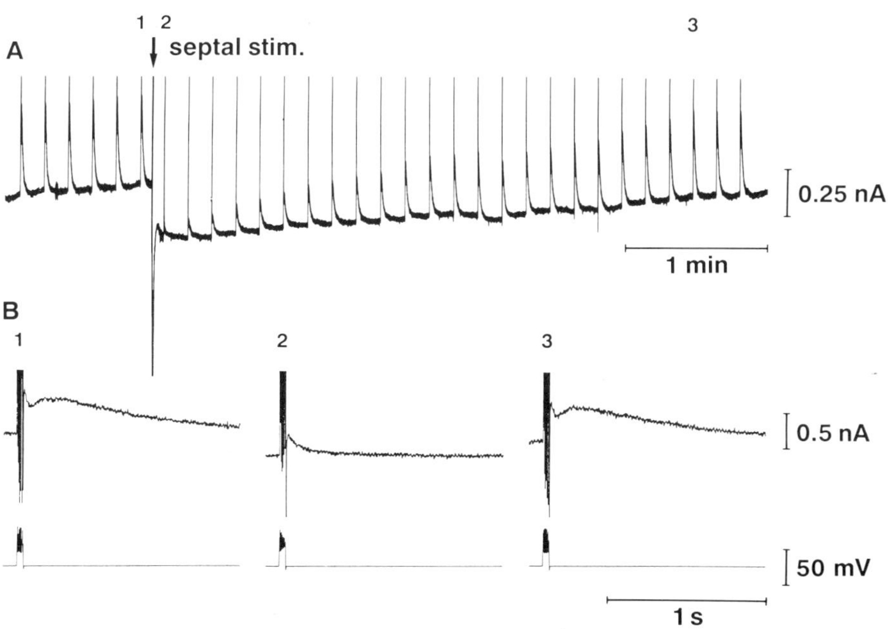

Figure 17 Effect of septal stimulation on CA3 pyramidal cell in septohippocampal co-culture: induction of slow inward current and reduction of I_{AHP}. (A) Continuous chart record shows the influence of septal stimulation (1-s train of stimuli at 40 Hz; 1 ms, 10 μA) on the slowly decaying outward currents which were induced by short command-voltage pulses (40 ms, 0.1 Hz, from -49 to -19 mV). (B) Single tail currents (upper traces) and corresponding voltage records (lower traces) monitored before (1), immediately after (2) and 225 s after (3) the septal stimulation. (Reproduced from Gähwiler *et al*., 1989, with permission)

The locus coeruleus is the major origin of noradrenergic projections in the brain. It consists of bilateral small nuclei in the brainstem which connect with almost all brain areas, including the hippocampus (Bloom *et al.*, 1972). To determine the survival of noradrenergic neurons *in vitro* and the fiber projections formed by these neurons, we have co-cultured slices from the locus coeruleus and the hippocampus (Knöpfel *et al.*, 1989). Slices derived from the locus coeruleus consistently contain tyrosinehydroxylase (TH)-immunoreactive (presumed noradrenergic) neurons. Dark-field illumination of such a co-culture reveals that a large number of TH-positive fibers reach the adjacent hippocampal target tissue (Figure 16B). The co-culture of locus coeruleus and hippocampus differ in several aspects from the septohippocampal cultures. First, NGF has no effect on the morphological development of locus coeruleus neurons, and second, noradrenergic fibers differ from the cholinergic ones in the distribution patterns as observed after several weeks *in vitro*.

VII CONCLUSIONS

The data discussed in this chapter demonstrate that slice cultures have a number of properties that make them an attractive preparation for a number of applications.

(1) Nerve cells in slice cultures continue to differentiate to a remarkable degree both functionally and morphologically.
(2) Important cytoarchitectural relationships are retained even during long-term cultivation.
(3) Immunocytochemical and histological techniques have demonstrated that there is no evidence for loss of a particular cell type due to prolonged denervation *in vitro*.
(4) Intrinsic fiber connections either form as *in situ* or according to principles known from lesion and transplant studies.
(5) Nerve cells in culture display normal chemosensitivity to putative neurotransmitters and to drugs which can, due to the minimal diffusion barriers of the quasi-monolayer preparation, be easily applied directly to individual cells.
(6) A whole spectrum of electrophysiological techniques ranging from recordings of single ion channels by the patch-clamp technique to the use of optical recording techniques can be applied to the slice culture preparation.
(7) Optical monitoring techniques appear to be particularly attractive since they allow simultaneous monitoring of membrane potential and intracellular ion concentrations from several cells.
(8) As demonstrated by the studies using NGF, the preparation is well suited for developmental studies since the short- and long-term effects of trophic substances or drugs can be studied under well-controlled *in vitro* conditions.
(9) The co-culture approach offers a unique opportunity to get further insight into

the basic mechanisms underlying specificity of axonal fiber growth and target recognition, and into the functional properties of neuronal projection between anatomically remote brain areas.

ACKNOWLEDGEMENTS

We thank L. Rietschin for excellent technical assistance, Mr W. Bauer for designing the electrophysiological set-up, including the microchamber, and Drs Goldstein, J. Mariani, M. Thomasset and C.O. Parks for generous supplies of antibodies. This work was supported by grant No. 3.354.86 from the Swiss National Science Foundation and the Dr. Eric Slack-Gyr Foundation.

REFERENCES

Bloom, F.E., Hoffer, B.J., and Siggins, G.R. (1972). 'Norepinephrine mediated cerebellar synapses: a model system for neuropsychopharmacology', *Biol. Psychiatry*, **4**, 157–177.

Cole, A.E., and Nicoll, R.A. (1984). 'Characterization of a slow cholinergic post-synaptic potential recorded *in vitro* from rat hippocampal pyramidal cells', *J. Physiol. (Lond.)*, **352**, 173–188.

Frotscher, M., and Gähwiler, B.H. (1988). 'Synaptic organization of intracellularly stained CA3 pyramidal neurons in slice cultures of rat hippocampus', *Neuroscience*, **24**, 541–551.

Grynkiewicz, G., Poenie, M., and Tsien, R.Y. (1985). 'A new generation of Ca^{2+} indicators with greatly improved fluorescence properties', *J. Biol. Chem.*, **260**, 3440–3450.

Gähwiler, B.H. (1975). 'The effects of GABA, picrotoxin and bicuculline on the spontaneous bioelectric activity of cultured cerebellar Purkinje cells', *Brain Res.*, **99**, 85–95.

Gähwiler, B.H. (1976). 'Inhibitory action of noradrenaline and cyclic AMP in explants of rat cerebellum', *Nature*, **259**, 483–484.

Gähwiler, B.H. (1980). 'Excitatory action of opioid peptides and opiates on cultured hippocampal pyramidal cells', *Brain Res.*, **194**, 193–203.

Gähwiler, B.H. (1981). 'Organotypic monolayer cultures of nervous tissue', *J. Neurosci. Methods*, **4**, 329–342.

Gähwiler, B.H. (1984a). 'Slice cultures of cerebellar, hippocampal and hypothalamic tissue', *Experientia*, **40**, 235–243.

Gähwiler, B.H. (1984b). 'Development of the hippocampus *in vitro*: cell types, synapses and receptors', *Neuroscience*, **11**, 751–760.

Gähwiler, B.H. (1988). 'Organotypic cultures of neural tissue', *TINS*, **11**, 484–489.

Gähwiler, B.H., and Brown, D.A. (1985a). '$GABA_B$-receptor-activated K^+ current in voltage-clamped CA3 pyramidal cells in hippocampal cultures', *Proc. Natl Acad. Sci. USA*, **82**, 1558–1562.

Gähwiler, B.H., and Brown, D.A. (1985b). 'Functional innervation of cultured hippocampal neurones by cholinergic afferents from co-cultured septal explants', *Nature*, **313**, 577–579.

Gähwiler, B.H., and Brown, D.A. (1987a). 'Effects of dihydropyridines on calcium currents in CA3 pyramidal cells in slice cultures of rat hippocampus', *Neuroscience*, **20**, 731–738.

Gähwiler, B.H., and Brown, D.A. (1987b). 'Muscarine affects calcium-currents in rat hippocampal pyramidal cells *in vitro*', *Neurosci. Lett.*, **76**, 301–306.

Gähwiler, B.H., and Dreifuss, J.J. (1980). 'Transition from random to phasic firing induced in neurons cultured from the hypothalamic supraoptic area', *Brain Res.*, **193**, 415–425.

Gähwiler, B.H., and Dreifuss, J.J. (1982). 'Multiple actions of acetylcholine on hippocampal pyramidal cells in organotypic explant cultures', *Neuroscience*, **7**, 1243–1256.

Gähwiler, B.H., and Hefti, F. (1984). 'Guidance of acetylcholinesterase-containing fibres by target tissue in co-cultured brain slices', *Neuroscience*, **13**, 681–689.

Gähwiler, B.H., Enz, A., and Hefti, F. (1987). 'Nerve growth factor promotes development of the rat septo-hippocampal cholinergic projection *in vitro*', *Neurosci. Lett.*, **75**, 6–10.

Gähwiler, B.H., Brown, D.A., Enz, A., and Knöpfel, T. (1989). 'Development of the septo-hippocampal projection *in vitro*', in *Central Cholinergic Synaptic Transmission* (Eds M. Frotscher and U. Misgeld), pp. 236–251, Birkhäuser Verlag, Basel.

Geneser-Jensen, F.A., and Blackstad, T.W. (1972). 'Distribution of acetylcholinesterase in the hippocampal region of the guinea pig. I. Entorhinal area, parasubiculum, and presubiculum', *Z. Zellforsch. Mikrosk. Anat.*, **114**, 460–481.

Halliwell, J.V., and Adams, P.R. (1982). 'Voltage-clamp analysis of muscarinic excitation in hippocampal neurons', *Brain Res.*, **250**, 71–92.

Knöpfel, T., and Fromherz, P. (1987). 'Fluorescence monitoring of membrane potentials: the spatio-temporal resolution in isolated neurons of helix pomatia', *Z. Naturforschung*, **42**, 986–990.

Knöpfel, T., Kasper, H.J., Kohler, B., Zglinski, Z., Zeller, L., and Gähwiler, B.H. (1988). 'Optical recording of neuronal activity in organotypic slice cultures', *Soc. Neurosci. Abstr.*, **14**, 247.

Knöpfel, T., Rietschin, L., and Gähwiler, B.H. (1989).'Organotypic co-cultures of rat locus coeruleus and hippocampus', *Eur. J. Neurosci.*, **1**, 678–689.

Llano, I., Marty, A., Johnson, J.W., Ascher, P., and Gähwiler, B.H. (1988). 'Patch-clamp recording of amino acid-activated responses in "organotypic" slice cultures', *Proc. Natl Acad. Sci. USA*, **85**, 3221–3225.

Loew, L.M. (1988). 'How to choose a potentiometric membrane probe', in *Spectroscopic Membrane Probes*, Vol. 2 (Ed. L.M. Loew) pp. 139–152, CRC Press, Boca Raton.

Matute, C., and Streit, P. (1986). 'Monoclonal antibodies demonstrating GABA-like immunoreactivity', *Histochemistry*, **86**, 147–157.

Mesulam, M.-M., Mufson, E.J., Wainer, B.H., and Levey, A.I. (1983). 'Central cholinergic pathways in the rat: an overview based on an alternative nomenclature (Ch1-Ch6)', *Neuroscience*, **10**, 1185–1201.

Milner, T.A., Loy, R., and Amaral, D.G. (1983). 'An anatomical study of the development of the septohippocampal projection in the rat', *Dev. Brain Res.*, **8**, 343–371.

Streit, P., Thompson, S.M., and Gähwiler, B.H. (1989). 'Anatomical and physiological properties of GABAergic neurotransmission in organotypic slice cultures of rat hippocampus', *Eur. J. Neurosci.*, **1**, 603–615.

Wray, S., Gähwiler, B.H., and Gainer, H. (1988). 'Slice cultures of LHRH neurons in the presence and absence of brainstem and pituitary', *Peptides*, **9**, 1151–1175.

Zimmer, J., and Gähwiler, B.H. (1987). 'Growth of hippocampal mossy fibers: a lesion and coculture study of organotypic slice cultures', *J. Comp. Neurol.*, **264**, 1–13.

Zimmer, J., Sunde, N., Sörensen, T., Jensen, S., Möller, A.G., and Gähwiler, B.H. (1985). 'The hippocampus and fascia dentata. An anatomical study of intracerebral transplants and intraocular and *in vitro* cultures', in *Neural Grafting in the Mammalian CNS* (Eds A. Björklund and U. Stenevi), pp. 285–299.

Part II: *En Bloc* Preparations

Preparations of Vertebrate Central Nervous System *In Vitro*
Edited by H. Jahnsen
Published by John Wiley & Sons Ltd

CHAPTER 5

The Lamprey CNS *In Vitro*, an Experimentally Amenable Model for Synaptic Transmission and Integrative Functions

LENNART BRODIN and STEN GRILLNER
The Nobel Institute for Neurophysiology, Karolinska Institute, Box 60400, S-10401 Stockholm, Sweden

I INTRODUCTION

Why do not neuroscientists work only on humans or at least mammals? The reasons are two-fold. (1) Many general questions about neural functions cannot be answered by studies on the complex nervous systems of mammals – the number of neurons and the number of alternative interpretations are too large when it comes to understanding the cellular basis of simple or more complex patterns of behaviour. Simplified versions of most basic patterns of behaviour can as a rule be elicited in lower vertebrates, in which the number of neurons, and in general the degree of complexity, is much less. These facts increase the chances of obtaining conclusive results and fundamental insights. Subsequently, these can be tested, to see whether they can explain the findings in more complex nervous systems. (2) In a scientific perspective, each species is also of interest in its own right.

In this chapter we will deal with jawless vertebrates (cyclostomes), the first group to evolve the general features of the vertebrate nervous system with its different subdivisions of the brainstem, basal ganglia, spinal cord, the different cranial nerves, dorsal and ventral roots, etc. The cyclostomes separated from the main vertebrate line as far back as approximately 450 million years ago (see Hotton, 1976). They are today represented by two main groups, hagfishes and lampreys. Lampreys comprise a number of species, some of which have been used extensively in neuroscience. They include *Petromyzon marinus*, *Lampetra fluviatilis*, *Lampetra planeri*, *Lampetra lamottei*, *Lampetra japonica*, *Entosphenus tridentata,* and *Ichthyomyzon unicuspis*. Since the last century, cyclostomes have attracted much attention due to their critical position in phylogeny. The classic works of Müller (1840), Retzius (1891, 1893), Ahlborn (1883), Kolmer (1905), Tretjakoff (1909a,b) and Whiting (1948) deserve special attention (see Nieuwenhuys, 1977; Rovainen, 1979b). As the CNSs of the different species of lampreys have a similar organization, we will use the term 'lamprey' for this entire group. Our own studies have been performed mainly on adult specimens of *Ichthyomyzon unicuspis* and *Lampetra fluviatilis*. The lamprey has both a larval and an adult form. During the larval stage the lamprey is a filter feeder and spends most of its time buried with the head sticking up filtering the contents of the water. After metamorphosis the lamprey instead feeds by sucking at other fishes and by means of its sharp tongue penetrates the skin and sucks in tissue from the host's body. The mouthpart changes during metamorphosis, but otherwise the body shape and movement repertoire remain unchanged (cf. Rovainen, 1982). The early development until hatching has been described by Piavis (1971), but modern methods have not yet been applied.

II *IN VITRO* PREPARATION

The lamprey spinal cord (Figures 1 and 2) is thin and ribbon-like with a dorsoventral thickness of only about 0.15–0.30 mm. It has no intrinsic blood vessels (Sterzi,

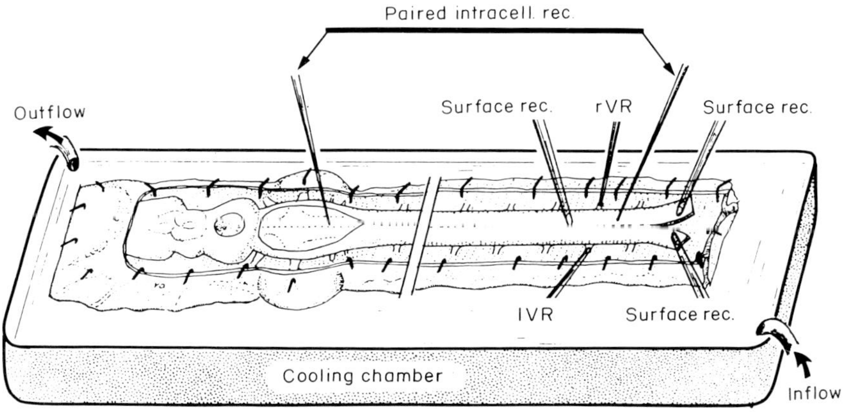

Figure 1 *In vitro* preparation of the lamprey CNS. The brain and spinal cord have been exposed by removal of the dorsal cranium and the tissue overlying the spinal cord. The preparation is fixed with small pins in the experimental chamber which contains oxygenated physiological solution. The chamber can be divided with barriers into separate pools allowing, for example, drug application to a restricted part of the preparation. The preparation is transilluminated from below which permits a direct identification of many types of neurons, and microelectrodes can be inserted under visual guidance. Suction electrodes are used for extracellular recording or stimulation of, for example, cranial nerves, ventral roots, or the surface of the spinal cord

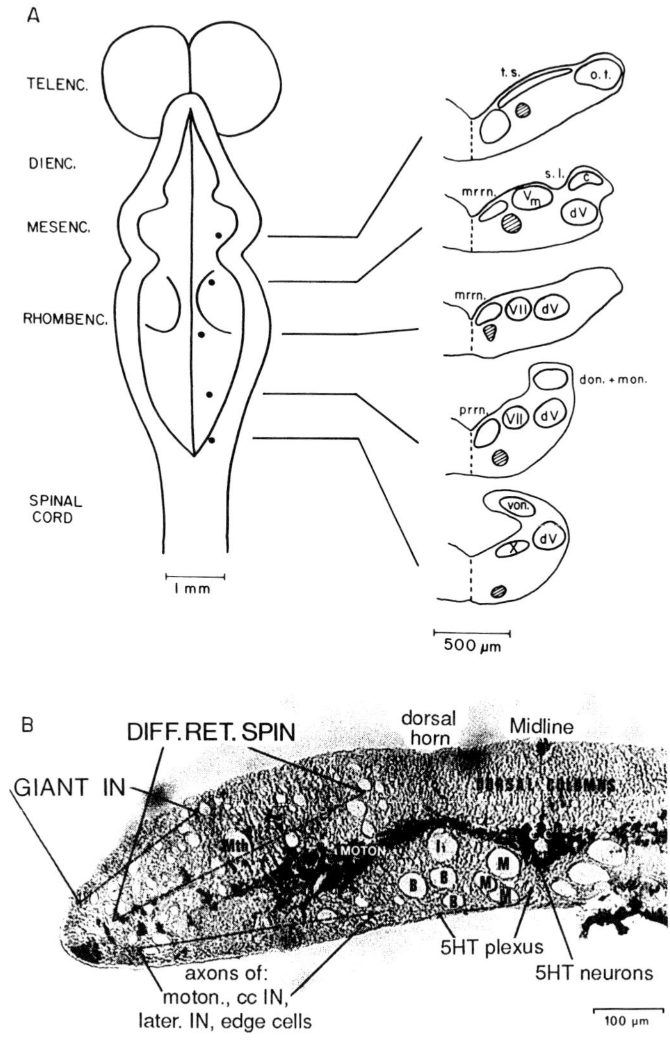

Figure 2 The lamprey CNS can generate fictive locomotion *in vitro*. (A) Schematic representation of the brain from a dorsal view, and transverse sections from different levels of the brainstem. The hatched areas indicate the regions from which fictive locomotion can be induced by microstimulation of low intensity (below 3 A). For abbreviations see Table 1. (B) Transverse section from the spinal cord stained with cresyl violet. The regions where axons of different neurons are located have been indicated with sectors (solid and dashed lines). Individual bulbar (B), isthmic (I) and mesencephalic (M) Müller cell axons, and the Mauthner (Mth) axon are indicated according to Rovainen (1979) (cc, central canal; diff. ret. spin, different reticulospinal axons)

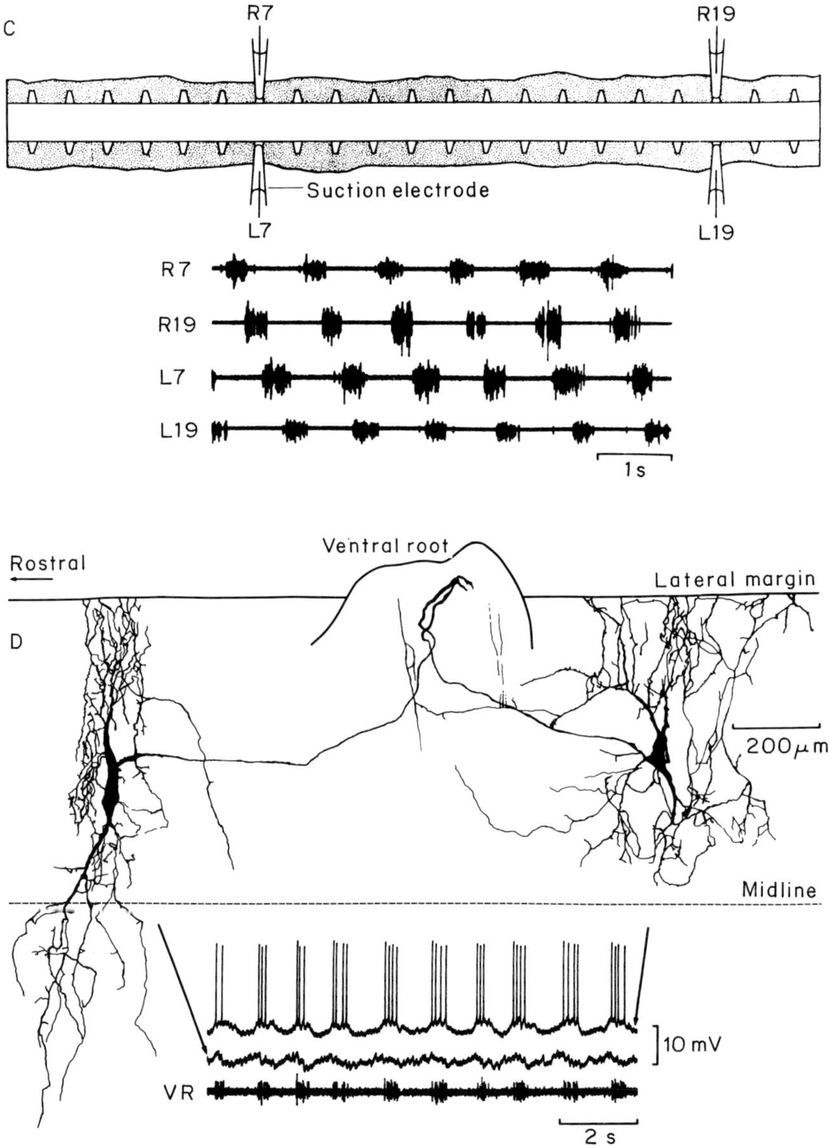

Figure 2 (*continued*) (C) Fictive locomotion in the lamprey spinal cord–notochord preparation induced by 0.6 mM D-glutamate applied to the bathing solution. The record traces show the ventral root activity in ventral roots number 7 and 19 on the right (R) and left (L) sides. (D) Two motoneurons reconstructed after intracellular injection with Lucifer yellow. The left cell projects to the ventral myotome and the right cell to the dorsal myotome. Note the difference in morphology. The record traces show the oscillatory activity in the neurons during fictive locomotion, and the burst activity in the ventral root

Table 1 Abbreviations

Individual neurons	
Spinal cord:	
CCIN	interneuron with crossed caudal axon
EIN	excitatory interneuron
GI	giant interneuron
LIN	lateral interneuron
MN	motoneuron
Brainstem:	
M	Müller cell
Mth	Mauthner cell
Nuclei	
aon	anterior olfactory nucleus
arrn	anterior rhombencephalic reticular nucleus
cho	chiasma opticum
cs	corpus striatum
di	diencephalon
dh	dorsal hypothalamic nucleus
dt	dorsal thalamic nucleus
hy	hypothalamus
ipn	interpeduncular nucleus
mes	mesencephalon
nmlf	nucleus of the medial longitudinal fascicle
mrn	mesencephalic reticular nucleus
mrrn	middle rhombencephalic reticular nucleus
ob	olfactory bulb
ot	optic tectum
otr	optic tract
pc	posterior commissure
pcn	posterior commissural nucleus
ph	primordium hippocampi
pn	preoptic nucleus
pocn	postinfundibular commissural nucleus
pp	primordium pallii
prcn	preinfundibular commissural nucleus
prrn	posterior rhombencephalic reticular nucleus
ptn	posterior tuberal nucleus
pon	postoptic commissural nucleus
rh	rhombencephalon
sd	saccus dorsalis
shl	subhippocampal lobe
sn	septal nucleus
tel	telencephalon
ts	torus semicircularis
V	Vth motor nucleus
vh	ventral hypothalamic nucleus
vt	ventral thalamic nucleus
vc	ventral commissure
vtc	ventral tegmental commissure

1904) and is oxygenated directly from the cerebrospinal fluid. The brainstem (Figure 1) has intrinsic blood vessels but is probably also oxygenated to a large extent from the cerebrospinal fluid. Rovainen (1967a,b), was the first to take advantage of these favourable conditions by developing an *in vitro* preparation of the lamprey nervous system (Figures 1 and 2).

The thin spinal cord resting on the notochord (Figures 1 and 2; Rovainen, 1967b, 1974a) or in isolation (Wallén *et al*., 1985) can be maintained *in vitro* for periods of 2–3 days at a temperature around 7–10°C. The physiological solution is similar to that of other vertebrates (Wickelgren, 1977). The brainstem (Figure 1) is somewhat more sensitive but can often be maintained during a similar period of time.

Transillumination from fibre optics or a mirror system will allow a visual identification of a number of cell types or individual cells, such as the different Müller and Mauthner cells in the brainstem, and sensory dorsal cells, intraspinal stretch receptor neurons (edge cells), giant interneurons (spinobulbar ascending cells) and lateral interneurons in the spinal cord (see below). In addition, a number of cells can be identified by simultaneous recording of the axonal action potential, such as motoneurons, certain inhibitory premotor interneurons, and reticulospinal neurons, whereas others only can be identified by recording the synaptic potential between pre- and postsynaptic cells such as excitatory and small inhibitory interneurons. The axonal spikes can be recorded with suction electrodes on the ascending or descending spinal halves or on the ventral roots. With slow-conducting axons, averaging can sometimes be very helpful. Depending on cell type, pharmacological criteria of synaptic transmission, membrane properties etc. may be of additional help.

Intracellular dye injection (Lucifer yellow, horseradish peroxidase) is useful for morphological identification, and detailed structural information of, for example, different subtypes within a class of neurons can be obtained. Since the spinal cord and medulla are thin, the stained neurons can be analysed in whole-mount preparations. Reconstructions can be obtained by camera lucida drawings, or photography in different focal planes (Figures 2D and 5; see, for example, Wallén *et al*., 1985). More recently a new technique for three-dimensional reconstruction of fluorescence-labelled neurons has been introduced, which involves confocal laser scanning of whole-mount preparations in sequential focal planes, followed by computerized reconstructions (Figure 7; Wallén *et al*., 1988). This technique can also be combined with immunofluorescence histochemistry (Brodin *et al*., 1988b), which allows studies of the structural relations between intracellularly stained neurons, and neurons containing a certain transmitter marker.

In this chapter we will consider the neuronal properties, the neurotransmitters, and the overall organization of the lamprey brain and spinal cord. We will emphasize (1) the physiology and biophysics of single identified neurons, still in their physiological environment, (2) the synaptic interaction between identified individual cells with regard to function and structure, and (3) the operation of behaviourally relevant networks of neurons, which are used to coordinate different motor acts or to process sensory information. The cyclostome CNS is also an attractive model for studies of a variety of subjects, such as regeneration, ageing and endocrine regulation.

III CELLULAR PHYSIOLOGY – SYNAPTIC TRANSMISSION

Several types of large and medium-sized neurons and axons can reliably be impaled with one or two microelectrodes, under visual inspection (Figure 1). These cells are intact and in their normal condition, rather than damaged by dissociation or slicing procedures, which usually cause damage to some neuronal processes of a cell. In the lamprey CNS, the biophysical properties of intact cells can thus be analysed, as well as the different types of membrane channels. The somata of most spinal neurons generate action potentials by means of Na^+ channels (blocked by tetrodotoxin), Ca^{2+} channels (blocked by inorganic Ca^{2+} blockers), and a repolarizing early K^+ current (blocked by tetraethylammonium and 4-aminopyridine). The late postspike afterhyperpolarization is generated preferentially by Ca^{2+}-dependent K^+ channels (Selzer, 1978a; Wickelgren, 1977; Hill *et al*., 1985; Leonard and Wickelgren, 1986; Wallén *et al*., 1989). Switch or two-electrode voltage clamp can be applied, with the unavoidable complications of a branched structure (Gold and Martin, 1984; Moore and Christensen, 1986; Moore *et al*., 1987, Wallén *et al*., 1989). The transmitter gated channels studied so far are similar to those of higher vertebrates (see below). Single-channel data will presumably be available in the near future, from recordings of dissociated cells, and perhaps also from surface recordings of cells *in situ* after cleaning of the tissue surface.

The properties of single synapses can without difficulty be analysed by recording single small and large unitary EPSPs or IPSPs. For instance, the large glutamatergic Müller axons can be impaled as well as their postsynaptic neurons, and recorded for long periods of time. The Müller axons make synapses *en passant* and the presynaptic active zones are located on the main membrane. It is thus possible to manipulate the presynaptic as well as the postsynaptic sites comparatively easily (Wickelgren *et al*., 1985; Alford *et al*., 1989). The ultrastructural changes in identifiable synapses can be analysed, for instance, at different levels of activity (Pfenninger and Rovainen, 1974; Wickelgren *et al*., 1985).

IV NEUROTRANSMITTERS

The lamprey CNS is well suited for studies of general aspects of vertebrate chemical transmission, since the transmitter systems are basically the same as in mammals. Detailed knowledge has accumulated on the amino acid transmitters, which appear to be the main mediators of fast excitatory and inhibitory synaptic transmission. Certain monoaminergic mechanisms are also fairly well understood. A characterization of the peptidergic systems has begun, but as yet little information is available on their functions. The summary below has focused on more recent studies. For earlier reviews see Rovainen (1979b) and Homma (1983).

IV.1 Excitatory Amino Acids

The classification of excitatory amino acid receptors into *N*-methyl-D-aspartate (NMDA), kainate and quisqualate receptors, initially established in amphibians and mammals (Watkins and Evans, 1981), is also applicable to lampreys. The pharmacology and ion selectivity of the different excitatory amino acid (EAA) receptor channels in lampreys are similar to those of higher vertebrates (Figure 3; Dale and Grillner, 1986; Grillner *et al*., 1988b; Watkins and Evans, 1981; Mayer and Westbrook, 1987). The NMDA and kainate/quisqualate gated channels are permeated by Na^+ and K^+, and the NMDA channels in addition by Ca^{2+} (Wallén and Grillner, 1987; Moore *et al*., 1987; Hill *et al*., in preparation; cf. Mayer and Westbrook, 1987). The NMDA gated conductance is blocked by Mg^{2+} in a voltage-dependent manner (Moore *et al*., 1987; cf. Mayer and Westbrook, 1987), which results in a region of negative-slope conductance in the I–V relationship (see also section VIII.3). In spinal neurons, NMDA receptor activation can elicit pacemaker-like membrane potential oscillations (Sigvardt *et al*., 1985; Wallén and Grillner, 1987).

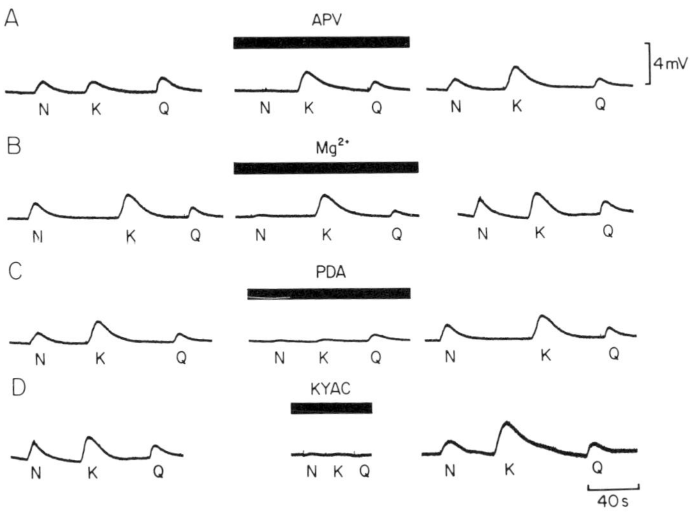

Figure 3 Effect of excitatory amino acid antagonists on agonist-evoked depolarizations of lamprey spinal neurons. The depolarizations evoked by pressure ejection of the agonists NMDA (N), kainate (K) and quisqualate (Q) are shown before the application of antagonist (left column), after maximal effect of the antagonist (middle column), and after recovery (right column). (A) Effect of bath application of 100 μM 2-amino-5-phosphonovalerate (APV). (B) 1.8 mM Mg^{2+}. (C) 2 mM *cis*-2,3-piperidine dicarboxylate (PDA). (D) 2 mM kynurenic acid (KYAC). (Modified from Dale and Grillner, 1986, with permission)

Three types of synaptic potentials mediated by EAA receptors have been recorded in spinal neurons. Brief EPSPs with a fast rise time (about 6–8 ms) and a short duration (half-decay time 20–25 ms) are due to an activation of kainate/quisqualate receptors. Slow EPSPs, with a slow rise time (25–30 ms) and a long duration (half-decay time >100 ms) are mediated by NMDA receptors. Furthermore, there are mixed synaptic potentials due to kainate/quisqualate as well as NMDA receptor activation (see Dale and Grillner, 1986; Ohta and Grillner, 1989). Some of the EAA synapses may also include an electrotonic component (Rovainen, 1974b; Buchanan *et al.*, 1987a). Several classes of EAA neurons have been identified in the brainstem and spinal cord with paired intracellular recordings (Figures 1 and 8) and/or with D-[^{3}H]aspartate tracing (Figure 11, see below). Sensory afferents, spinal premotor interneurons and reticulospinal neurons from all reticular nuclei are examples of EAA neurons (see below).

The compounds released in EAA synapses have not been established in direct experiments. However, a blockade of the EAA uptake causes a marked increase in the overflow of endogenous glutamate, whereas the efflux of aspartate is only slightly elevated (Brodin *et al.*, 1988d). Furthermore, such an uptake blockade potentiates the motor output evoked by sensory synaptic activity (Brodin and Grillner, 1985).

IV.2 Inhibitory Amino Acids

IV.2.a Glycine

The properties of lamprey glycine receptors appear similar to those of mammals. Glycine activates a Cl^- conductance in spinal cord and brainstem neurons with a reversal potential of about −80 to −85 mV (−83 mV in Müller cells (Homma and Rovainen, 1978; Homma, 1983; Gold and Martin, 1984)). The glycine effect is antagonized by low doses of strychnine, and a large proportion of the IPSPs in the brainstem and spinal cord is depressed by similar levels of strychnine. Three types of spinal glycinergic interneurons have been characterized physiologically and morphologically (see below).

The amount of glycine in the spinal cord is somewhat higher than in the brain, and is comparable to that in other vertebrates (Homma, 1983). The levels of taurine and β-alanine, which can both activate glycine receptors, are low (Homma, 1983). The presence of a Na^+-dependent uptake system for glycine has been suggested by experiments with Na^+ substitution (Homma and Rovainen, 1978).

IV.2.b GABA

γ-Aminobutyric acid induces a postsynaptic Cl^- conductance, which is antagonized by $GABA_A$ receptor blockers such as bicuculline (Homma and Rovainen, 1978). A presynapic depolarization, mediated by both $GABA_A$ and $GABA_B$ receptors, has been recorded in axons of spinal interneurons (Alford *et al*., 1989). The depolarizing action is probably due to the resting membrane potential of these axons, which is around −90 mV. A phasic GABAergic modulation of interneuronal axons occurs during fictive locomotion. The transmitter release from primary afferents is also subject to a presynaptic modulation by GABA (Christenson and Grillner, 1987). Two types of GABAergic interneurons, as well as GABA-immunoreactive (ir) liquor contacting cells, have been identified in the spinal cord by GABA immunohistochemistry (see below; Brodin *et al*., in preparation). The levels of GABA and the GABA-synthesizing enzyme glutamate decarboxylase are several times higher in the brain than in the spinal cord (Homma, 1983; Wald *et al*., 1981).

IV.3 Acetylcholine

As in other vertebrates, the motoneurons in lamprey are cholinergic and activate postsynaptic nicotinic receptors. The muscle membrane is sensitive to acetylcholine, and end-plate potentials are blocked by curare and α-bungarotoxin (see Rovainen, 1979b). Acetylcholinesterase is present in motoneurons, and the level of choline acetyltransferase is high in the spinal cord and in the cranial nerve motor nuclei (Wächtler, 1974, 1978). The corpus striatum shows a relatively high acetylcholinesterase activity indicating that it contains cholinergic neurons (Wächtler, 1974).

IV.4 Monoamines

IV.4.a 5-Hydroxytryptamine

Several groups of 5-HT neurons have been identified with the Falck–Hillarp method and/or with immunohistochemistry (Figures 4 and 10; Baumgarten, 1972; Steinbusch *et al*., 1981; Van Dongen *et al*., 1985a; Harris-Warrick *et al*., 1985; Brodin *et al*., 1986, 1988a). Ultrastructural studies of the spinal cord have suggested that the 5-HT-containing terminals do not form synaptic specializations, but instead release transmitter diffusely in the extracellular space (Christenson *et al*., 1989c). A dense 5-HT plexus occurs around spinal motoneurons, certain spinal interneurons (Van Dongen *et al*., 1985a), and certain reticulospinal neurons (Dubuc, 1989). In these cell types, 5-HT causes a depression of the late phase of the postspike afterhyperpolarization (Figure 4) which is due to Ca^{2+}-dependent K^+ channels (Van Dongen *et al*., 1986a; Dubuc, 1989; Wallén *et al*., 1989). This effect will lead to an increase in the firing rate in response to a given excitatory stimulus. The 5-HT effect is probably exerted directly on the Ca^{2+}-dependent K^+ channels, or on the intracellular handling of Ca^{2+}, and does not appear to involve a second-messenger system. No antagonism by metysergide or cinanserin has been observed, suggesting a possible involvement of 5-HT_3 receptors (Wallén et al., 1989). A selective blockade of the 5-HT uptake mimics the effect of exogenous 5-HT, and causes an increase in the overflow of endogenous 5-HT (Christenson *et al.*, 1989b).

IV.4.b Catecholamines

The lamprey brain contains a relatively high level of dopamine but only small amounts of noradrenaline (Baumgarten, 1972). Catecholamine-fluorescent cells have been identified in the diencephalon with the Falck–Hillarp method (see below). The brainstem and spinal cord appear to lack catecholaminergic neurons, apart from a small system of liquor-contacting cells near the central canal (see below; Ochi and Hosoya, 1974).

IV.4.c Histamine

Neurons containing histamine have recently been identified by immunohistochemistry using an antiserum raised against conjugated histamine. The histaminergic neurons are restricted to the diencephalon and caudal mesencephalon (Figure 12N; Brodin *et al.*, in preparation). Their location and projections resemble those found in other vertebrates (Panula and Airaksinen, 1989).

IV.5 Neuropeptides

IV.5.a Tachykinins

Chromatographic and immunohistochemical studies of the spinal cord have suggested the presence of at least three different forms of tachykinin-related peptides (Van Dongen *et al.*, 1986b). These peptides probably differ from previously known tachykinins, including substance P. Several groups of tachykinin-ir neurons have been detected in the brain (Nozaki and Gorbman, 1986).

IV.5.b Somatostatin

Three forms of somatostatin, containing 14, 34 and 37 residues respectively, have recently been isolated from lamprey pancreas (Andrews *et al.*, 1988). Their common C-terminal peptide is identical to mammalian somatostatin in 13 out of 14 amino acids. Somatostatin from the CNS has not been characterized, but immunohistochemical studies have demonstrated the presence of several groups of somatostatin-ir neurons in the brain and spinal cord (see below; Wright, 1986; Buchanan *et al.*, 1987b). Mammalian somatostatin has a hyperpolarizing effect on edge cells in the spinal cord which is not due to an increased Cl^- conductance (Christenson *et al.*, 1989a).

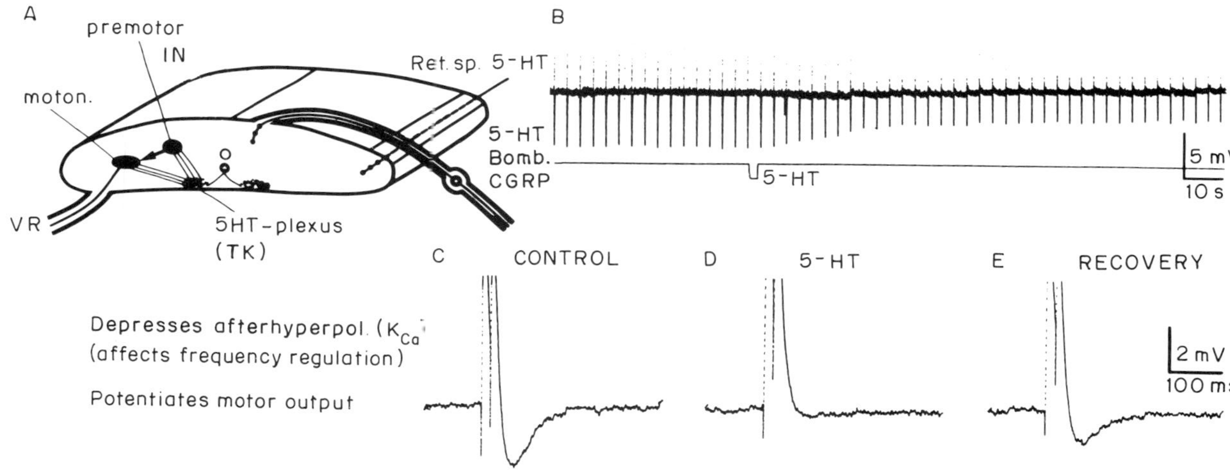

Figure 4 5-Hydroxytryptamine causes a depression of the afterhyperpolarization (AHP) in motoneurons and premotor interneurons. (A) Schematic representation of the spinal cord. The 5-HT innervation has three different origins: intraspinal 5-HT neurons situated ventral to the central canal, descending neurons from the medullary reticular formation, and afferent 5-HT cells in the dorsal root ganglia. The latter also contain calcitonin gene-related peptide (CGRP) and bombesin-like peptides. (B) A continuous recording from a motoneuron during repeated stimulation at 0.3 Hz. At each stimulation three pulses with an interval of 20 ms were given. Note the reduction of the AHP after application of 5-HT (pressure ejection). (C) and (D) show the AHPs with higher time resolution before and after 5-HT application, respectively

IV.5.c Neurotensin

Neurotensin-like material has been detected immunochemically in the lamprey gut (Carraway *et al*., 1982). As in other lower vertebrates, this peptide(s) resembles mammalian neurotensin in its C-terminal end. The brain and spinal cord contain several populations of neurotensin-ir neurons as detected immunohistochemically with both N- and C-terminal specific neurotensin antisera (Brodin *et al*., in preparation).

IV.5.d Pancreatic polypeptides (PP)

Radioimmunoassay and chromatographic analyses of CNS extracts have shown the presence of multiple forms of peptides, which react with antiserum towards the common C-terminal of pancreatic polypeptides (PP) (Brodin *et al*., 1989b). Immunohistochemically, two populations of PP-ir cells were recognized. One of these, containing neuropeptide Y- (NPY)-like immunoreactivity, is located in the spinal dorsal horn, the alar plate, and the retina, and another, showing peptide YY- (PYY)-like immunoreactivity, is present in the reticular motor area, including a reticulospinal projection from the anterior rhombencephalon (see below).

IV.5.e Cholecystokinin (CCK)

A CCK-like compound which resembles the mammalian CCK octapeptide has been detected in extracts of brain, spinal cord and gut (Holmquist *et al*., 1979; Brodin *et al*., 1988a). Low levels of extended precursor forms were also detected (Brodin *et al*., 1988a). Several populations of CCK neurons have been detected immunohistochemically in the CNS using antisera directed to the C- as well as the N-terminus of the octapeptide, including a reticulospinal projection from the posterior rhombencephalon (see below; Brodin *et al*., 1988a; Ohta *et al*., 1988).

IV.5.f Thyrotropin hormone releasing hormone (TRH)

Immunoreactivity to TRH has been detected in brain and spinal cord extracts, with a four-fold higher level in the brain (Jackson and Reichlin, 1974; Young *et al*., 1985). The TRH level in the spinal cord was similar 4 weeks after a spinal transection, suggesting that there is no significant contribution from descending neurons. It should be noted, however, that a post-lesion survival period of 2–3 months was required to obtain a complete degeneration of the descending CCK-ir pathway (Brodin *et al*., 1988a).

IV.5.g Gonadotropin releasing hormone (GnRH)

The structure of GnRH from lamprey brain has recently been established (Sherwood *et al.*, 1986). The peptide is eleven amino acids long, and differs in as much as five amino acids compared to the mammalian and avian peptides. It has no or little biological effect in birds or mammals (Sower *et al.*, 1987). The distribution of GnRH in the lamprey brain has been studied immunohistochemically with antibodies to mammalian luteinizing hormone releasing hormone (LHRH) (Crim *et al.*, 1979).

IV.5.h Vasotocin

Chromatographic studies of brain extracts have suggested the presence of a peptide resembling arginine–vasotocin (Sawyer, 1965). Vasotocin-ir has been localized in the preoptico-neurohypophysial system (Goossens *et al.*, 1977).

IV.5.i Other peptides

Immunohistochemical studies have demonstrated immunoreactivity to several other peptides in neural elements. These include growth hormone (GH), prolactin (Wright, 1986), β-endorphin, met-enkephalin (Nozaki and Gorbman, 1986), bombesin, calcitonin, calcitonin gene-related peptide (CGRP), corticotropin releasing factor (CRF), FMRFamide (phe-met-arg-phe-amide) and metorphamide (Buchanan *et al.*, 1987b; cf. also Negishi *et al.*, 1986). A further characterization of these compounds is as yet lacking.

IV.5.j Coexistence of transmitters and peptides

Immunohistochemical double-labelling experiments have demonstrated neuronal co-localization of different types of transmitter immunoreactivity. The following combinations have been observed: 5-HT–tachykinin (Van Dongen *et al.*, 1985a); 5-HT–CGRP–bombesin; 5-HT–CCK (Brodin *et al.*, 1988a); GABA–somatostatin; GABA–neurotensin; acetylcholine–CGRP (Brodin *et al.*, in preparation).

V ORGANIZATION OF THE SPINAL CORD

The spinal cord (Figure 1) comprises approximately 100 segments. It is flattened throughout its extent, except for the rostralmost segments, which have an ellipsoid shape. Dorsal and ventral roots are present in each segment, but they do not exit at the same transverse level and do not unite to form mixed spinal nerves (Tretjakoff, 1927). The 'grey matter' is composed of a ventral cell column containing the perikarya of both motoneurons and interneurons, and a 'dorsal horn' with predominantly small perikarya. The cell column is surrounded by tracts of unmyelinated fibres ranging from very thin to 40 μm or more, divided, as in other vertebrates, into dorsal, lateral and ventral funiculi.

V.1 Cell Types in the Spinal Cord and Spinal Ganglia

The motoneurons (Figure 2) are arranged as a column of cells in the grey matter with dendrites extending into all areas of the white matter, except into the dorsal columns (Tretjakoff, 1909a; Teräväinen and Rovainen, 1971; Wallén *et al*., 1985). Their physiological properties are similar to those of other vertebrate motoneurons. Fast and slow motoneurons have been distinguished by the speed of contraction of their motor units. The latter are smaller and more numerous (Teräväinen and Rovainen, 1971). The orientation of the motoneuronal dendrites is fan-like (Figure 2) and varies somewhat with which part of the myotome they supply. Motor units supplying muscle fibres near the ventral or dorsal midline may have crossed dendritic processes (Wallén *et al*., 1985), and appear to resemble fin motoneurons (Rovainen and Birnberger, 1971; Rovainen and Dill, 1984). The motoneuronal axons from the rostral and caudal hemisegment converge towards the same ventral root. They have no recurrent collaterals.

The *dorsal cell* is a round intraspinal sensory neuron (7–9/hemisegment) clearly visible in the dissection microscope, with a branch in the dorsal root, which innervates the skin (Freud, 1878; Rovainen, 1967b; Martin and Wickelgren, 1971). Dorsal cells are of two types, rapidly (touch; T) and slowly (pressure; P) adapting (Christenson *et al*., 1988a,b). Their central axons ramify in an ascending and a descending branch entering the dorsal column. They activate EAA receptors and are thus probably glutamatergic (Brodin *et al*., 1987). Dorsal cells are similar to, and have been thought to correspond to, the intraspinal Rohon–Beard cells of fish and amphibians (Rovainen, 1979b), but whether this is so or not is at present unclear (Nakao and Ishizawa, 1987).

Dorsal root ganglion cells, altogether 15–30 in each ganglion, give rise to the majority of the fibres in the dorsal roots (Freud, 1878; Van Dongen *et al*., 1985a). Morphologically, they are of two main types, large cells with a diameter of about 50 μm, and small cells with a diameter of 10–20 μm. The latter neurons co-contain two types of peptide-ir (CGRP and bombesin), as well as serotonin (Brodin *et al*., 1988a). Their function has not been studied, but it should be expected that some of these neurons convey nociception and perhaps the temperature modality.

Edge cells (Figure 5; Retzius, 1891; Rovainen, 1974a) are intraspinal stretch receptor neurons (Grillner *et al*., 1984) which are located along the lateral edge of the spinal cord, with specialized dendritic structures along the lateral margin (Figure 5). They sense the normal lateral bending which occurs when the body and thus also the spinal cord is bent during swimming (see section VIII.3). Edge cells are of two types, an excitatory type, which produces monosynaptic EPSPs in ipsilateral motoneurons and interneurons of the locomotor network, and an inhibitory type, which provides glycinergic IPSPs in contralateral neurons (Viana di Prisco *et al*., 1989; see section VIII.3). Edge cells resemble crustacean stretch receptor neurons ultrastructurally.

Giant interneurons are large relay cells located in the caudal spinal cord

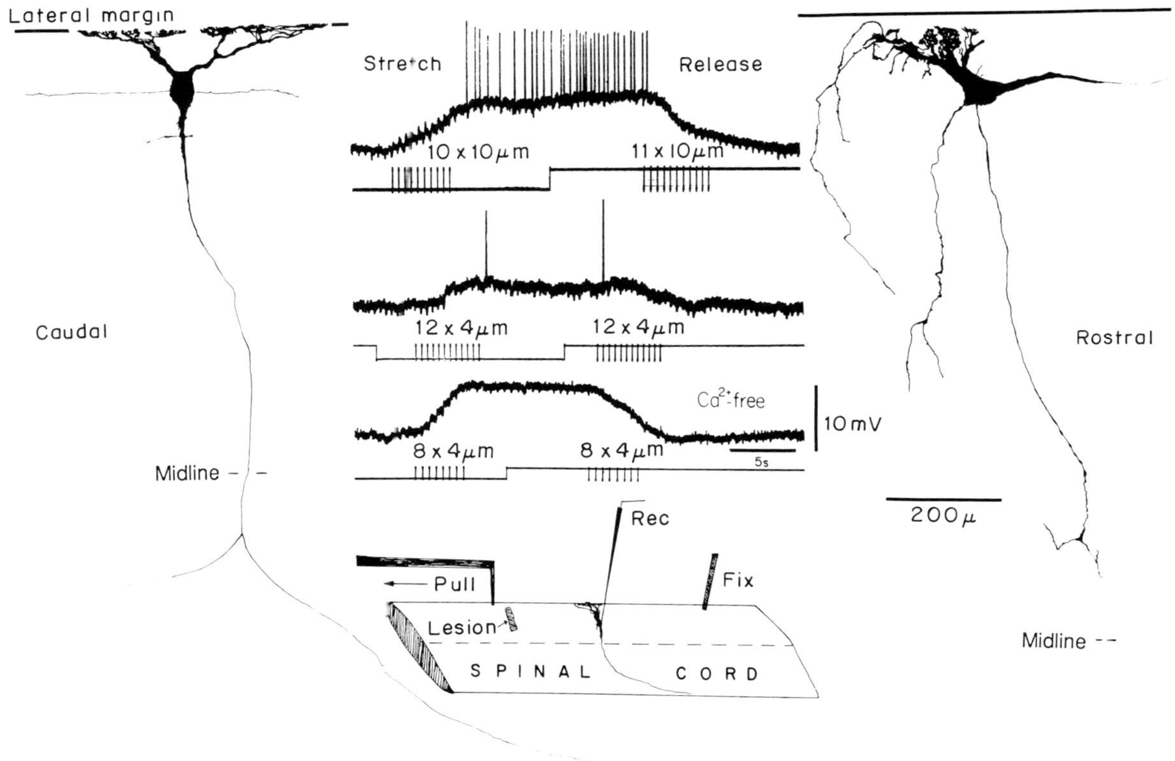

Figure 5 Edge cells – intraspinal mechanosensory neurons. Two edge cells were reconstructed after injection with Lucifer yellow. Note the extensive dendritic ramifications along the lateral margin of the spinal cord. The middle panel shows recordings of their membrane potential during stretch. The two upper traces are from the right cell, and the bottom trace is from the left cell. The top trace shows depolarization during ten repeated stretches of 10 μm. The middle trace shows depolarization during twelve stretches of 4 μm. In the lower trace, stretches were applied after removal of Ca^{2+} from the solution. The experimental set-up is shown beneath. (Modified from Grillner *et al*., 1984, *Science*, **223**, 500–503, with permission. Copyright 1984 by the AAAS)

(Rovainen, 1974a) which are monosynaptically excited from dorsal cells of the touch type (Christenson *et al.*, 1988c). They have very large cell bodies with dendrites extending bilaterally above the central canal, ramifying within the dorsal column (Rovainen, 1974a), and large axons which extend to the contralateral side to ascend to the brainstem level within the dorsolateral funiculus (Rovainen, 1974a). Caudal giant interneurons excite more rostral giant interneurons monosynaptically by a presumed EAA synapse (Homma, 1981), presumably thereby providing an amplification of the ascending signal. They constitute the most rapid conduction pathway to the brain for signals from the caudal part of the body, including the dorsal fins. Such signals could be important in producing rapid corrections of body position.

Dorsal interneurons (Figure 6) are small and identified only to a limited degree by their tonic inhibition during bouts of swimming (Brodin and Grillner, 1986b). Presumably they can gate sensory signals (cf. Sillar and Roberts, 1988).

Excitatory interneurons (EIN) are fairly small interneurons (Figure 7) which activate postsynaptic EAA receptors. They have a descending axon extending up to 6–7 segments and may have a shorter ascending branch (Dale, 1986). The EINs receive monosynaptic reticulospinal and polysynaptic sensory input. During locomotion they are phasically active, and they monosynaptically excite motoneurons and certain inhibitory premotor interneurons (see section VIII.3; Buchanan and Grillner, 1987a; Buchanan *et al.*, 1989).

Small ipsilateral inhibitory interneurons (sIIIN) mediate glycinergic IPSPs in motoneurons. They have a similar morphology to that of the excitatory interneurons, and they are phasically active during locomotion (Buchanan and Grillner, 1987b).

Crossed caudal interneurons (CCIN) have a medium-sized motoneuron-like cell body, with a contralateral axon with a long descending branch extending over 19 segments or more. Most of these cells are inhibitory (glycinergic) to motoneurons and lateral interneurons (Buchanan, 1982, 1986). They receive direct input from reticulospinal and excitatory interneurons (Buchanan and Grillner, 1987a), and they are phasically active during locomotion (see section VIII.3). One subgroup is excitatory.

Lateral interneurons (LIN) have an ipsilateral long descending axon. The cell bodies (Figure 8) are large and apparent. They occur in the rostral part of the spinal cord, but they may have smaller counterparts in the caudal part of the spinal cord (Rovainen, 1974a). They are located in the lateral part of the grey matter with a very large process extending to the lateral margin (Figure 8; Rovainen, 1974a; Wallén *et al.*, 1988). They are inhibitory to CCINs (Buchanan, 1982) and may at least partially subserve the role of a long ipsilateral propriospinal interneuron. The LINs receive monosynaptic reticulospinal input and they are phasically modulated during locomotion.

GABAergic interneurons have been identified with immunohistochemistry using antisera to conjugated GABA (Brodin *et al.*, 1990). These include segmental medium-sized multipolar cells located in the lateral cell column, and

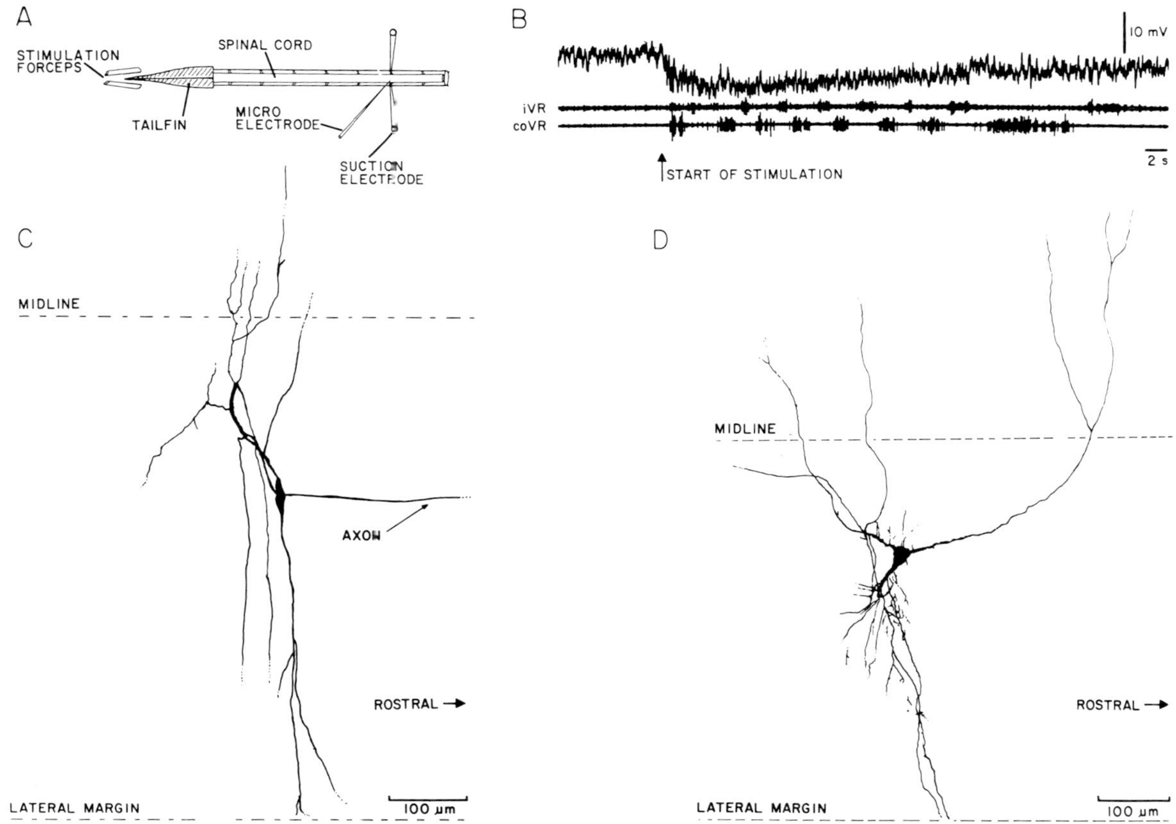

Figure 6 Dorsal interneurons receive tonic inhibitory input during bouts of fictive locomotion. (A) shows a schematic illustration of the tailfin–caudal spinal cord preparation used. (B) Intracellular recording from a dorsal interneuron during a bout of locomotor activity evoked by mechanical stimulation of the tailfin. The traces below show the efferent activity in the ipsilateral (iVR) and contralateral (coVR) ventral roots. (C) and (D) show reconstructions of two dorsal interneurons after intracellular injection with Lucifer yellow. (Modified from Brodin and Grillner, 1986b, with permission)

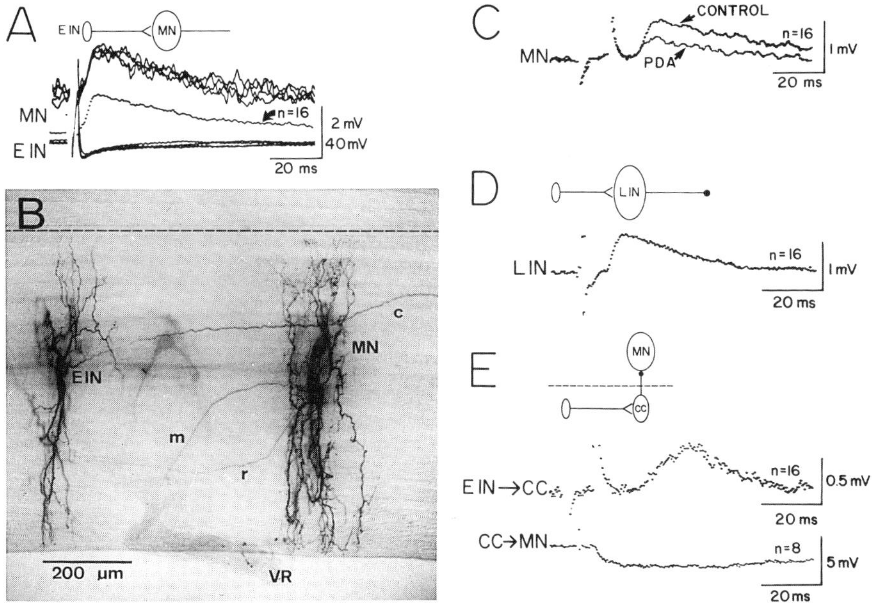

Figure 7 Synaptic output and morphology of excitatory interneurons (EINs) in the spinal cord. (A) The EPSP elicited in a motoneuron by intracellular stimulation of an EIN. The top trace shows several superimposed EPSPs, and the middle trace shows an average of 16 sweeps. (B) A whole-mount photograph of an EIN and a postsynaptic motoneuron injected intracellularly with horseradish peroxidase. The axon of the EIN projects caudally (c) through the motoneuron dendrites and also has a rostrally projecting axon (r) which branches from the caudal axon. The motoneuron axon (m) can be followed to the ventral root (VR). (C) The effect of *cis*-2,3-piperidine dicarboxylate (PDA) on an EPSP in a motoneuron elicited by an EIN. Averages of sixteen sweeps. (D,E) Averaged EPSPs evoked in a lateral interneuron (D; LIN), and a CC interneuron (E; CCIN) by an EIN. (Modified from Buchanan and Grillner, 1987a, *Science*, **236**, 312–314, with permission. Copyright 1987 by the AAAS)

small bipolar cells in the ventral aspect of the dorsal horn. These two types probably give rise to the dense GABAergic innervation, which is present throughout the lateral, ventromedial and, to a lesser extent, the dorsal columns. These neurons might mediate the presynaptic $GABA_A$ and $GABA_B$ receptor-mediated inhibition of transmitter release from primary afferents and interneuronal axons (Christenson and Grillner, 1987, Alford *et al*., 1989).

Intraspinal 5-HT neurons (Figure 4) identified by the Falck–Hillarp technique (Honma, 1970; Baumgarten, 1972) and by immunohistochemistry (Van Dongen *et al*., 1985a,b; Harris-Warrick *et al*., 1985), are located ventral to the central canal. These neurons give rise to a dense serotonergic plexus in the ventromedial part of the spinal cord. Some of these cells co-contain a tachykinin-like peptide (van Dongen *et al*., 1985a). In this 5-HT plexus, motoneurons and a number of premotor

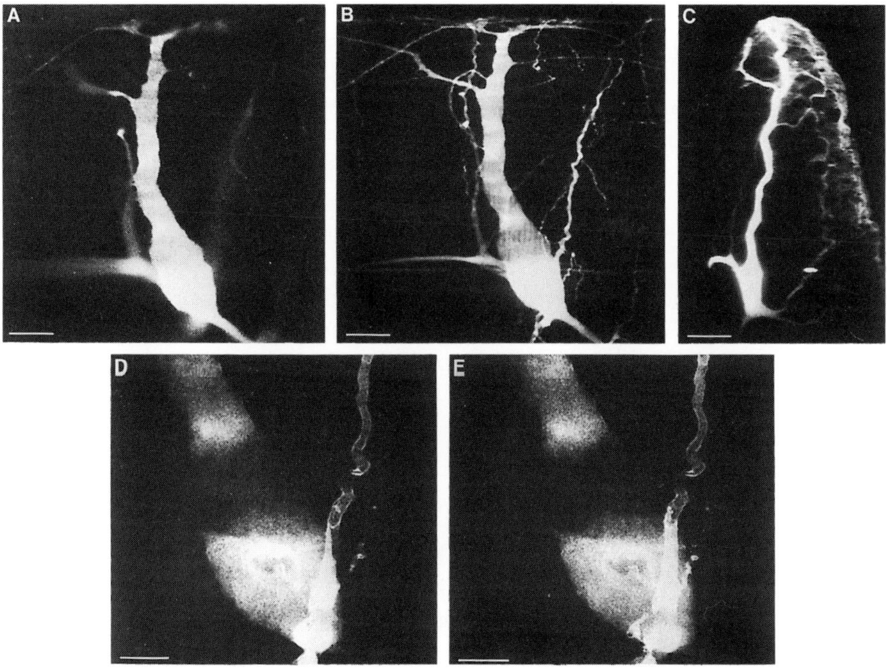

Figure 8 Lateral interneuron injected with Lucifer yellow in a whole-mount preparation of the spinal cord. (A) shows a photomicrograph taken in a conventional fluorescence microscope. Many processes are out of focus. (B) and (C) show three-dimensional reconstructions of the same neuron seen from a dorsal (B) and caudal (C) view. The preparation was scanned at sequential focal planes using a confocal laser scanning system. The digitized data were used to calculate three-dimensional images. Note the pronounced depth resolution compared to (A). The large process extending to the lateral spinal margin is characteristic of LINs. (Modified from Wallén *et al.*, 1988, with permission)

interneurons have dendrites in close apposition to 5-HT varicosities (Van Dongen *et al.*, 1985b). However, 5-HT varicosities do not form synaptic specializations with postsynaptic cells even if they are in close contact (Christenson *et al.*, 1989c; see also section IV).

Peptidergic neurons – small NPY- and neurotensin-ir cell bodies occur in the ventral aspect of the dorsal horn (Van Dongen *et al.*, 1985a; Brodin *et al.*, in preparation). Their function is unknown, but NPY-ir cells form close appositions with processes of sensory dorsal cells (Bongianni *et al.*, 1989). Furthermore, immunohistochemical studies have indicated the presence of additional peptidergic neurons in the spinal cord (see section IV). Immunoreactivity to several peptides (see Van Dongen *et al.*, 1986b; Buchanan *et al.*, 1987b) has been observed in spinal fibres, which remain unaffected after a chronic spinal transection. This is in contrast to the two systems of reticulospinal peptidergic fibres which disappear below a lesion (see below).

Liquor-contacting neurons are located around the central canal in lampreys as well as in other vertebrates (Vigh and Vigh-Teichmann, 1973; Vigh *et al.*, 1977; Dale *et al.*, 1987) investigated. They have a kinocilium, which extends into the central canal, and axon-like processes, which extend to the lateral margin of the spinal cord where they form a dense plexus near edge-cell dendrites. These liquor-contacting neurons contain GABA immunoreactivity and they may in addition contain either somatostatin or neurotensin immunoreactivity, and physealamine immunoreactivity (Brodin *et al.*, unpublished; Van Dongen *et al.*, 1986b), and also catecholamines (Ochi and Hosoya, 1974). Somatostatin and GABA exert a hyperpolarizing action on edge cells by effects on different types of ionic channels (Christenson *et al.*, 1989a). Despite this finding, the role of these neurons remains enigmatic even if the presence of a kinocilium might suggest mechanosensitivity and the ramifications on the edge cells might support such a notion.

V.2 Fibre Tracts in the Spinal Cord

The unmyelinated fibre tracts of the lamprey spinal cord (Figure 2) can be subdivided into a dorsal, lateral and ventral funiculus divided by the entry zones for the dorsal and ventral roots as in other vertebrates.

The *dorsal column* contains the central axons of dorsal root ganglion cells and spinal dorsal cells (see above; Tretjakoff, 1909a; Ronan, 1983; Dubuc and Grillner, 1987, unpublished). They appear to be organized in a somatotopic fashion with caudal afferents running in the medial part and rostral afferents in the lateral part of the column. The dorsal column pathway projects mainly to the area of the dorsal column nuclei, which have recently been identified in the alar plate (Dubuc *et al.*, in preparation), even if some fibres continue to the cerebellum.

The *propriospinal* and *ascending axons* of the excitatory and inhibitory neurons mentioned above and motoneurons run mainly in the ventral funiculus, whereas the ascending axons of giant interneurons are located in the dorsal part of the lateral funiculus. Numerous peptidergic propriospinal fibres (see above) are located in the lateral and ventral funiculi.

The *reticulospinal projections* run in the lateral and ventral columns (Rovainen *et al.*, 1973; Rovainen, 1982). The prominent Müller axons (diameters of 20–40 μm) are the largest. They are located in the ventral column, and the large Mauthner axons in the lateral column. Many large and medium-sized reticulospinal axons from the posterior rhombencephalic reticular nucleus run in the dorsal part of the lateral column. All of the larger axons, with conduction velocities of 1 m/s or more, activate EAA receptors and are thus probably glutamatergic (Buchanan *et al.*, 1987a; Ohta and Grillner, 1989). Small reticulospinal axons, with diameters down to less than 1 μm, including 5-HT- and CCK-ir axons, are distributed throughout the lateral and ventral columns (Brodin *et al.*, 1988a), whereas PYY-ir reticulospinal axons are restricted to the lateral column (Brodin *et al.*, 1989b).

VI ORGANIZATION OF THE BRAINSTEM

The rhombencephalon is about 3–4 mm long (Figures 9 and 10). It consists of a vertically oriented alar plate, and a wide horizontal basal plate, which is particularly accessible for *in vitro* recordings. The basal plate (Figure 10) can roughly be divided into a medial reticular area and a lateral region containing the cranial nerve motor nuclei. The rostral rhombencephalon contains a small structure which is considered as a primordial cerebellum (see Nieuwenhuys, 1977). The mesencephalon (Figures

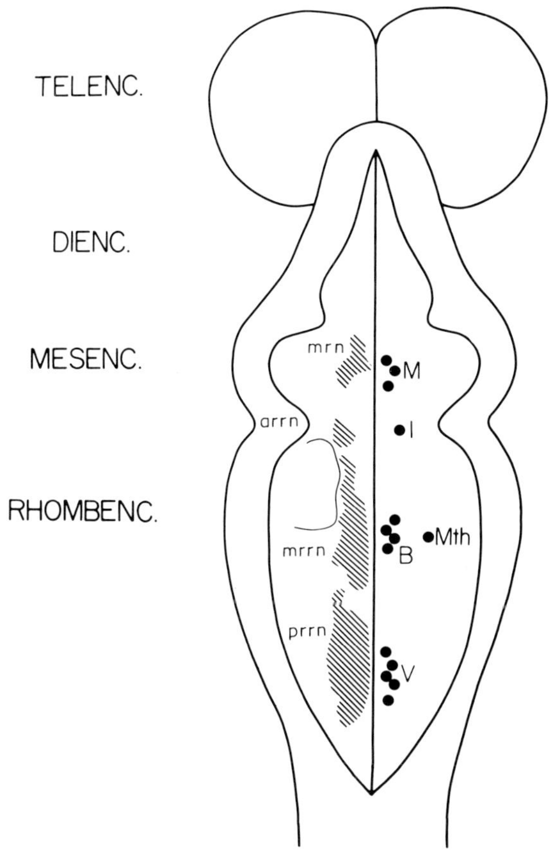

Figure 9 Schematic drawing of the lamprey brain from a dorsal view. The reticular area (left side) occupies the medial brainstem and is subdivided into the posterior (prrn), middle (mrrn), and anterior reticular nuclei, and the mesencephalic reticular nucleus (mrn). Within the reticular nuclei there are several large, identifiable reticulospinal neurons (right side). These are termed mesencephalic (M), isthmic (I), and bulbar (B) Müller cells, Mauthner cells (Mth), and large prrn cells (or vagal cells (V), due to their proximity to the vagal nucleus)

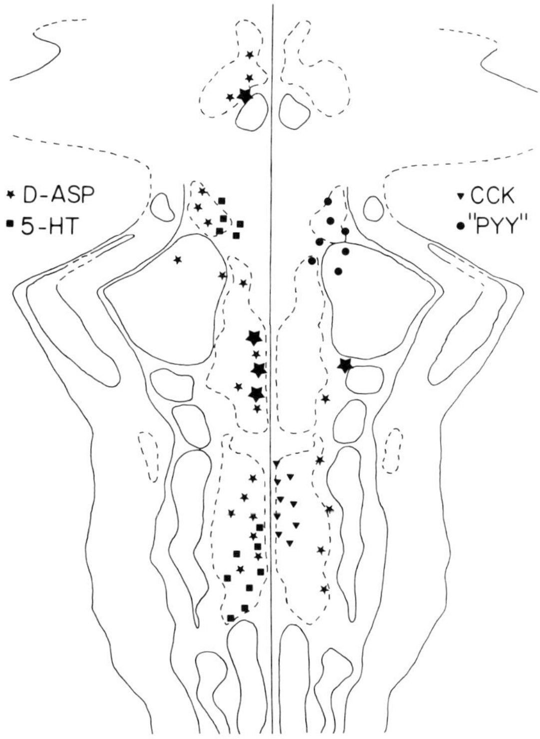

Figure 10 Schematic representation of different types of reticulospinal neurons classified according to transmitter markers. [^{3}H]D-aspartate labelled cells following spinal injection (stars), including Müller and Mauthner cells (large stars), are present in all four reticular nuclei (cf. Figure 8). Descending 5-HT cells are located in the prrn and arrn. Cells containing CCK immunoreactivity are located in the medial aspect of the prrn, and PYY/PP-immunoreactive cells are present in the arrn. The 5-HT, CCK, and PYY/PP cells were identified by retrograde fluorescence tracing from the spinal cord

10 and 12) is approximately 1.5–2 mm long, and comprises a ventral part, the tegmentum mesencephali, and a dorsal part, which includes the optic tectum and the torus semicircularis.

VI.1 The Alar Plate

The alar plate extends from the obex to the caudal border of the mesencephalon (Figure 10). It contains the main sensory nuclei of the brainstem, including the dorsal column termination area, the vestibular nuclei, the octavolateralis complex, and the Vth sensory nucleus. The sensory projections of the remaining cranial nerves are less well explored.

VI.1.a The dorsal column termination area

This is for the most part located in the caudal alar plate in two groups of neurons, the axons of which project towards the midline (Dubuc and Grillner, 1987, unpublished) and may cross to the contralateral side.

VI.2.b The sensory trigeminal system

The Vth nerve comprises an ophthalmic nerve (ganglion V1) innervating the surface of the head, a suboptic nerve (ganglion V2) with three branches supplying the mouth region, and a mandibular nerve containing mainly motor fibres (Tretjakoff, 1927; Rovainen, 1979b). The response properties of trigeminal afferents have been characterized by intracellular recordings from the trigeminal ganglia (Matthews and Wickelgren, 1978). The axons of some medullary and even spinal dorsal cells (see above) also exit the Vth nerve (Finger and Rovainen, 1982). The major component of the central trigeminal projection is a descending ipsilateral root.

The neurons within the boundaries of the descending trigeminal tract do not form a distinct nucleus, but they are often referred to as the sensory trigeminal nucleus. The specific areas of termination of the different trigeminal nerve branches have been characterized with retrograde tracing (Koyama *et al*., 1987). The second-order neurons of the Vth sensory nucleus are thought to project to the Vth and VIIth motor nuclei, the reticular nuclei, and the tectum (Heier, 1948, Rovainen, 1982). Some neurons project to the spinal cord (Ronan, 1989).

VI.1.c The lateral line system

The mechanoreceptive neuromasts of the head and body are innervated by the anterior and posterior lateral line nerves, which project to the octavolateralis area in the rostral alar plate (Ronan and Northcutt, 1987; cf. also Bodznick and Northcutt, 1981). The anterior nerve innervates the ipsilateral dorsal octavolateralis nucleus. The posterior nerve projects to the ipsilateral medial nucleus, continues rostrally to the cerebellum, and crosses to the contralateral medial nucleus. Fibres originating from the octavolateral region have been traced to the cerebellum and into the medial longitudinal fascicle (Heier, 1948).

VI.1.d The vestibular (octavomotor) nuclei

After entering the brain the vestibular nerve fibres (cf. section VIII.1) branch rostrally and caudally, and reach the vestibular or octavomotor nuclei, the cerebellum, the octavolateralis complex, and the Vth sensory nucleus (see Nieuwenhuys, 1977; Northcutt, 1979; Koyama *et al*., 1989). The vestibular complex comprises an anterior, an intermediate, and a posterior octavomotor nucleus with secondary vestibular neurons. The octavomotor neurons project to extraocular motoneurons, reticulospinal neurons, and directly to the spinal cord (see below; Rovainen, 1982). *Vestibulospinal axons* descend in the ventral column and, to some extent, in the lateral column (Rovainen *et al*., 1973; Rovainen, 1979a), but not beyond the level of the gills. They originate from the ipsilateral intermediate octavomotor nucleus and the contralateral posterior octavomotor nucleus (Ronan, 1983). Efferents to the labyrinth originate from a group of cells located caudal to the ipsilateral Vth motor nucleus (Fritzsch *et al*., 1989; Koyama *et al*., 1989).

VI.1.e The cerebellum

This is situated in the rostralmost part of the alar plate, and is very small in comparison with that of other vertebrates. It consists of a periventricular cell layer, and a group of migrated cells which may correspond to the cerebellar nuclei (see Nieuwenhuys, 1977). The cerebellum receives input from the vestibular and lateral line nerves, the octavolateral nuclei, and probably also from the tectum and the spinal cord (see Larsell, 1947; Nieuwenhuys, 1977). The cerebellar efferents appear to reach the mesencephalic and rhombencephalic reticular formation, the nucleus III, and the tectum (Heier, 1948; Nieuwenhuys, 1977).

VI.2 The Basal Plate and Ventral Mesencephalon

The basal plate and the ventral mesencephalon, which contain the reticular nuclei and the cranial nerve motor nuclei, will be considered together (see Figures 9, 10 and 12).

The *reticular formation* occupies the medial part of the basal plate and extends from the caudal rhombencephalon into the mesencenphalon. By analogy with other lower vertebrates, the rhombencephalic reticular formation (Figure 9) has been subdivided into a posterior, middle and anterior rhombencephalic reticular nucleus (prrn, mrrn and arrn, respectively) (Nieuwenhuys, 1972; cf. Opdam and Nieuwenhuys, 1976; Opdam *et al*., 1976). The corresponding area in the mesencephalic tegmentum is usually called the mesencephalic reticular nucleus (mrn) (Nieuwenhuys, 1972), and its rostral part may correspond to the nucleus of the medial longitudinal fascicle. The reticular nuclei receive input from several regions, including the mesencephalic tectum (Heier, 1948), the cranial nerves, and the spinal cord (see below; Rovainen, 1978, 1982; Dubuc *et al*., in preparation).

The reticulospinal cells, which include several giant Müller neurons (Tretjakoff, 1909a), have attracted particular interest, partly due to their role in motor control (see below; for reviews see Rovainen, 1978; Brodin *et al*., 1988c). The ascending and intrinsic neurons of the reticular formation are less well characterized.

All four reticular nuclei contain reticulospinal cells which activate *EAA receptors* (Figures 1 and 10; Buchanan *et al*., 1987a; Ohta and Grillner, 1989), and accumulate [^{3}H]D-aspartate after a spinal injection (Figure 11; Brodin *et al*., 1989a). These presumed glutamatergic cells range in size from small to very large. They include the giant Müller and Mauthner cells, which can be identified individually from animal to animal. The Müller cells have ipsilateral axons whereas the Mauthner cells project to the contralateral spinal cord (Müller; 1840, Rovainen, 1967a; Rovainen *et al.*, 1973). Smaller ipsilateral glutamatergic cells are present in all four reticular nuclei, and a crossed projection originates from the lateral border of the prrn (Figures 10 and 11; Brodin *et al.*, 1989a). Monosynaptic reticulospinal EPSPs (from Müller cells and prrn cells) have been recorded in motoneurons, EINs, CCINs, and LINs (Rovainen, 1974b; Buchanan, 1982; Ohta and Grillner, 1989; Brodin et al., 1988b). These synapses frequently utilize both electrical and EAA receptor-mediated EPSPs (via both kainate/quisqualate and NMDA receptors).

Serotonergic neurons are present in two discrete groups (Figure 10) in the caudal part of the prrn, and in the isthmic area near the arrn, respectively. These groups have been proposed to correspond to the serotonergic raphe system of higher vertebrates (Baumgarten, 1972). The majority of the neurons in the caudal group and some neurons in the rostral group project to the spinal cord (Brodin *et al.*, 1986). Serotonergic fibres are widely distributed throughout the brainstem, particularly in the alar plate.

Several groups of *peptidergic neurons* have been identified immunohistochemically in the reticular nuclei. Small *somatostatin-ir* neurons are present in the mrn and the rostral aspect of the arrn (Wright, 1986). They give rise to a dense fibre plexus in the ventral tegmentum and isthmus region.

Three groups of *cholecystokinin-ir* cells are present in the prrn, mrrn, and mrn, respectively (Brodin *et al.*, 1988a). The medium-sized prrn cells (Figure 10) project to the spinal cord, whereas the other two groups of smaller cells provide a local innervation in the brainstem. Neurons containing a *pancreatic polypeptide/peptide YY-like compound* are present in the mrrn, mrn, and arrn (Brodin *et al.*, 1989b). The latter group of medium-sized neurons (Figure 10) have a spinal projection. *Tachykinin-ir* neurons have been detected in the mrn and arrn (Nozaki and Gorbman, 1986).

The *trigeminal motor nucleus*, which occupies the rostral portion of the lateral basal plate (Figure 10), is large and readily visible with the dissection microscope. Its large cholinergic (Wächtler, 1974) motoneurons innervate the mouth and the rasping organ (see section VIII.5). These motoneurons also contain CGRP immunoreactivity (Brodin *et al.*, unpublished observation), similar to mammalian motoneurons.

The *VIIth, IXth and Xth nuclei* are located in a row extending caudally from the Vth nucleus (Figure 10). They contain respiratory motoneurons (see section VIII.4; Rovainen, 1974c), and presumably visceral efferents (Nieuwenhuys, 1977).

The *oculomotor nuclei (III, IV and VI)* – the locations of the motoneurons supplying the different eye muscles have recently been determined with selective retrograde tracing (cf. section VIII.1; Fritzsch and Sontag, 1988; Fritzsch *et al.*, in preparation; cf. Finger and Rovainen, 1978). The oculomotorius nucleus (III) is situated in the caudal aspect of the mesencephalic reticular formation at the level of

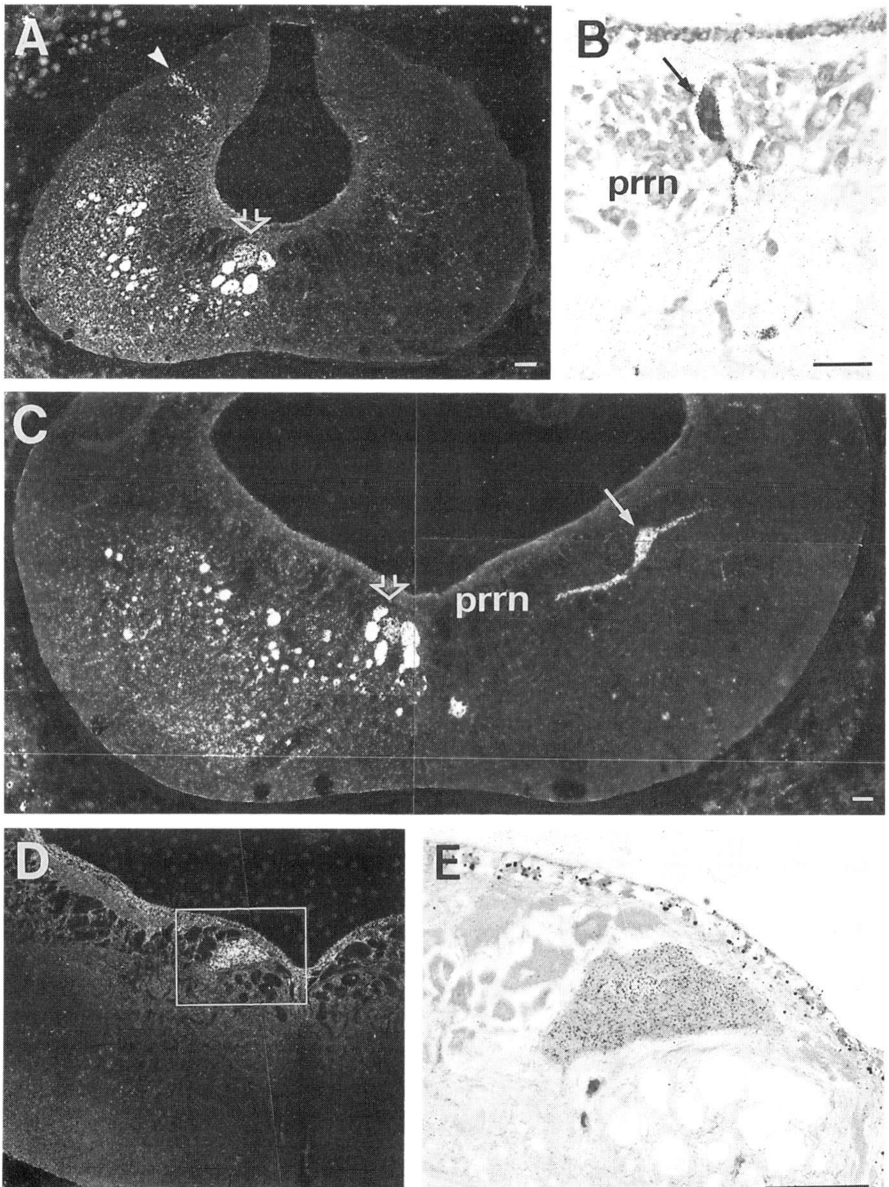
A
B
prrn
C
prrn
D
E

the arrn, and includes motor nuclei to three groups of eye muscles (see below). The trochlear nucleus (IV) is situated in the dorsal part of the isthmus region near the mrn, caudal to the IIIrd nucleus. The abducens nucleus (VI) is located in the lateral and ventral aspect of the mrrn.

The *interpeduncular nucleus* is an elongated cell accumulation situated in the midline at the junction of mesencephalon and rhombencephalon (see Nieuwenhuys, 1977). It receives input from the habenulae, and its efference passes caudally via the diffuse interpedunculobulbar tract.

VI.3 The Dorsal Mesencephalon

The mesencephalic tectum is a relatively large, distinctly laminated structure (Figure 10) similar to that of other lower vertebrates. The basal tectal cell layer is directly continuous with the torus semicircularis. Apart from the optic nerve input (see below), both of these structures receive input from the vestibular and lateral line nerves, the Vth sensory nerve, and probably also from the spinal cord (Heier, 1948, Fritzsch *et al*., 1989). The tectum also receives input from the epithalamus and dorsal thalamus. Efferents from the tectum and the torus semicircularis reach the mesencephalic reticular formation via the toro-tecto tegmental tract, and tectal efferents descend to the rhombencephalic reticular nuclei via the tectobulbar tract (Heier, 1948). A small tectospinal projection may reach the rostral spinal cord (Brodin *et al*., 1989a; Ronan, 1989). The tectum is most likely of importance for integration between visual and other sensory modalities and the motor system of the eyes and the body including orienting responses.

The tectum receives a prominent optic nerve input, whereas the projection to the torus semicircularis is relatively small (Heier, 1948). Most fibres are crossed but some ipsilateral fibres have been observed (Northcutt and Przybylski, 1972; Reperant *et al*., 1985). In addition the optic nerve projects to the outer layer of the dorsal thalamus ('lateral geniculate nucleus', see below). Cells giving rise to

---

Figure 11 Putative EAA neurons descending from the reticular formation. The micrographs show autoradiographically labelled neurons in the caudal rhombencephalon following unilateral injection of [^{3}H]D-aspartate in the rostral spinal cord (left side). (A) Dark-field micrograph showing accumulated [^{3}H]D-aspartate in axons at the level of obex. Large Müller axons near the midline are indicated with an open arrow. The arrowhead indicates anterogradely labelled dorsal column fibres. Note the lack of labelling contralateral to the site of [^{3}H]D-aspartate injection. (B) Light-field micrograph showing intense radiolabelling of a neuron (arrow) in the posterior rhombencephalic reticular nucleus (prrn), ipsilateral to the injection site. (C) A section from the caudal medulla at the level of the caudal pole of the prrn. Axonal labelling is present ipsilateral to the site of injection, Müller axons are indicated with an open arrow. A labelled cell body (arrow) is present on the contralateral side. (D) A large [^{3}H]D-aspartate-labelled neuron in the prrn ipsilateral to the injection site. The framed area is shown at high magnification in light-field illumination in (E) (A–C) are cryostat sections (14 μm), (D) and (E) show a thin section (1–2 μm) from a brain embedded in araldite. All bars = 50 μm. (Modified from Brodin *et al*., 1989a, with permission)

retinofugal fibres are present in the mesencephalic nucleus M5 of Schober (Reperant *et al*., 1985).

The retina of lampreys is similar to that of other vertebrates with a complexity intermediate between the primate and the frog retina (Öhman, 1976; Rovainen, 1982). The photoreceptors are of two types, but it is unclear whether they represent rods and cones. Two types of horizontal cells have been defined physiologically and morphologically (Teranishi *et al*., 1982). Bipolar, ganglion and amacrine cells occur together in the inner nuclear layer. Several putative transmitters present in the gnathostome retina have been detected with immunohistochemistry (Negishi *et al*., 1986).

VII ORGANIZATION OF THE FOREBRAIN

In comparison with the brainstem and spinal cord, the forebrain (Figure 12) is less amenable to *in vitro* studies, since it is comparatively small (approximately 1–1.5 mm long) and it contains smaller neurons. Knowledge of the forebrain organization is limited and primarily based on anatomical studies performed with conventional techniques (silver impregnation, Nissl and Golgi methods), but modern tracing techniques and transmitter histochemistry have been used to some extent. The following brief survey is focussed on more recent studies. For earlier reviews see Ariens Kappers *et al*. (1936), Nieuwenhuys (1977) and Parent (1986).

VII.1 The Diencephalon

The diencephalon (Figure 12) is divisible into the hypothalamus, the ventral and dorsal thalamus, and the epithalamus.

VII.1.a Hypothalamus

This consists of the dorsal and ventral hypothalamic nuclei, and the pre- and postinfundibular commissural nuclei (Figure 12; Schober, 1964). These nuclei are believed to receive input from several telencephalic nuclei, the dorsal thalamus and the mesencephalon (Heier, 1948). The hypothalamic neurons are located subependymally and many cells send liquor-contacting processes into the third ventricle. By analogy with other vertebrates, numerous transmitter markers have been detected in hypothalamic neurons. They include 5-HT, catecholamines (Baumgarten, 1972), histamine (Brodin *et al*., in preparation), CCK (Brodin *et al*., 1988a), somatostatin (Wright, 1986), substance P/tachykinin (Nozaki and Gorbman, 1986), β-endorphin (Nozaki and Gorbman, 1984) and neurotensin (Brodin *et al*., in preparation).

The hypothalamic neurons are thought to project to the ventral mesencephalon and rhombencephalon, thalamus, and telencephalon (Johnston, 1912; Heier, 1948). Serotonergic fibres ascend from the hypothalamus to arborize within the

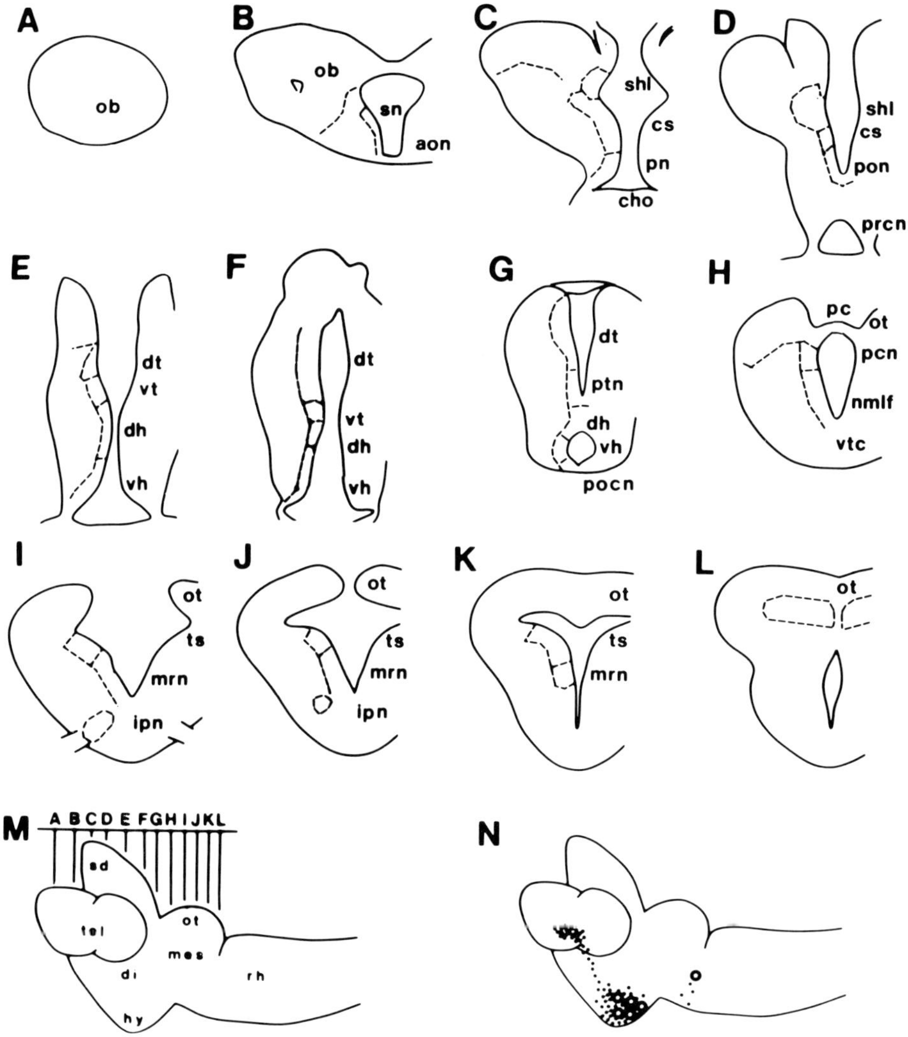

Figure 12 Schematic map of the forebrain of *Lampetra fluviatilis*. The positions of the sections are indicated in (M). For abbreviations see Table 1. (N) shows the distribution of histaminergic neurons (star-on-spot) and fibres (spots) seen from a lateral view (Brodin *et al*., in preparation). (cf. Schober, 1964)

telencephalon (Steinbusch and Nieuwenhuys, 1979, Steinbusch *et al*., 1981). Histaminergic (Figure 12N) and neurotensin-ir neurons project to the corpus striatum and the septal nucleus (Brodin *et al*., in preparation).

The posterior tuberculum, in the junction between the hypothalamus and tegmentum, contains a group of catecholaminergic cells (Baumgarten, 1972), which have been suggested to correspond to the mesolimbic dopamine system in mammals (Baumgarten, 1972), but as yet only descending projections have been found from this area (Ronan, 1989; cf. Parent, 1986).

VII.1.b The dorsal thalamus

This comprises a dense central zone and an external zone. The latter may correspond to the lateral geniculate nucleus (Nieuwenhuys, 1977). The dorsal thalamus contains both serotonin (Steinbusch *et al*., 1981) and somatostatin-ir perikarya (Wright, 1986), and it receives input from the retina (see above), and probably also from the tectum, hypothalamus, epithalamus, and telencephalon. The efference from the dorsal thalamus includes projections to the telencephalon, ventral thalamus, mesencephalon and epithalamus (see Heier, 1948; Nieuwenhuys, 1977).

VII.1.c The ventral thalamus

This contains both 5-HT- and noradrenaline-ir neurons (Steinbusch *et al*., 1981). It appears to receive input from the dorsal thalamus, tectum, hypothalamus, habenulae, and telencephalon. Most of its efferent fibres descend to the hypothalamus and the brainstem (Heier, 1948).

VII.1.d Epithalamus

This comprises the pineal organ and the habenular ganglia. The pineal organ contains photoreceptors connected to the pineal ganglia. Axons of the ganglion cells reach the habenulae, the posterior commissural nucleus and the mesencephalic tectum. The habenular ganglia are asymmetrical due to a marked hypertrophy of the superficial cell layer in the right ganglion. They receive fibres from the telencephalon and the dorsal thalamus, and project to the interpeduncular nucleus and probably also to other areas of the brainstem (Johnston, 1912; Heier, 1948).

VII.2 Telencephalon

The telencephalon (Figure 12) is small and consists of the olfactory bulbs, and the subpallial and pallial cell groups forming the true cerebral hemispheres.

VII.2.a The olfactory bulb

This has a similar organization and projections to those of higher vertebrates (see Nieuwenhuys, 1977; Northcutt and Puzdrowski, 1988). Two olfactory tracts have been identified with HRP tracing: a medial tract projecting to the septal nucleus, the preoptic area and, to some extent, the rostral striatum, and a lateral tract projecting to the lateral and dorsal pallium, and the posterior diencephalon. Efferent neurons to the olfactory bulb are located in the dorsal and lateral pallium, the preoptic area, and the mesencephalic tegmentum (Northcutt and Puzdrowski, 1988).

VII.2.b Subpallium

This consists of the corpus striatum, the septal and preoptic nuclei. The septum, and to some extent the rostral striatum, receives input from the olfactory bulb (Northcutt and Puzdrowski, 1988). The striatum probably also receives input from the dorsal thalamus (Heier, 1948). Both nuclei are innervated by 5-HT (Steinbusch *et al*., 1981), histamine, and neurotensin-ir fibres (Brodin *et al*., in preparation) originating from the hypothalamus. In the caudal striatum, catecholaminergic fibres and cell bodies are present (Baumgarten, 1972). Striatal efferents are thought to reach the dorsal thalamus and the mesencephalic tegmentum (Johnston, 1912).

The preoptic nucleus is located ventral to the striatum and septal nucleus. It contains the cell bodies of a preopticoneurohypophysial system, containing immunoreactivities to vasotocin (Goossens *et al*., 1977), LHRH (Crim *et al*., 1979), growth hormone, and prolactin (Wright, 1986).

VII.2.c Pallium

This can be divided into medial (primordium hippocampi), submedial (subhippocampal lobe), dorsal, and lateral (primordium piriforme) cellular zones (see Schober, 1964; Nieuwenhuys, 1977). These structures may primarily be concerned with olfaction (see above; Northcutt and Puzdrowski, 1988), but the dorsal and lateral pallium probably also receive input from the hypothalamus (Johnston, 1912). It is not clear to what extent the above-mentioned telencephalic centres are homologous to the corresponding structures of jawed vertebrates (see Northcutt, 1981; Parent, 1986).

VIII MOTOR SYSTEMS

VIII.1 Eye Movements and Vestibular Compensatory Movements

If the head of a lamprey is partially moved back and forth, the eyes are rotated in a direction opposite to that of the movement as if to stabilize the gaze. Consequently the eye movements on one side are in an opposite direction to those on the other side. If the rotation is continued in the same direction, the eyes will stop in a maximal rostral or caudal position until the head is moved in a different direction. This is in contrast to other vertebrates, which display a fast reset of eye position in the form of the rapid phase of nystagmus. If both labyrinths are removed, these compensatory eye movements disappear and the animal also loses the ability to swim with the head and the body oriented with the dorsal side up (de Burlet and Versteegh, 1930). If only one labyrinth is removed the animal swims in spirals (Rubinson, 1974) with the ipsilateral eye rotated downwards.

The eye bulb is controlled by six eye muscles, a dorsal and a rostral rectus innervated from separate motor nuclei within the oculomotor (N III) complex, a caudal and a ventral rectus innervated from the abducens nucleus (N VI), and finally a rostral and a caudal oblique muscle innervated from the IIIrd and IVth motor nuclei, respectively (Fritzsch *et al.*, in preparation; Finger and Rovainen, 1978). The anterior canal (nerve) innervates eye motor nuclei disynaptically (Rovainen, 1976a), rotating both eyes in the appropriate direction (ipsilateral eye upwards and backwards; contralateral eye anterior and downwards). These three ocular motor nuclei on the contralateral side are activated together (dorsal rectus (motoneuronal axon crosses back to the ipsilateral side), rostral rectus and oblique) with the ipsilateral abducens for the caudal rectus and the contralateral for the ventral rectus.

The posterior canal, on the other hand, produces an ipsilateral upward and forward eye movement and a contralateral movement in the opposite direction by combining an activation of the ipsilateral oculomotor nucleus for the rostral rectus and the contralateral for the dorsal rectus and rostral oblique and the contralateral abducens nuclei for both the ventral and caudal recti. From these findings it is clear that the anterior and posterior vestibular projections can excite, separately, specific motor nuclei within the oculomotor and abducens complexes, respectively. A more detailcd analysis of angular movements of different types would probably provide evidence of a much more refined control.

The compensatory eye movements serve presumably to stabilize the gaze during any type of head movement. The vestibular input also elicits compensatory movements of head and trunk, which serve to counteract 'unintended' movements away from the 'derived' orientation such as the dorsal side up during swimming. The vestibular input has a very powerful excitatory electric and chemical synapse. The anterior nerve projects onto the ipsilateral intermediate nucleus, which contains ipsilateral vestibulospinal neurons. The latter monosynaptically excite motoneurons supplying the dorsal myotomes of the most rostral segments (Rovainen, 1979a) and in addition also very powerfully reticulospinal neurons of the posterior rhombencephalon (prrn). The prrn cells excite all ipsilateral premotor interneurons and motoneurons of the locomotor network (Ohta and Grillner, 1989). The anterior vestibular nerve in addition excites, presumably disynaptically, the contralateral mesencephalic Müller cells in the mrn, which are located around the oculomotor nuclei, which are also excited (see above), and the contralateral isthmic Müller cell in the arrn, while the ipsilateral Müller cells in the mrrn and arrn are inhibited. The bulbar Müller cells in the mrrn are bilaterally excited as are the abducens motor nuclei, ventral (ipsilateral) and caudal (contralateral) recti. The abducens nuclei are located in the same area as the mrrn and their dendrites overlap to a large extent. It is interesting to note that the smaller reticulospinal cells in the mrrn are inhibited, while the large Müller cells are excited (Rovainen, 1979a). With regard to the trochlear motoneurons, which have a bilateral representation, less is known. The cell bodies are located at the same level as the arrn, but more dorsally. Their dendrites partially overlap, however (Fritzsch *et al.*, in preparation).

The posterior vestibular nerve projects to the posterior vestibular nucleus, which has contralateral descending projections to the rostral segments exciting motoneurons, lateral interneurons, and very powerfully contralateral reticulospinal cells in the prrn. The posterior vestibular nerve affects Müller cells in different nuclei and different combinations of eye muscle nuclei with a preference for contralateral excitation of juxtaposed reticulospinal and eye motor nuclei. The pattern of eye muscle control is, however, quite specific and the linkage of different motor nuclei to different reticulospinal projections may relate to a variety of conditions, such as the orientation of the body in a given instant. For instance, during locomotion the reticulospinal activity is modulated via ascending projections from the spinal locomotor circuitry (Kasicki *et al*., 1989; Dubuc and Grillner, 1989).

The spinal overall effect of an activation of the different fast-conducting Müller cells and other reticulospinal cells remains to be fully clarified. It appears clear that they may be used in compensatory movements such as those discussed here and perhaps for steering commands during locomotion. This presupposes that although different cells have partially overlapping functions, they must show a preference for certain target cells, for instance dorsal and ventral myotome motoneurons (Wallén *et al*., 1985). For example, Rovainen's (1974b, 1979a, 1982) results suggest that the mesencephalic M_3 Müller cell in the mrn and the I1 cell in the arrn should preferentially control dorsal fin movements in a way which would counteract a rotating movement along the long lateral axis, that is a compensatory body movement.

VIII.2 Startle and Escape Reactions

Tactile or nociceptive stimuli as well as vibration transmitted through the water can elicit escape reactions. An adult lamprey will initially withdraw from the stimulus and subsequently swim away. A stimulus applied to the left part of the head will for instance cause an initial bending of the head to the right away from the stimulus and subsequently either a swimming movement or a rapid C- or S-shaped flexure of the body, moving the head backwards (McClellan, 1984). Conversely, a tail-fin stimulus will cause a tail bending and a body flexure that will move the tail forwards and often a subsequent bout of swimming (McClellan and Grillner, 1983). The situation is somewhat different for the young ammocoetes (larvae), which spend most of their time burrowed in the sand, with only the mouthpiece sticking out for the purpose of filter feeding. Vibration in the sand or any stimulus to the oral parts causes a rapid flexure of the body, pulling the head in under the surface of the sand (Currie and Carlsen, 1987b).

Cutaneous input from the trigeminal nerve or via dorsal roots (McClellan, 1984; McClellan and Grillner, 1983; Martin and Wickelgren, 1971) can elicit startle reactions. Vibration gives a powerful activation of certain labyrinthine afferents, which monosynaptically activate the large crossed reticulospinal Mauthner cells and via excitatory relay interneurons the contralateral (to the vibration) Mauthner and Müller cells (see above). Ipsilateral Müller cells in the midbrain and anterior rhombencephalon are, however, inhibited. A vibratory stimulus to the left labyrinth will tend to cause a more powerful activation on the right side of the body (Currie and Carlsen, 1987a,b), but the initial position of the body determines to a large extent the response. The change in position amplifies the initial body position. For instance, if the body has a slight S-shape the curvature will be amplified further. The cellular mechanisms for this reinforcement of the existing posture are not clear. An apparent possibility is that the body shape under resting conditions is actively maintained by an increased excitability in the motoneurons which create the inner curvature of the body whenever it is not straight. A descending signal exciting motoneurons will then most likely exert the most powerful effect on the side which is already excited (to a varying degree) compared to the contralateral side. Sensory input may also contribute but if so in a hitherto unknown way (see also section VIII.3).

VIII.3 Locomotion

Lampreys swim with alternating contractions of the two sides at any given segment. The frequency of the contractions can vary approximately between 0.3 and 10 Hz. The intersegmental coordination is characterized by a delay between each consecutive segment resulting in a wave propagated along the spinal cord. This delay is not a fixed time delay but rather a constant phase lag. Accordingly the delay is always a certain proportion (1% per segment) of the cycle duration, whether the latter is 2 s or 0.1 s (Wallén and Williams, 1984). A phase lag of 1% per segment means that there will be a phase lag of approximately a whole cycle between the most rostral segment and the tip of the tail 100 segments further down.

The motor pattern underlying locomotion can be produced in the brainstem–spinal cord *in vitro* preparation, and conveniently be recorded in the ventral roots along the spinal cord. The full frequency range and the intersegment coordination can be retained under *in vitro* conditions. In the brainstem–spinal cord preparation, locomotor activity can be initiated by stimulation of reticulospinal pathways (McClellan and Grillner, 1984, Brodin *et al*., 1988c), by sensory stimuli (if the tailfin is left innervated) (McClellan and Grillner, 1983), and by the application of EAA agonists in the bath (Poon, 1980; Cohen and Wallén, 1980; Grillner *et al*., 1981b; Brodin *et al*., 1985) or blockade of their uptake (Brodin *et al*., 1988d). From an experimental point of view, bath application of EAA agonists is very convenient. It acts by elevating the excitability of the interneuronal pattern-generating circuitry. This circuitry is distributed throughout the spinal cord since any small part of the cord down to two segments can be made to generate alternating burst activity. The intersegmental coordination is achieved by short and long propriospinal neurons (cf. Grillner *et al*., 1988b).

The prrn is most likely the source for the major descending drive of the spinal locomotor circuitry, although EAA neurons in the more rostral reticular nuclei may contribute (Kasicki *et al*., 1989; McClellan and Grillner, 1984; Brodin *et al*.,

1988c; McClellan, 1988b). Neurons in the prrn are turned on during locomotion and they provide NMDA and kainate receptor-mediated EPSPs to all interneurons identified in the segmental network underlying locomotion (Figure 13; Ohta and Grillner, 1989).

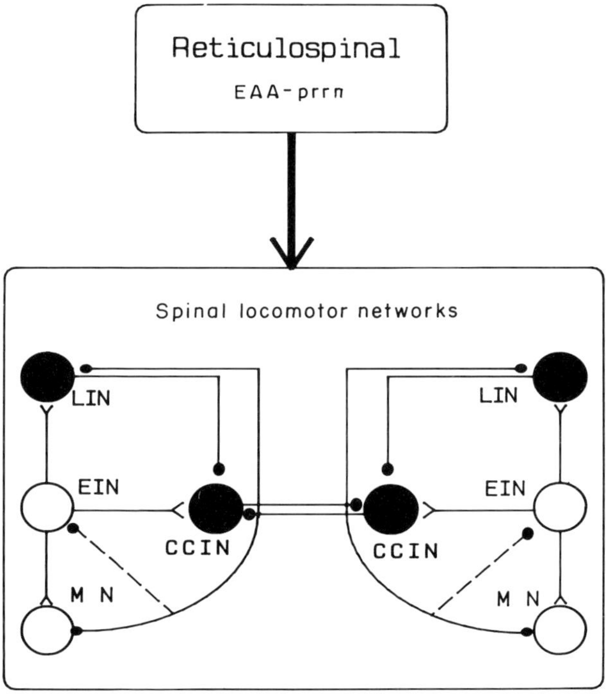

Figure 13 Segmental network of premotor interneurons in the lamprey spinal cord. The network has a left and right symmetrical side. The connections indicated with solid lines have been established with paired intracellular recordings. All types of premotor interneurons, as well as motoneurons, receive EAA receptor-mediated excitation from reticulospinal neurons in the prrn. For abbreviations see Table 1

The segmental locomotor interneurons (Figure 13) include one excitatory (EIN) (Buchanan and Grillner, 1987a; Buchanan *et al*., 1989) and two inhibitory (CCIN, LIN) types (Rovainen, 1974a; Buchanan, 1982; cf. Buchanan and Grillner, 1987b), and their synaptic interactions have been demonstrated by paired recordings. EINs excite all ipsilateral neurons, CCINs inhibit all contralateral neurons and LINs discharge only at the peak of the depolarization and inhibit CCINs and thereby disinhibit the contralateral side. Provided that the background excitation is sufficient, the neurons on the contralateral side will start discharging, thereby inhibiting neurons on the initially active side. Simulation experiments (Figure 14; Grillner *et al*., 1988a, 1989, unpublished) have corroborated this interpretation. In addition to the connectivity, additional mechanisms are important.

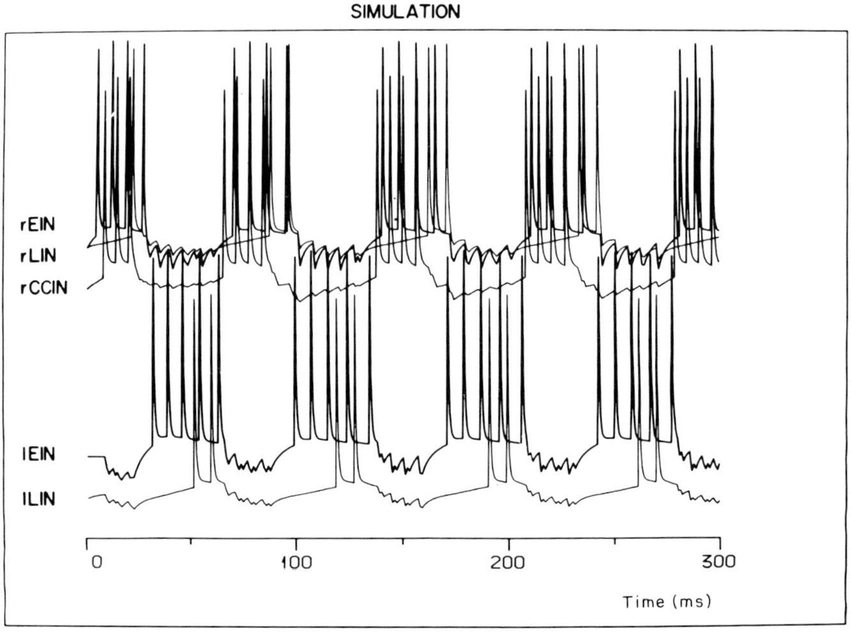

Figure 14 Simulation of the neural network of Figure 13. The network comprised three neurons of each type for each half-segment. Motoneurons were not simulated but can be assumed to be co-active with EIN. The pattern of activity of each type of neuron is shown on the right (r) side, and only EIN and LIN on the left (l) side. The network was turned on by a constant depolarizing current applied to all neurons. (Modified from Grillner *et al*., 1988a, with permission)

(1) The burst termination is achieved not only by the lateral interneuron activity (see above), but also by summation of the postspike afterhyperpolarization (Grillner *et al*., 1988b; Wallén *et al*., 1989).
(2) NMDA–receptor activation leads to a generation of pacemaker-like oscillations in several types of neurons (Grillner and Wallén, 1985; Wallén and Grillner, 1987; Sigvardt *et al*., 1985) and these properties appear to be of particular importance to maintain long bursts during slow-rate swimming (Brodin and Grillner, 1986a).
(3) The excitatory drive from reticulospinal axons is not tonic but phasic with bursts of spikes when the ipsilateral ventral roots are active (Kasicki *et al*., 1989). This modulation originates from the spinal locomotor circuitry and results from an ascending phasic excitation alternating with an active Cl^--dependent inhibition (Dubuc and Grillner, 1989).

The cellular mechanisms underlying the intersegmental coordination are still enigmatic. The constant intersegmental phase lag is always 1% per segment and consequently it will vary between 1 ms at fast swimming (10 Hz) up to 30 ms at very slow swimming (0.3 Hz). Such a 30-fold difference can obviously not be due to fixed axonal conduction delays (see Grillner *et al.*, 1988b). It could be explained, however, if the excitation of the different segmental networks varied, with the most caudal segments having the lowest excitability. Actually, the known interconnections with long propriospinal fibres would tend to create such an excitability gradient (Grillner *et al.*, 1988b). The analogy can be made with a chain of coupled mutually interacting oscillators, each of which would have, if isolated, a basic tendency to oscillate. In isolation the most rostral would be faster and the caudalmost the slowest. If they are hooked up together, they will, under the appropriate conditions, become coordinated, in such a way that the frequency will be the same along the chain, but the rostral oscillator will be ahead of the others and a phase lag will most likely be obtained between successive oscillators.

Movement-related feedback interacts with the central network underlying locomotion in all species investigated (see Grillner, 1981). In the lamprey the lateral undulation of the body activates sensory stretch receptors (edge cells), which are located in the lateral margin of the spinal cord itself (Figure 5; Grillner *et al.*, 1984). When the body or spinal cord is bent passively as if to simulate the locomotor movements, the stretch receptor neurons will very effectively entrain ongoing fictive locomotion. Depending on the imposed rhythm, the frequency may become either higher or lower than rest activity (Grillner and Wallén, 1984; Grillner *et al.*, 1981a; McClellan and Sigvardt, 1988). The edge cells (see above) are of two types: an excitatory type, which excites ipsilateral motoneurons and interneurons of the locomotor network, and an inhibitory type, which inhibits contralateral neurons (Viana di Prisco *et al.*, 1989). The connection of the neurons and their pattern of activity during locomotion can explain the entrainment. This means that the cellular mechanisms underlying the sensory regulation of the locomotor circuitry can now be explained.

VIII.4 Respiration

The isolated lamprey brainstem can spontaneously produce fictive respiration over periods of several days. This robust motor pattern is transmitted via motoneurons in the facial (VII), glossopharyngeal (IX) and vagal motor nuclei (X) (Rovainen, 1974c, 1982). The respiratory motor pattern is simple, and consists of essentially bilateral, almost synchronous, bursts with a free burst interval. The actual respiration in the adult lamprey is produced by seven gillpouches consisting essentially of elastic bags around which muscle fibres contract to empty the bags and expire the water. After the contraction, the elastic recoil will fill the gillpouch with fresh water. The most rostral gillpouch is controlled by cranial nerve VII and IX motoneurons and the remaining gillpouches by vagal motoneurons arranged rostrocaudally within the vagal motor nucleus (Rovainen, 1985; Russell, 1986).

During each respiratory cycle the motoneurons are activated by at least three different sets of interneurons. They receive a slow EPSP, followed by a large phasic EPSP and an IPSP in the interburst interval (Thompson, 1985). The large EPSP often gives rise to an action potential in the motoneurons followed by a large afterhyperpolarization (Rovainen, 1985; Thompson, 1985; Russell, 1986). The premotor interneurons are not well characterized, although some crossed interneurons in the rostral rhombencephalon appear to elicit monosynaptic EPSPs and other IPSPs. Stimulation of their axonal bundle can entrain the locomotor rhythm (Rovainen, 1985). Other putative respiratory interneurons have been located by Russell (1986).

Lesion studies have been used extensively. The full respiratory pattern can be produced by the isolated rhombencephalon. After a longitudinal midline section, each half can produce a respiratory burst pattern. Under ordinary conditions there appear to be respiratory interneurons at the level of the Vth motor nucleus, although a slower burst pattern can under some conditions be produced after a transverse section caudal to the Vth motor nucleus (Rovainen, 1983, 1985; Thompson, 1985; Kawasaki, 1979; Russell, 1986). Fibre bundles coordinating the two respiratory centres on each side are located in both the rostral and the caudal brainstem.

Rovainen (1983) has concluded that the respiratory pattern generator circuitry depends to a very small degree on glycinergic or $GABA_A$ receptor-mediated postsynaptic inhibition. The motor patterns seem to be influenced only to a limited degree by a combination of strychnine and bicuculline or picrotoxin.

The respiratory motor pattern may thus normally be produced by interaction between groups of excitatory interneurons at different levels of the brainstem, each of which in isolation under some conditions can produce a respiratory burst pattern. This network would appear to be a good candidate for elucidating a respiratory vertebrate pattern generation.

VIII.5 Oral Motor Control

In the larval stage the mouth has a shape compatible with filter feeding and after metamorphosis a new elaborate oral apparatus develops with a sucker and a rasper (tongue), and the central pharynx muscles are also modified (Rovainen and Schieber, 1975; Homma, 1975, 1978). The motoneurons supplying the different muscles are located in different parts of the trigeminal motor nucleus (Homma, 1978), but little is known about the interneuronal control of this behaviour. The feeding behaviour has recently been described in some detail by Kawasaki and Rovainen (1988).

IX REGENERATION

After a complete transection of the spinal cord in larval lampreys, they recover swimming movements which to the eye appear quite normal, and EMG activity can be coordinated across the lesion (Hubbard, 1963; Rovainen, 1976b; Selzer, 1978b; Currie and Ayers, 1983). Reticulospinal axons are known to grow across the lesion over a distance of several millimetres (1–15 mm) and to make synaptic contacts (Wood and Cohen, 1979; Yin and Selzer, 1983; Hill *et al*., 1987). Moreover, giant interneurons have been shown to make synaptic contacts on appropriate types of cells (Mackler and Selzer, 1985). To ascertain whether the regenerating fibres make contacts which are sufficient to be able to coordinate motor activity, experiments have been performed *in vitro*, under conditions when no movements occur. This is important since sensory feedback can reflexively coordinate the two parts (Wallén, 1982; Hill *et al*., 1987). Such experiments in curarized preparations have demonstrated a functional regeneration with coordinated activity across the lesion (Cohen *et al*., 1986; McClellan, 1988a).

X CONCLUDING REMARKS

A number of central neuronal mechanisms can be studied in the 'intact' lamprey CNS under favourable *in vitro* conditions. In particular, the sensorimotor systems of the brainstem and spinal cord are well suited for an analysis of the cellular bases of behaviour. Rapidly accumulating morphological, physiological and neurochemical data are compatible with a view that the basic features of the CNS are similar in vertebrates ranging from cyclostomes to mammals, but the present comparative neurobiological knowledge is, however, as yet insufficient (cf. Nieuwenhuys, 1977; Northcutt, 1981, 1984; Parent, 1986).

REFERENCES

Ahlborn, F. (1883). 'Untersuchungen uber das Gehirn der Petromyzon', *Z. Wiss. Zool.*, **39**, 191–294.

Alford, S., and Grillner, S. (1989). 'CNQX and DNQX block non NMDA synaptic transmission but not NMDA evoked fictive locomotion in lamprey spinal cord', *Brain Res.*, **506**, 297–302.

Alford, S., Christenson, J., and Grillner, S. (1989). 'Presynaptic $GABA_A$ and $GABA_B$ receptor-mediated phasic modulation in axons of spinal interneurones', *Eur. J. Neurosci.* (in press).

Andrews, P.C., Pollock, H.G., Elliot, W.M., Youson, J.H., and Plisetskaya, E.M. (1988). 'Isolation and characterization of a variant somatostatin-14 and two related somatostatins of 34 and 37 residues from lamprey (*Petromyzon marinus*)', *J. Biol. Chem.*, **263**, 15809–15814

Ariens Kappers, C.U., Huber, G.C., and Crosby, E.C. (1936). *The Comparative Anatomy of the Nervous System of Vertebrates, Including Man,* Vols 1 and 2, Macmillan, New York.

Baumgarten, H.G. (1972). 'Biogenic monoamines in the cyclostome and lower vertebrate brain', *Prog. Histochem. Cytochem.*, **4**, 1–90.

Bodznick, D., and Northcutt, R.G. (1981). 'Electroreception in lampreys: evidence that the earliest vertebrates were electroreceptive', *Science*, **212**, 465–467.

Bongianni, F., Christenson, J., Grillner, S., and Hökfelt, T. (1989). 'Relation between identified primary sensory neurons and neuropeptide Y- or GABA-immunoreactive neurons and fibers in the spinal cord', *Eur. J. Neurosci.* **73** (Suppl. 2), 6.

Brodin, L., and Grillner, S. (1985). 'The role of putative excitatory amino acid neurotransmitters in the initiation of locomotion in the lamprey spinal cord. II. The effects of amino acid uptake inhibitors', *Brain Res.*, **360**, 149–158.

Brodin, L., and Grillner, S. (1986a). 'Effects of magnesium on fictive locomotion induced by activation of N-methyl-D-aspartate (NMDA) receptors in the lamprey spinal cord *in vitro*', *Brain Res.*, **380**, 244–252.

Brodin, L., and Grillner, S. (1986b). 'Tonic inhibition of a new type of spinal interneurone during fictive locomotion in lamprey', *Acta Physiol. Scand.*, **128**, 327–329.

Brodin, L., Grillner, S., and Rovainen, C.M. (1985). 'NMDA, kainate and quisqualate receptors and the generation of fictive locomotion in the lamprey spinal cord', *Brain Res.*, **325**, 302–306.

Brodin, L., Buchanan, J.T., Hökfelt, T., Grillner, S., and Verhofstad, A.A.J. (1986). 'A spinal projection of 5-hydroxytryptamine neurons in the lamprey brainstem', *Neurosci. Lett.*, **67**, 53–57.

Brodin, L., Christenson, J., and Grillner, S. (1987). 'Single sensory neurons activate excitatory amino acid receptors in the lamprey spinal cord', *Neurosci. Lett.*, **75**, 75–79.

Brodin, L., Buchanan, J.T., Hökfelt, T., Grillner, S., Rehfeld, J.F., Frey, P., Verhofstad, A.A.J., Dockray, G., and Walsh, J.H. (1988a). 'Immunohistochemical studies of cholecystokinin (CCK)-like peptides and their relation to CGRP, 5-HT and bombesin immunoreactivities in the brainstem and spinal cord of lampreys', *J. Comp. Neurol.*, **271**, 1–18.

Brodin, L., Ericsson, M., Mossberg, K., Hökfelt, T., Grillner, S., and Ohta, Y., (1988b). 'Three-dimensional reconstruction of transmitter-identified neurons by en bloc immunohistochemistry and confocal scanning microscopy', *Exp. Brain Res.*, **73**, 441–446.

Brodin, L., Grillner, S., Dubuc, R., Ohta, Y., Kasicki, S., and Hökfelt, T. (1988c). 'Reticulospinal neurons in lamprey: Transmitters, synaptic interactions and their role during locomotion', *Arch. Ital. Biol.*, **126**, 317–345.

Brodin, L., Tossman, U., Ungerstedt, U., and Grillner, S. (1988d). 'The effect of an uptake inhibitor (dihydrokainate) on endogenous excitatory amino acids in the lamprey spinal cord as revealed by microdialysis', *Brain Res.*, **458**, 166–169.

Brodin, L., Ohta, Y., Grillner, S., and Hökfelt, T. (1989a). 'Further evidence for excitatory amino acid transmission in lamprey reticulospinal neurons: selective retrograde labeling with (3H)-D-aspartate', *J. Comp. Neurol.*, **281**, 225–233.

Brodin, L., Rawitch, A., Taylor, T., Ohta, Y., Ring, H., Hökfelt, T., Grillner, S., and Terenius, L. (1989b). 'Multiple forms of pancreatic polypeptide related compounds in the lamprey CNS: partial characterization and immunohistochemical localization in the brainstem and spinal cord', *J. Neurosci.*, **9**, 3428–3442.

Brodin, L., Dale, N., Christensen, J., Storm-Mathisen, J., Hökfelt, T., and Grillner, S. (1990) 'Three types of GABA-immunoreactive cells in the lamprey spinal cord', *Brain Res.*, **508**, 172–175.

Buchanan, J.T. (1982). 'Identification of interneurons with contralateral, caudal axons in the lamprey spinal cord: synaptic interactions and morphology', *J. Neurophysiol.*, **47**, 961–975.

Buchanan, J.T. (1986). 'Premotor interneurons in the lamprey spinal cord: morphology, synaptic interactions and activities during fictive swimming', in *Neurobiology of Vertebrate Locomotion* (Eds S. Grillner, P.S.G. Stein, D.G. Stuart, H. Forssberg and R.M. Herman), pp. 321–334, Macmillan, London.

Buchanan, J.T., and Cohen, A. (1982). 'Activities of identified interneurons, motoneurons and muscle fibers during fictive swimming in the lamprey and the effects of reticulospinal and dorsal cell stimulation', *J. Neurophysiol.*, **47**, 948–960.

Buchanan, J.T., and Grillner, S. (1987a). 'Newly identified "glutamate interneurons" and their role in locomotion in the lamprey spinal cord', *Science*, **236**, 312–314.

Buchanan, J.T., and Grillner, S. (1987b). 'A new class of small inhibitory interneurones in the lamprey spinal cord', *Brain Res.*, **438**, 404–407.

Buchanan, J.T., Brodin, L., Dale, N., and Grillner, S. (1987a). 'Reticulospinal neurons activate excitatory amino acid receptors', *Brain Res.*, **408**, 321–325.

Buchanan, J.T., Brodin, L., Hökfelt, T., Van Dongen, P.A.M., and Grillner, S. (1987b). 'Survey of neuropeptide-like immunoreactivity in the lamprey spinal cord', *Brain Res.*, **408**, 299–302.

Buchanan, J.T., Grillner, S., Cullheim, S., and Risling, M. (1989). 'Identification of excitatory interneurons contributing to generation of locomotion in lamprey: structure, pharmacology, and function', *J. Neurophysiol.*, **62**, 1–11.

de Burlet, H.M., and Versteegh, C. (1930). 'Uber Bau und Funktion des Petromyzon-labyrinthes', *Acta Otolaryngol.* (Suppl.), **13**, 5–58.

Carraway, R., Ruane, S.E., and Kim, H.-R. (1982). 'Distribution and immunochemical character of neurotensin-like material in representative vertebrates and invertebrates: apparent conservation of the COOH-terminal during evolution', *Peptides*, **1**, 115–123.

Christenson, J., and Grillner, S. (1987). 'Modulation of synaptic transmission in primary sensory neurons in the lamprey spinal cord', *Neuroscience* Suppl. **22**, 1022P.

Christenson, J., Bohman, A., Lagerbäck, P.Å., and Grillner, S. (1988a). 'The dorsal cell, one class of primary sensory neurone in the lamprey spinal cord. I. Touch, pressure but no nociception – A physiological study', *Brain Res.*, **440**, 1–8.

Christenson, J., Lagerbäck, P.Å., and Grillner, S. (1988b). 'The dorsal cell, one class of primary sensory neurone in the lamprey spinal cord. A light and electron-microscopical study', *Brain Res.*, **440**, 9–17.

Christenson, J., Alford, S., and Grillner, S. (1989a). 'Co-localized somatostatin and GABA activate different conductances in intraspinal mechanoreceptor neurons', *Proc. XXXI Int. Congr. Physiol. Sci.* (Eds L. Hirvonen *et al.*), p. 4424, Helsinki.

Christenson, J., Franck, J., and Grillner, S. (1989b). 'Increase in endogenous 5-hydroxytryptamine levels modulates the central network underlying locomotion in the lamprey spinal cord', *Neurosci. Lett.*, **100**, 188–192.

Christenson, J., Cullheim, S., Grillner, S., and Cullheim, S. (1989c). '5-hydroxytryptamine immunoreactive varicosities in the lamprey spinal cord have no synaptic specializations – an ultrastructural study', *Brain Res.* (in press).

Cohen A., and Wallén, P. (1980). 'The neuronal correlate of locomotion in fish. "Fictive swimming" induced in an *in vitro* preparation of the lamprey spinal cord', *Exp. Brain Res.*, **41**, 11–18.

Cohen A.H., Mackler, S.A., and Selzer, M.E. (1986). 'Functional regeneration following spinal transection demonstrated in the isolated spinal cord of the larval sea lamprey', *Proc. Natl Acad. Sci. USA*, **83**, 2763–2766.

Crim, J.W., Urano, A., and Gorbman, A. (1979). 'Immunocytochemical studies of LHRH in brains of agnathan fishes. I. Comparisons of adult pacific lamprey (*Entosphenus tridentata*) and the pacific hagfish (Eptatretus stouti)', *Gen. Comp. Endocrinol.*, **37**, 294–305.

Cuénod, M., and Streit, P. (1983). 'Neuronal tracing using retrograde migration of transmitter-related compounds', in *Methods in Chemical Neuroanatomy* (Eds A. Björklund and T. Hökfelt), pp. 365–397, Elsevier, Amsterdam.

Currie, S.N., and Ayers, J. (1983). 'Regeneneration of locomotor command systems in the lamprey', *Brain Res.*, **279**, 238–240.

Currie, S.N., and Carlsen, R.C. (1987a). 'Modulated vibration-sensitivity of lamprey Mauthner neurons' *J. Exp. Biol.*, **129**, 41–51.

Currie, S.N., and Carlsen, R.C. (1987b). 'Functional significance and neural basis of larval lamprey startle behavior', *J. Exp. Biol.*, **133**, 121–135.

Dale, N. (1986). 'Excitatory synaptic drive for swimming mediated by excitatory amino acid receptors in the lamprey', *J. Neurosci.*, **6**, 2662–2675.

Dale, N., and Grillner, S. (1986). 'Dual-component synaptic potentials in the lamprey mediated by excitatory amino acid receptors', *J. Neurosci.*, **6**, 2653–2661.

Dale, N., Roberts, A., Ottersen, O.P., and Storm-Mathisen, J. (1987). 'The morphology and distribution of "Kolmer-Agdur" cells, a class of cerebrospinal contacting neurons revealed in frog embryo by GABA immunocytochemistry', *Proc. R. Soc. Lond.*, **232**, 193–203.

Dubuc, R., and Grillner, S. (1987). 'Spinal cord input to reticulospinal neurons in lamprey', *Acta Physiol. Scand.*, **129**(3), 28A.

Dubuc, R., and Grillner, S. (1989). 'The role of spinal cord inputs in modulating the activity of reticulospinal neurons during fictive locomotion in the lamprey', *Brain Res.*, **483**, 196–200.

Dubuc, R., Viana di Prisco, G., Wallén, P., and Grillner, S. (1989). 'Serotonin modulates membrane properties of brainstem neurons in the lamprey', *Proc. XXXI Int. Congr. Physiol. Sci.* (Eds L. Hirvonen *et al.*), p. 4458, Helsinki.

Finger, T.E., and Rovainen, C.M. (1978). 'Retrograde HRP labeling of the oculomotoneurons in adult lampreys', *Brain Res.*, **154**, 123–127.

Finger, T.E., and Rovainen, C.M. (1982). 'Spinal and medullary dorsal cell axons in the trigeminal nerve in lampreys', *Brain Res.*, **240**, 331–333.

Freud, S. (1878). 'Über Spinalganglien und Rückenmark des Petromyzon', *Sitzungsber. Akad. Wiss. Wien.*, **75**, 81–167.

Fritzsch, B., Dubuc, R., Ohta, Y., and Grillner, S. (1989). 'Efferents to the labyrinth of the river lamprey (*Lampetra fluviatilis*) as revealed with retrograde tracing techniques', *Neurosci. Lett.*, **96**, 241–246.

Fritzsch, B., and Sontag, R. (1988). 'The troclear motoneurons of lampreys (*Lampetra fluviatilis*): location, morphology and numbers as revealed with horseradish peroxidase', *Cell Tissue Res.*, **252**, 223–229.

Gold, M.R., and Martin, A.R. (1984). 'γ-Aminobutyric acid and glycine activate Cl^- channels with different characteristics in CNS neurones', *Nature*, **308**, 639–641.

Goossens, N., Dierickx, K., and Vandesande, F. (1977). 'Immunocytochemical demonstration of the hypothalamo–hypophysial vasotocinergic system of *Lampetra fluviatilis*', *Cell Tissue Res.*, **177**, 317–323.

Grillner, S. (1981). 'Control of locomotion in bipeds, tetrapods and fish', in *Handbook of Physiology, Motor Control* (Ed. V. Brooks), pp. 1179–1236, American Physiological Society, Bethesda, MD.

Grillner, S., and Wallén, P. (1980). 'Does the central pattern generator for locomotion in lamprey depend on glycine inhibition?', *Acta Physiol. Scand.*, **110**, 103–105.

Grillner, S., and Wallén, P. (1984). 'How does the lamprey CNS make the lamprey swim?' *J. Exp. Biol.*, **112**, 337–357.

Grillner, S., and Wallén, P. (1985). 'The ionic mechanisms underlying NMDA receptor induced, TTX-resistant membrane potential oscillations in lamprey neurones active during locomotion', *Neurosci. Lett.*, **60**, 289–294.

Grillner, S., McClellan, A. and Perret, C. (1981a). 'Entrainment of the spinal pattern generators for swimming by mechanosensitive elements in the lamprey spinal cord *in vitro*', *Brain Res.*, **217**, 380–386.

Grillner, S., McClellan, A., Sigvardt, K., Wallén, P., and Wilén, M. (1981b). 'Activation of NMDA receptors elicits "fictive locomotion" in lamprey spinal cord *in vitro*', *Acta Physiol. Scand.*, **113**, 549–551.

Grillner, S., Williams, T.L., and Lagerbäck, P.Å. (1984). 'The edge cell, a possible intraspinal mechanoreceptor', *Science*, **223**, 500–503.

Grillner, S., Buchanan, J.T., and Lansner, A. (1988a). 'Simulation of the segmental burst generating network for locomotion in lamprey', *Neurosci. Lett.*, **89**, 31–35.

Grillner, S., Buchanan, J.T., Wallén, P., and Brodin, L. (1988b). 'Neural control of locomotion in lower vertebrates – From behavior to ionic mechanisms', in *Neural Control of Rhythmic Movements in Vertebrates* (Eds A.H. Cohen, S. Rossignol and S. Grillner), pp. 1–40, John Wiley and Sons, New York.

Grillner, S., Christenson, J., Brodin, L., Wallén, P., Hill, R.H., Lansner, A., and Ekeberg, Ö. (1989). 'Locomotor system in lamprey: Neuronal mechanisms controlling spinal rhythm generation', in *Neuronal and Cellular Oscillators* (Ed. J.W. Jacklet), pp. 407–434, Dekker, New York.

Harris-Warrick, R.M., and Cohen, A.H. (1985). 'Serotonin modulates the central pattern generator for locomotion in the isolated lamprey spinal cord', *J. Exp. Biol.*, **116**, 27–46.

Harris-Warrick, R.M., McPhee, J.C., and Philler, J.A. (1985b). 'Distribution of serotonergic neurons and processes in the lamprey spinal cord', *Neuroscience*, **14**, 1127–1140.

Heier, P. (1948). 'Fundamental principles in the structure of the brain. A study of the brain of *Petromyzon fluviatilis*', *Acta Anat.* Suppl. **8**, 1–213.

Hill, R.H., Århem, P., and Grillner, S. (1985). 'Ionic mechanisms of three types of functionally different neurons in the lamprey spinal cord', *Brain Res.*, **358**, 40–52.

Hill, R.H., Wallén, P., Carlstedt, T., Grillner, S., and Aguayo, A.J. (1987). 'A re-examination of regeneration and recovery of swimming after spinal transection in lamprey', *Acta Physiol. Scand.*, **129**, 22.

Holmquist, A.L., Dockray, G.J., Rosenquist, G.L., and Walsh, J.H. (1979). 'Immunochemical characterization of cholecystokinin-like peptides in lamprey brain and gut', *Gen. Comp. Endocrinol.*, **37**, 474–481.

Homma, S. (1970). 'Presence of monoaminergic neurons in the spinal cord and intestine of the lamprey, *Lampetra japonica*', *Arch. Histol. Jpn, Niigata, Jpn*, **32**, 383–393.

Homma, S. (1975). 'Velar motoneurons of lamprey larvae', *J. Comp. Physiol.*, **104**, 175–183.

Homma, S. (1978). 'Organization of the trigeminal motor nucleus before and after metamorphosis in lampreys', *Brain Res.*, **140**, 33–42.

Homma, S. (1981). 'Effects of DL-aminoadipate on synaptic transmission in spinal interneurones of the lamprey', *J. Comp. Physiol.*, **143**, 423–426.

Homma, S. (1983). 'Physiology and pharmacology of putative transmitters in lamprey central nervous system', *Prog. Neurobiol.*, **20**, 287–311.

Homma, S., and Rovainen, C.M. (1978). 'Conductance increases produced by γ-aminobutyric acid and glycine in lamprey interneurones', *J. Physiol. Lond.*, **279**, 231–252.

Hotton, N. (1976). 'Origin and radiation of poikilothermous vertebrates', in *Evolution of Brain and Behavior in Vertebrates* (Eds R.B. Masterton, C.B.G. Campbell, M.E. Bitterman and N. Hotton), pp. 1–24, Erlbaum Publishers, New Jersey.

Hubbard, E. (1963). 'Regeneration in the severed spinal cord of chordate larvae of Petromyzon marinus', *Exp. Neurol.*, **7**, 175–185.

Jackson, I.M.D., and Reichlin, S. (1974). 'Thyrotropin releasing hormone (TRH): distribution in hypothalamic and extrahypothalamic brain tissues of mammalian and submammalian vertebrates', *Endocrinology*, **95**, 854–862.

Jansen, J. (1930). 'The brain of Myxine glutinosa', *J. Comp. Neurol.*, **49**, 359.

Johnston, J.W. (1912). 'The telencephalon in cyclostomes', *J. Comp. Neurol.*, **22**, 341–404.

Kasicki, S., Grillner, S., Ohta, Y., Dubuc, R., and Brodin, L. (1989). 'Phasic modulation of reticulospinal neurones during fictive locomotion and other types of motor activity in lamprey', *Brain Res.*, **484**, 203–216.

Kawasaki, R. (1979). 'Breathing rhythm generation in the adult lamprey, *Entosphenus japonicus*', *Jap. J. Physiol.*, **29**, 327–338.

Kawasaki, R., and Rovainen, C.M. (1988). 'Feeding behavior by parasitic phase lampreys, *Ichthyomyzon unicuspis*', *Brain Behav. Evol.*, **32**, 317–329.

Kolmer, W. (1905). 'Zur kenntnis der Ruckenmarks von Ammocoetes', *Anat. Hefte*, **29**, 165–214.

Koyama, H., Kishida, R., Goris, R., and Kusunoki, T. (1987). 'Organization of sensory and motor nuclei of the trigeminal nerve in lampreys', *J. Comp. Neurol.*, **264**, 437–448.

Koyama, H., Kishida, R., Goris, R., and Kusunoki, T. (1989). 'Afferent and efferent projections of the VIIIth cranial nerve in lampreys', *J. Comp. Neurol.*, **280**, 663–671.

Larsell, O. (1947). 'The nucleus of the IVth nerve in petromyzontz', *J. Comp. Neurol.*, **86**, 447–466.

Leonard, J.P., and Wickelgren, W.O. (1986). 'Prolongation of calcium action potentials by γ-aminobutyric acid in primary sensory neurones of lamprey', *J. Physiol.*, **375**, 481–497.

Mackler, S.A., and Selzer, M.E. (1985). 'Regeneration of functional synapses between individual recognizable neurons in the lamprey spinal cord', *Science*, **229**, 774–776.

Martin, A.R., and Wickelgren, W.O. (1971). 'Sensory cells in the spinal cord of the sea lamprey', *J. Physiol. Lond.*, **212**, 65–83.

Martin, R.J. (1979). 'A study of the morphology of the large reticulospinal neurons of the lamprey ammocoete by intracellular injection of Procion yellow', *Brain Behav. Evol.*, **16**, 1–18.

Martin R.J., and Bowsher, D. (1977). 'An electrophysiological investigation of the projection of the intramedullary primary afferent cells of the lamprey ammocoete', *Neurosci. Lett.*, **5**, 39–43.

Matthews, G., and Wickelgren, W.O. (1978). 'Trigeminal sensory neurons of the sea lamprey', *J. Comp. Physiol.*, **123**, 329–333.

Mayer, M.L., and Westbrook, G. (1987). 'The physiology of excitatory amino acids in the vertebrate central nervous system', *Prog. Neurobiol.*, **28**, 197–276.

McClellan, A. (1984). 'Descending control and sensory gating of "fictive" swimming and turning responses elicited in an *in vitro* preparation of the lamprey brainstem/spinal cord', *Brain Res.*, **302**, 151–162.

McClellan, A. (1986). 'Command systems for initiating locomotion in fish and amphibians: Parallels to initiation systems in mammals', in *Neurobiology of Vertebrate Locomotion* (Eds S. Grillner, P. Stein, D. Stuart, H. Forsberg and R. Herman), pp. 3–20, Macmillan, London.

McClellan, A. (1988a). 'Functional regeneration of descending brainstem command pathways for locomotion demonstrated in the *in vitro* lamprey CNS', *Brain Res.*, **448**, 339–345.

McClellan, A. (1988b). 'Brainstem command systems for locomotion in the lamprey: localization of descending pathways in the spinal cord', *Brain Res.*, **457**, 338–349.

McClellan, A., and Grillner, S. (1983). 'Initiation and sensory gating of "fictive" swimming and withdrawal responses in an *in vitro* preparation of the lamprey spinal cord', *Brain Res.*, **269**, 237–250.

McClellan, A., and Grillner, S. (1984). 'Activation of "fictive swimming" by electrical microstimulation of brainstem locomotor regions in an *in vitro* preparation of the lamprey central nervous system', *Brain Res.*, **300**, 357–361.

McClellan, A., and Sigvardt, K. (1988). 'Features of entrainment of spinal pattern generators for locomotor activity in the lamprey spinal cord', *J. Neurosci.*, **8**, 133–145.

Moore, L.E., and Christensen, B.G. (1986). 'White noise analysis of cable properties of neuroblastoma cells and lamprey central neurons', *J. Neurophysiol.*, **53**, 636–651.
Moore L.E., Hill, R.H., and Grillner, S. (1987). 'Voltage clamp analysis of lamprey neurons – role of N-methyl-D-aspartate receptors in fictive locomotion', *Brain Res.*, **419**, 397–402.
Müller, J. (1840). 'Vergleichende Neurologie der Myxinoiden', *Berlin Acad. Wissenschaften*, 37–40.
Nakao, T., and Ishizawa, A. (1987). 'Development of the spinal nerves in the lamprey: I. Rohon-Beard cells and interneurons', *J. Comp. Neurol.*, **256**, 342–355.
Negishi, K., Kiyama, H., Kato, S., Teranishi, T., Hatakenaka, S., Katayama, Y., Miki, N., and Tohyama, M. (1986). 'An immunohistochemical study of the river lamprey retina', *Brain Res.*, **362**, 389–393.
Nieuwenhuys, R. (1972). 'Topological analysis of the brainstem of the lamprey *Lampetra fluviatilis*', *J. Comp. Neurol.*, **145**, 165–178.
Nieuwenhuys, R. (1977). 'The lamprey brain in comparative perspective', *Ann. N.Y. Acad. Sci.*, **299**, 97–145.
Northcutt, R.G. (1979). 'Central projections of the eighth cranial nerve in lampreys', *Brain Res.*, **167**, 163–167.
Northcutt, R.G. (1981). 'Evolution of the telencephalon in nonmammals', *Annu. Rev. Neurosci.*, **4**, 301–350.
Northcutt, R.G. (1984). 'Evolution of the vertebrate central nervous system: patterns and processes', *Am. Zool.*, **24**, 701–716.
Northcutt, R.G., and Przyblski, R.J. (1973). 'Retinal projections in the lamprey *Petromyzon marinus*', *Anat. Rec.*, **175**, 400.
Northcutt, R.G., and Puzdrowski, R.L. (1988). 'Projections of the olfactory bulb and nervus terminalis in the silver lamprey', *Brain Behav. Evol.* **32**, 96–107.
Nozaki, M., and Gorbman, A. (1984). 'Distribution of immunoreactive sites for several components of pro-opiocortin in the pituitary and brain of adult lampreys', *Gen. Comp. Endocrinol.*, **53**, 335–352.
Nozaki, M., and Gorbman, A. (1986). 'Occurrence and distribution of substance P-related immunoreactivity in the brain of adult lampreys', *Gen. Comp. Endocrinol.*, **62**, 217–229.
Ochi, J., and Hosoya, Y. (1974). 'Fluorescence microscopic differentiation of monoamines in the hypothalamus and spinal cord of the lamprey, using a new filter system', *Histochemistry*, **40**, 263–266.
Öhman, P. (1976). 'Fine structure of photoreceptors and associated neurons of *Lampetra* fluviatilis', *Vision Res.*, **16**, 659–662.
Ohta, Y., and Grillner, S. (1989). 'Monosynaptic excitatory amino acid transmission from the posterior rhombencephalic reticular nucleus to spinal neurones involved in the control of locomotion in lamprey', *J. Neurophysiol.*, **62**, 1079–1089.
Ohta, Y., Brodin, L., Grillner, S., Hökfelt, T., and Walsh, J.H. (1988). 'Possible target neurons of the reticulospinal cholecystokinin (CCK) projection to the lamprey spinal cord:
immunohistochemistry combined with Lucifer yellow staining', *Brain Res.*, **445**, 400–403.
Opdam, P., and Nieuwenhuys, R. (1976). 'Topological analysis of the brainstem of the urodele *Ambystoma mexicanum*', *J. Comp. Neurol.*, **165**, 285–305.
Opdam, P., Kemali, M., and Nieuwenhuys, R. (1976). 'Topological analysis of the brainstem of the frogs *Rana esculenta* and *Rana catesbiana*', *J. Comp. Neurol.*, **165**, 285–305.
Panula, P., and Airaksinen, M.S. (1989). 'The histaminergic neuronal system as revealed with antisera against histamine', in *Histaminergic Neurons: Morphology and Function*, (Eds T. Watanabe and H. Wada), CRC Press Florida. In press.

Parent, A. (1986). '*Comparative Neurobiology of the Basal Ganglia*', pp. 1–335, John Wiley, New York.

Pfenninger, K.H., and Rovainen, C.M. (1974). 'Stimulation- and calcium-dependence of vesicle attachment sites in the presynaptic membrane; a freeze-cleave study in the lamprey spinal cord', *Brain Res.*, **72**, 1–23.

Piavis, G.W. (1971). 'Embryology', in *Biology of Lampreys*, Vol. I (Eds M.W. Hardisty and I.C. Potter), pp. 405–432, Academic Press, London.

Poon, M. (1980). 'Induction of swimming in lamprey by L-DOPA and amino acids', *J. Comp. Physiol.*, **136**, 337–344.

Reperant, J., Vesselkin, N.P., Miceli, D., Kenigfest, N.B., and Rio, J.P. (1985). 'A comparative radioautographic study of the bidirectional axonal and transcellular transport of different amino acids and sugars in the lamprey visual system', *Brain Res.*, **348**, 348–354.

Retzius, G. (1891). 'Zür Kenntnis der Zentrale Nervensystem von *Myxine glutinosa*', *Biol. Unters.*, **2**, 47–53.

Retzius, G. (1893). 'Ependym und Neuroglia beiden Cyclostomen', *Biol. Unters.*, **5**, 15–18.

Ronan, M. (1983). 'Ascending and descending spinal projections in petromyzontid and myxinoid agnathans', Doctoral thesis, University of Michigan.

Ronan, M. (1989). 'Origins of the descending spinal projections in petromyzontid and myxinoid agnathans' *J. Comp. Neurol.*, **281**, 54–68.

Ronan, M., and Northcutt, R.G. (1987). 'Primary projections of the lateral line nerves in adult lampreys', *Brain Behav. Evol.*, **30**, 62–81.

Rovainen, C.M. (1967a). 'Physiological and anatomical studies on large neurons of central nervous system of the sea lamprey (*Petromyzon marinus*). I. Müller and Mauthner cells', *J. Neurophysiol.*, **30**, 1000–1023.

Rovainen, C.M. (1967b). 'Physiological and anatomical studies on large neurons of central nervous system of the sea lamprey (*Petromyzon marinus*). II. Dorsal cells and giant interneurons', *J. Neurophysiol.*, **30**, 1024–1042.

Rovainen, C.M. (1974a). 'Synaptic interactions of identified nerve cells in the spinal cord of the sea lamprey', *J. Comp. Neurol.*, **154**, 189–206.

Rovainen, C.M. (1974b). 'Synaptic interactions of reticulospinal neurons and nerve cells in the spinal cord of the sea lamprey', *J. Comp. Neurol.*, **154**, 207–223.

Rovainen, C.M. (1974c). 'Respiratory motoneurones in lampreys', *J. Comp. Physiol.*, **94**, 57–68.

Rovainen, C.M. (1976a). 'Vestibulo-ocular reflexes in the adult sea lamprey', *J. Comp. Physiol.*, **112**, 159–164.

Rovainen, C.M. (1976b). 'Regeneration of Müller and Mauthner axons after spinal transection in larval lampreys', *J. Comp. Neurol.*, **168**, 545–554.

Rovainen, C.M. (1978). 'Müller cells, "Mauthner" cells, and other identified reticulospinal neurons in the lamprey', in *Neurobiology of the Mauthner Cell* (Eds D.S. Faber and H. Korn), pp. 245–269, Raven Press, New York.

Rovainen, C.M. (1979a). 'Electrophysiology of vestibulospinal and vestibuloreticulospinal systems in lampreys', *J. Neurophysiol.*, **42**, 745–766.

Rovainen, C.M. (1979b). 'Neurobiology of lampreys', *Physiol. Rev.*, **59**, 1007–1077.

Rovainen, C.M. (1982). 'Neurophysiology', in *Biology of Lampreys*, (Eds M.W. Hardisty and I.C. Potter), pp. 1–136, Academic Press, London.

Rovainen, C.M. (1983). 'Generation of respiratory activity by the lamprey brain exposed to picrotoxin and strychnine, and weak synaptic inhibition in motoneurons', *Neuroscience*, **10**, 875–882.

Rovainen, C.M. (1985). 'Respiratory bursts at the midline of the rostral medulla of the lamprey', *J. Comp. Physiol.*, **157**, 303–309.

Rovainen, C.M., and Birnberger, K.L. (1971). 'Identification and properties of motoneurons to fin muscle of the sea lamprey', *J. Neurophysiol.*, **34**, 974–982.

Rovainen, C.M., and Dill, D.A. (1984). 'Counts of axons in electron microscopic sections of ventral roots in lampreys', *J. Comp. Neurol.*, **225**, 433–440.

Rovainen, C.M., and Schieber, M.H. (1975). 'Ventilation in larval lampreys', *J. Comp. Physiol.*, **104**, 185–203.

Rovainen, C.M., Johnson, P.A., Roach, E.A., and Mankovsky, J.A. (1973). 'Projections of individual axons in lamprey spinal cord determined by tracing through serial sections', *J. Comp. Neurol.*, **149**, 193–201.

Rubinson, K. (1974). 'The central distribution of VIIIth nerve afferents in larval Petromyzon marinus', *Brain Behav. Evol.*, **10**, 121–129.

Russell, D.F. (1986). 'Respiratory pattern generation in adult lampreys (*Lampetra fluviatilis*): interneurons and burst resetting', *J. Comp. Physiol.*, **158**, 91–102.

Sawyer, W.H. (1965). 'Active neurohypophysial principles from a cyclostome (*Petromyzon marinus*) and two cartilaginous fishes (*Squalus acanthias* and *Hydrolagus collei*)', *Gen. Comp. Endocrinol.*, **5**, 427–439.

Schober, W. (1964). 'Vergleichend-anatomische Undersuchungen am Gehiren der Larven und adulten Tiere von *Lampetra fluvialis* (Linné 1758) und *Lampetra planeri* (Bloch, 1784)', *J. fur Hirnforschung*, **7**, 9–209.

Selzer, M.E. (1978a). 'The effect of phenytoin on the action potential of a vertebrate spinal neuron', *Brain Res.*, **171**, 511–521.

Selzer, M.E. (1978b). 'Mechanisms of functional recovery and regeneration after spinal cord transection in larval sea lampreys', *J. Physiol.*, **277**, 395–408.

Sherwood, N.M., Sower, S.A., Marshak, D.R., Fraser, B.A., and Brownstein, M.J. (1986). 'Primary structure of gonadotropin-releasing hormone from lamprey brain', *J. Biol. Chem.*, **261**, 4812–4820.

Sigvardt, K.A., Grillner, S., Wallén, P., and Van Dongen, P.A.M. (1985). 'Activation of NMDA receptors elicits fictive locomotion and bistable membrane properties in the lamprey spinal cord', *Brain Res.*, **336**, 390–395.

Sillar, K.T., and Roberts A. (1988). 'A neuronal mechanism for sensory gating during locomotion in a vertebrate', *Nature*, **331**, 262–265.

Sower, S.A., King, J.A., Millar, R.P., Sherwood, N.M., and Marshak, D.R. (1987). 'Comparative biological properties of gonadotropin-releasing hormone in vertebrates', *Endocrinology*, **120**, 773–779.

Starck, D. (1979). *Vergleichende Anatomie der Wirbeltiere, auf Evolutionsbiologisher Grundlage*, Vol. 3, Springer-Verlag, Heidelberg.

Stefanelli, A. (1933). 'Le cellule le fibre di Müller dei petromizonti', *Arch. Ital. Anat. Embriol.*, **31**, 519–548.

Steinbusch, H.W.M., and Nieuwenhuys, R. (1979). 'Serotonergic neuron systems in the brain of the lamprey *Lampetra fluviatilis*', *Anat. Rec.*, **193**, 693–694.

Steinbusch, H.W.M., Verhofstad, A.A.J., Penke, B., Varga, J., and Joosten, H.W.J. (1981). 'Immunohistochemical characterization of monoamine-containing neurons in the central nervous system by antibodies to serotonin and noradrenaline. A study in rat and lamprey, *Lampetra fluviatilis*', *Acta Histochem. Cytochem.*, Suppl. **24**, 107–122.

Sterzi, G. (1904). 'Die Blutgefässe des Rückenmarks', *Anat. Hefte*, **24**, 3–364.

Teranishi, T., Negishi, K., and Kato, S. (1982). 'Two types of light-induced response recorded from horizontal cells in the river lamprey retina', *Neurosci. Lett.*, **33**, 41–46.

Teräväinen, H., and Rovainen, C.M. (1971). 'Fast and slow motoneurons to body muscle of the sea lamprey', *J. Neurophysiol.*, **34**, 990–998.

Thompson, K.J. (1985). 'Organization of inputs to motoneurons during fictive respiration in the isolated lamprey brain', *J. Comp. Physiol.*, **157**, 291–302.

Tretjakoff, D. (1909a). 'Das Nervensystem von Ammocoetes. I. Das Rückenmark.', *Arch. Mikroskop. Anat.*, **74**, 636–779.

Tretjakoff, D. (1909b). 'Das Nervensystem von Ammocoetes. II. Gehirn', *Arch. Mikroskop. Anat.*, **74**, 636–779.

Tretjakoff, D. (1927). 'Das Periphere Nervensystem des Flussneunauges', *Z. Wiss. Zool.*, **129**, 359–452.

Van Dongen, P.A.M., Hökfelt, T., Grillner, S., Verhofstad, A.A.J., Steinbusch, H.W.M., Cuello, A.C., and Terenius, L. (1985a). 'Immunohistochemical demonstration of some putative neurotransmitters in the lamprey spinal cord and spinal ganglia. 5-hydroxytryptamine-, tachykinin-, and neuropeptide γ-immunoreactive neurons and fibers', *J. Comp. Neurol.*, **234**, 501–522.

Van Dongen, P.A.M., Hökfelt, T., Grillner, S., Verhofstad, A.A.J., and Steinbusch, H.W.M. (1985b). 'Possible target neurons of 5-hydroxytryptamine fibers in the lamprey spinal cord', *J. Comp. Neurol.*, **234**, 523–525.

Van Dongen, P.A.M., Grillner, S., and Hökfelt, T. (1986a). '5-hydroxytryptamine (serotonin) causes a reduction in the afterhyperpolarisation following the action potential in lamprey motoneurons and premotor interneurons', *Brain Res.*, **366**, 320–325.

Van Dongen, P.A.M., Theodorsson-Norheim, E., Brodin, E., Hökfelt, T., Grillner, S., Peters, A., Cuello, A.C., Forssman, W.G., Reinecke, M., Singer E.A., and Lazarus, L.H. (1986b). 'Immunohistochemical and chromatographical studies of peptides with tachykinin-like immunoreactivity in the central nervous system of the lamprey', *Peptides*, **7**, 297–313.

Viana di Prisco, G., Wallén, P., and Grillner, S. (1989). 'Spinal target neurons and synaptic effects of sensory neurons responsible for entrainment of central locomotor pattern generators in lamprey', *Proc. XXXI Int. Congr. Physiol. Sci.* (Eds L. Hirvonen *et al.*), p. 4459, Helsinki.

Vigh, B., and Vigh-Teichmann, I. (1973). 'Comparative ultrastructure of the cerebrospinal fluid-contacting neurons', *Int. Rev. Cytol.*, **35**, 189–251.

Vigh, B., Vigh-Teichmann, I., and Aros, B. (1977). 'Special dendritic and axonal endings formed by the cerebrospinal fluid-contacting neurons of the spinal cord', *Cell Tissue Res.*, **183**, 541–552.

Wächtler, K. (1974). 'The distribution of acetylcholinesterase in the cyclostome brain: I. *Lampetra planeri*', *Cell Tissue Res.*, **152**, 259–270.

Wächtler, K. (1978). 'The regional distribution of choline acetyl transferase and acetylcholinesterase in the brain of the lamprey (*Petromyzon fluviatilis*)', *Comp. Biochem. Physiol.*, **61C**, 411–413.

Wald, U., Selzer, M.E., and Krieger, N.R. (1981). 'Glutamic acid decarboxylase in sea lamprey (*Petromyzon marinus*): characterization, localization, and developmental changes', *J. Neurochem.*, **36**, 363–368.

Wallén, P. (1982). 'Spinal mechanisms controlling locomotion in dogfish and lamprey', *Acta Physiol. Scand.*, Suppl. **503**, 1–45.

Wallén, P., and Grillner, S. (1987). 'N-methyl-D-aspartate receptor induced, inherent oscillatory activity in neurons active during fictive locomotion in the lamprey', *J. Neurosci.*, **7**, 2745–2755.

Wallén, P., and Williams, T.L. (1984). 'Fictive locomotion in the lamprey spinal cord *in vitro* compared with swimming in the intact and spinal animal', *J. Physiol.*, **347**, 225–239.

Wallén, P., Grillner, S., Feldman, J., and Bergelt, S. (1985). 'Dorsal and ventral myotome motoneurons and their input during fictive locomotion in lamprey', *J. Neurosci.*, **5**, 654–651.

Wallén, P., Carlsson, K., Liljeborg, A., and Grillner, S. (1988). 'Three-dimensional reconstruction of neurons in the lamprey spinal cord in whole-mount, using a confocal laser scanning microscope', *J. Neurosci. Methods*, **24**, 91–100.

Wallén, P., Buchanan, J.T., Grillner, S., Christenson, J., and Hökfelt, T. (1989). 'The effects of 5-hydroxytryptamine on the afterhyperpolarisation, spike frequency regulation and oscillatory membrane properties in lamprey spinal cord neurons', *J. Neurophysiol.*, **61**, 759–768.

Watkins, J.C., and Evans, A.H. (1981). 'Excitatory amino acid transmitters', *Annu. Rev. Pharmacol. Toxicol.*, **21**, 165–204.

Whiting, H.P. (1948). 'Nervous structure of the spinal cord of the young larval brook lamprey', *Q. J. Microsc. Sci.*, **89**, 359–383.

Wickelgren, W.O. (1977). 'Physiological and anatomical characteristics of reticulospinal neurons in lamprey', *J. Physiol. Lond.*, **270**, 89–114.

Wickelgren, W.O., Leonard, J.P., Grimes, M.J., and Clark, R.D. (1985). 'Ultrastructural correlates of transmitter release in presynaptic areas of lamprey reticulospinal axons', *J. Neurosci.*, **5**, 1188–1201.

Wood, M., and Cohen, A.H. (1979). 'Synaptic regeneration in identified neurons of the lamprey spinal cord', *Science*, **206**, 344–347.

Wright, G.M. (1986). 'Immunocytochemical demonstration of growth hormone, prolactin and somatostatin-like immunoreactivities in the brain of larval, young adult and upstream migrant adult sea lamprey, Petromyzon marinus', *Cell Tissue Res.*, **246**, 23–31.

Yin, H.S., and Selzer, M.E. (1983). 'Axonal regeneration in lamprey spinal cord', *J. Neurosci.*, **3**, 1135–1144.

Young, L.J., Winokur, A., and Selzer, M.E. (1985). 'Thyrotropin-releasing hormone in lamprey central nervous system', *Brain Res.*, **338**, 177–180.

Preparations of Vertebrate Central Nervous System *In Vitro*
Edited by H. Jahnsen
Published by John Wiley & Sons Ltd

CHAPTER 6

The Isolated Turtle Brain and the Physiology of Neuronal Circuits

JØRN HOUNSGAARD
*Institute of Neurophysiology, Blegdamsvej 3,
DK-2200 Copenhagen, Denmark*

CHARLES NICHOLSON
*Departments of Physiology and Biophysics,
New York University Medical Center,
550 First Avenue, New York, NY 10016, USA*

I INTRODUCTION

Compared to other isolated brain preparations that often require elaborate surgery and perfusion techniques as well as careful control of temperature and medium, the isolated turtle brain is remarkably simple. It is merely necessary to anesthetize the animal, decapitate it, remove the brain parts required and place them in a chamber with flowing, oxygenated saline at ambient temperature. The preparation will then live for several days without further effort. In this chapter we shall first examine the merits of the preparation, the reasons why the brain is able to survive under such primitive conditions and the question of whether there is an optimum bathing medium. Then we shall go on to survey the electrophysiology that has been accomplished with this preparation.

The reader will note that the species of turtle is rarely specified in the following review. That is because most of the studies reported have used one of two North American pond turtles; either *Pseudemys scripta* (the red-eared slider) or *Chrysemys picta* (the painted turtle). Some anoxia studies have used other varieties including marine turtles and a number of anatomical studies (not reviewed here) have also used other species.

I.1 Merits of Isolated Whole-brain Preparations

There are at least three major advantages of the isolated brain preparation. (1) The bathing medium composition can be completely controlled. Generally speaking, in time the composition of the interstitial space of the brain will acquire the same composition as the bath. Two caveats must be made. First, the time for exchange may be quite long, since it depends on the diffusion properties of the brain. Nicholson and Hounsgaard (1983) have provided a quantitative discussion of this issue for brain slices and similar remarks apply to the whole brain. The second caveat is that complete equilibrium may never be achieved if the brain is actively consuming a compound (e.g. oxygen) or producing it (e.g. lactate). Usually, however, investigators are interested in changing ionic composition or applying a drug and in both cases the isolated preparation, in common with the slice, offers clear advantages. (2) The isolated preparation lacks the noise and vibration associated with heartbeat and respiration. This is obviously important for intracellular recording but is also valuable with other techniques including most forms of extracellular recording. Again, this advantage is shared with the slice. Where the isolated brain is demonstrably superior to the slice is in (3) its possession of entire neuronal circuits. This advantage applies both to intrinsic neuronal circuits, such as the inherently three-dimensional architecture of the cerebellum, and extrinsic circuits, as witnessed by the studies on the brainstem–cerebellar system and the eye–optic tectum (see below). These advantages can only be realized if the preparation can preserve its physiological integrity; consequently we must next address the fascinating issue of brain anoxia in the turtle.

I.2 The Rationale for Using the Turtle Brain: Resistance to Anoxia

The major reason for the rapidly increasing popularity of the isolated turtle brain as an isolated preparation is the phenomenal ability of this preparation to withstand anoxia. Numerous studies (e.g. Berkson, 1966; Jackson and Silverblatt, 1974; Robin *et al*., 1964; Penney, 1974) have documented that turtles routinely dive for several hours, while their supply of oxygen is exhausted within 60 min. In fact, at typical ambient temperatures (22°C) turtles can breathe N_2 for 12 h while other reptiles can only tolerate such conditions for about 45 min; typical mammals survive only 5 min (Belkin, 1963). It is also established that such diving feats are accompanied by a systemic lactic acidosis (Cserr *et al*., 1988). An extreme example of this was revealed in studies by Jackson and co-workers on turtles submerged for 3 months or more at 3°C in water gassed with N_2. These animals continued to survive but suffered severe metabolic acidosis with plasma lactate rising to over 200 mM (Jackson and Ultsch, 1982). The acid was buffered by changes in the strong ions, primarily a rise in Ca^{2+} from 3 to over 100 mEq/l and Mg^{2+} from 3 to 30 mEq/l (Jackson and Ultsch, 1982). Most of the lactate was bound to Ca^{2+} and Mg^{2+} (Jackson and Heisler, 1982), with these ions coming from the skeleton and shell (Jackson and Heisler, 1983). Jackson and Heisler also concluded that several unmeasured solutes increase during anoxia.

Controversy still exists about how the turtle brain withstands anoxia. Several mechanisms are probably involved so that different investigators, using different approaches, may emphasize one mechanism to the exclusion of others.

In a recent review Suarez (1988) has summarized the different abilities of mammals and turtles to withstand anoxia. Mammals and probably some anoxia-prone non-mammalian species demonstrate an inability to produce ATP using anaerobic pathways at rates sufficient to sustain ATP-requiring functions. This is true despite efforts to 'keep up' via the activation of glycolysis in response to anoxia, the so-called Pasteur effect. The glycolytic enhancement only lasts for about 15 min, however, and is Ca^{2+}-dependent. In another recent review, Lutz *et al.* (1985) drew attention to the fact that many investigators had shown that EEG suppression in mammals occurs before either ATP is significantly diminished or lactate (the major product of anaerobic glycolysis) is increased. Lutz *et al.* concluded that the main compensatory change in the mammalian brain during hypoxia is a decrease in energy consumption rather than an increase in anaerobic glycolysis.

In marked contrast to mammals, no Pasteur effect is seen in turtles (Suarez, 1988) and it has been claimed that the brains of turtles actually decrease their energetic requirements and, consequently, their glycolytic rates (Suarez, 1988). This reduced energy requirement has been attributed in part either to 'channel arrest' (Hochachka, 1986), whereby the leakage of K^+ from brain cells is curtailed, or 'neuronal dormancy' (Suarez, 1988), where synaptic activity may be reduced. The latter conjecture has received some support from a recent study of brain excitability during anoxia (Feng *et al.*, 1988a) but a related investigation (Feng *et al.*, 1988b) showed that inhibition is not enhanced during anoxia although it may increase during the recovery phase. Needless to say, if either channel arrest or neuronal dormancy was confirmed they would be of considerable experimental and clinical interest. The overall conclusion of Suarez (1988) regarding the behavior of the turtle brain during prolonged anoxia is corroborated by Lutz *et al.* (1985) but those authors also add a considerable amount of important detail based in large part on their own, continuing, studies.

What has been established experimentally is that, firstly, the obvious difference in body temperature between turtles and mammals is insufficient to account for the marked difference in ability to withstand anoxia (Lutz *et al.*, 1985; Suarez,

1988). A key observation in coming to terms with the underlying mechanisms is embodied in a series of experiments in which Sick *et al*. (1982) established that N_2 inspiration in the turtle led to only moderate increases in $[K^+]_o$ of less than 2 mM, in sharp contrast to the results in rats under similar conditions where $[K^+]_o$ elevates to as much as 80 mM. When iodoacetate (IAA), a potent blocker of glycolysis, is also present in the turtle, however, a dramatic rise in $[K^+]_o$ to about 30 mM was seen during anoxia. But in the normoxic turtle IAA had no such effect. These data indicate that the turtle does place increased reliance on glycolytic ATP production during anoxia which apparently conflicts with the conclusions summarized by Suarez (see above). Glucose availability is another important factor during anoxia in the turtle brain. During enforced 24-h dives in de-oxygenated water, blood glucose rises from 3.6 mM to 17.7 mM (Penney, 1974). It was also established that a level of at least 20 mM must be maintained in the bathing medium to sustain the anoxic brain, while levels of 10 mM or less are adequate during normoxia (Sick, personal communication). We have also discovered this empirically in the formulation of physiological salines (see below). Despite the evidence for increased glycolysis, Lutz *et al*. (1985) also concluded that a real reduction of metabolic rate was apparent as anoxia proceeded and that this may have been missed in earlier investigations using brain slices (in contrast to whole brain) (Robin *et al*., 1979) either because of differences in pH and the dependence of energy production on this parameter or because of other factors relating to intrinsic energy control in the intact preparation.

Where there are differences between the conclusions of Lutz *et al*. (1985) and Suarez (1988) for the turtle brain they may perhaps be reconciled by the summary Figure 1, derived from a similar one by Lutz *et al*. (1985), showing that the relative roles of different mechanisms change as anoxia proceeds. Thus during the early phase of a dive the animal still has some oxygen in its lungs and oxidative phosphorylation continues but diminishes. The deficit in ATP production is made up by increasing reliance on anaerobic glycolysis and ATP consumption continues at a normal rate. After about 20 min a transition phase occurs where PO_2 is close to zero and glycolysis is rapidly diminishing. Remarkably, beyond this time glycolysis levels off at only 5–10% of its previous value but the level of ATP is undiminished (Lutz *et al*., 1984). This can only mean that metabolic consumption is greatly reduced. Again Lutz *et al*. (1985) favor the idea that a decreased ion permeability is responsible for the reduction in metabolic need.

A final comment on these studies is that there are considerable differences between the brain of the intact diving turtle and the isolated brain bathed in flowing physiological saline. It is known that, in common with diving mammals, there is a redistribution of blood supply in the diving turtle to ensure improved brain oxygenation (Lutz *et al*., 1985) and that ventilation of the intact animal is under the control of chemoreceptors that are accessible from CSF in the ventricles (Hitzig and Jackson, 1978). Moreover the intact preparation maintains a perfused brain system, enabling movement of substances to and from the brain across the blood–

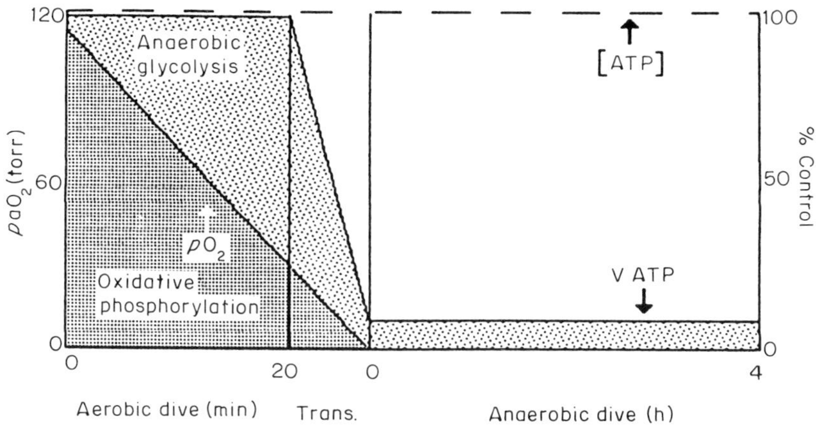

Figure 1 Schematic representation of brain adaptations to hypoxia during prolonged dive and anoxia in the freshwater turtle. (From Lutz *et al.* 1985, with permission)

brain barrier. In contrast the isolated brain is not only anoxic but also ischemic and all substrates and products must diffuse from and to the brain surfaces.

I.3 Choice of Bathing Media

The ideal physiological saline for a given preparation is a vexed and perhaps impossible question. In most species, different investigations use widely different compositions, often without any declared rationale (see Burton, 1975). The composition of the CSF of the turtle has been analyzed (Heisy, 1968; Cserr *et al.*, 1988) and is given in Table 1. It is commonly assumed that CSF is in equilibrium with the brain interstitial space so that a physiological saline might seek to mimic CSF. Other issues make this simple strategy less than optimal, however, and the results of necessary compromise are indicated in Table 1, where the actual salines used in several investigations are characterized. The major problem is that the circulatory system is inoperative in the isolated brain. This results in the following:

(1) A pH gradient from the surface of the brain to the interior (Figure 2). Since lactic acid is being produced, the interior is *always* more acidic than the surface, so it may be advantageous to adjust the pH of the surface to be more alkaline than normal so that the interior of the brain will be less acidic. Using a bicarbonate buffer and typical 95% O_2/5% CO_2 gas mixture, this implies that bicarbonate must be elevated somewhat beyond the natural 23 mM, unless special gas mixtures are used (Hounsgaard *et al.*, 1988). Other pH buffers are generally less effective (less buffering power) or introduce possible artifacts, although they are required by certain studies.
(2) In all isolated, unperfused, brain preparations (including the slice) the O_2

Table 1 Ringer for turtle

	Na^+	Cl^-	K^+	Ca^{2+}	Mg^{2+}	HCO_3^-	Other pH buffer	Gas	pH	Glu	Representative ref.
Plasma	133.1	87.4	3.2	3.6	1.85	30.1					Heisy (1968)
	121	76	2.1	3.5	2						Cserr *et al*. (1988)
CSF	128.5	100.3	2.6	1.8	1.7	22.9					Heisy (1968)
	129	91	1.9	1.3	1.2						Cserr *et al*. (1988)
Ringer	142	108.5	5	0.75	1	30	Na_2HPO_4 5; NaH_2PO_4 2		7.5	5	Robin *et al*. (1979)
	128	111.1	2.6	4	2	31.5		95/5	7.6	10	Mori *et al*. (1981a)
	114	122	2	3	0	0	Tris 10	100/0	7.2	2.5	Jahr and Nicoll (1982a)
	133	146.8	3.8	2.8	2.2	0	Tris 5	100/0	6.9–7.6	10	Keifer and Stein (1983)
	140	120	6	3.5	3.5	40		95/5	20	20	Chan and Nicholson (1986)
	140	119	5	3.5	3.5	40		95/5	7.6	20	Lauritzen *et al*. (1988)
	155	133	5	2	2	35		95/5	7.6	10	Chesler and Chan (1988)
	155	135	5	3	2	35		95/5		20	Chan *et al*. (1988a)
	135	135	5	3	2	15		98/2	7.6	10*	Hounsgaard *et al*. (1988)
	130	139	5	1	1	0	Hepes 20			25	Shen *et al*. (1988)

Notes: Concentrations in mM; gas is O_2/CO_2 as %; Glu = glucose.
* Glucose concentration is now 20 mM.

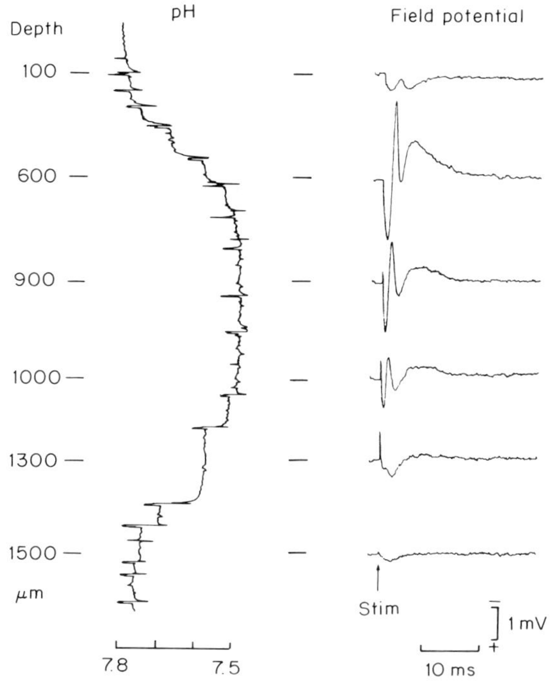

Figure 2 Depth profiles of pH and field potentials in the isolated turtle cerebellum. pH was measured with a double-barreled ion-selective microelectrode (ISM) at the depths indicated below the pial surface. A pH scale is shown beneath the record. The values above 100 μm and below 1500 μm represent the pH of the Ringer solution in which the cerebellum was bathed. The field potentials were recorded on the reference barrel of the ISM and were evoked by local surface stimulation (stim) of the parallel fibers. They are characteristic of cerebellar field potentials in many different species with a fast presynaptic volley representing the parallel fiber compound action potential and a later slower negativity representing the depolarization of Purkinje cell dendrites (at 600 μm). Note that the depths are arbitrary: the cerebellum extends from approximately 600 μm to 1500 μm. (C. Nicholson, unpublished data)

tension is too high at the surface and often too low in the center of the preparation. Of course in the turtle this is relatively unimportant but it may imply a gradient of metabolic processes. At present we do not know of studies of O_2 tension as a function of depth in the isolated turtle brain.

(3) It is evident from the discussion of anoxia that adequate glucose is essential if true anoxic conditions are established in any region of the turtle brain. Such conditions may occur even when well-oxygenated saline is supplied due to problems with chamber design or flow patterns. It would seem that 20 mM glucose is required for safety. Even then the glucose concentration in the center might be reduced depending on consumption rate.

(4) Most slice preparations of all species use an elevated $[K^+]_o$, usually approaching twice the *in vivo* value, because there is an impression that the physiological responses are enhanced. Our own observations support this in the turtle, where the locally evoked cerebellar responses appeared best with 5 mM $[K^+]_o$ (Nicholson, unpublished observations). Several investigators do use 2.6 mM, however, presumably derived from Heisy's measurements (Heisy, 1968).

(5) The best concentrations of Ca^{2+} and Mg^{2+} are not easy to specify. Synaptic activity demands Ca^{2+}, as do a vast number of other processes, so adequate Ca^{2+} is important. Mg^{2+} is implicated increasingly in the dynamics of NMDA receptors. Its omission can also make the cerebellum (and possibly other regions) greatly susceptible to spreading depression (Lauritzen *et al*., 1988; see below). Finally Ca^{2+} activity is quite dependent on the anions present so that if Cl^- is substituted or HCO_3^- is substantially altered then the effect of Ca^{2+} may be changed. These conflicting and complex issues probably account in part for the variety of salines in current use.

Finally it is worth noting that even with an excellent saline there can be noticeable variation from animal to animal. This can be reduced to some extent by screening incoming animals for leeches and other parasites and by keeping turtles in tanks where they can both swim and rest out of the water; regular feeding is also important. Some practical advice on turtles can be found in the book by Harless and Morlock (1979).

II REGIONAL SURVEY OF *IN VITRO* STUDIES OF TURTLE CNS

In this section we review how *in vitro* preparations are being used to investigate several regions of the turtle CNS. We shall concentrate on physiological investigations on the isolated brain but two points should be mentioned. There have been a number of physiological studies on the *in vivo* brain which obviously complement those to be described and there is a large number of neuroanatomical studies on the brain of the turtle covering all the regions to be described. No comprehensive anatomical review exists at this time and we are unable to provide one within the confines of this chapter but the reader should be aware that this literature exists.

II.1 Olfactory Bulb

The isolated olfactory bulb was the first *in vitro* preparation from the CNS of the turtle (Mori and Shepherd, 1979). The motivation for developing this preparation was the unique prospect of studying a well-defined local synaptic circuitry involving dendrodendritic synapses. The morphology of the dendritic arbor of the mitral cell – the principal neuron in the bulb – excluded the use of mammalian brain slices. The dendrites of a mitral cell occupy a cylindrical territory of the external plexiform layer with a diameter of more than 2000 μm. By isolating the entire bulb, Shepherd and his collaborators not only preserved the cells and the local circuitry but also included the extrinsic, afferent and efferent pathways, which proved highly advantageous for the experimental analysis.

In their first report Mori and Shepherd (1979) documented the viability of the preparation and its usefulness for intracellular recording from mitral cells. They also described the complex sequence of synaptic responses in mitral cells following activation of the lateral olfactory tract and the olfactory nerve. These initial observations formed the basis of subsequent work in this preparation. Structurally the dendrites of mitral cells engage in reciprocal synaptic connections with the dendrites of periglomerular cells and granule cells (see Shepherd, 1979). Exploitation of the *in vitro* preparation has served to establish the credibility of dendrodendritic synapses as functional entities and describe their properties. In agreement with a presynaptic function, a Ca^{2+}-mediated regenerative response was identified in mitral cell dendrites (Jahr and Nicoll, 1980, 1982a; Mori *et al*., 1981a). This response was shown to be bounded by a voltage-dependent K^+ conductance, a Ca^{2+}-dependent K^+ conductance and a Cl^- conductance (Jahr and Nicoll, 1982a). The Cl^- conductance was elicited in response to orthodromic and antidromic activation by stimulating the olfactory nerve and the lateral olfactory tract (Mori *et al*., 1981b; Jahr and Nicoll, 1982a). It was blocked when large anions were substituted for Cl^- in the bathing medium (Nowycky *et al*., 1981a) and by picrotoxin and bicuculline (Jahr and Nicoll, 1980, 1982a; Nowycky *et al*., 1981a). Although synaptic in nature, the Cl^- conductance in a mitral cell nevertheless strongly depended on the impulse activity in the cell itself. In response to orthodromic and antidromic nerve stimulation, the inhibitory postsynaptic potential (IPSP) increased dramatically in amplitude when accompanied by an action potential in the mitral cell recorded from Jahr and Nicoll (1982a). In fact a train of action potentials, generated by a depolarizing current pulse through the intracellular recording electrode, could elicit the IPSP in the mitral cell recorded from even in the presence of TTX (Jahr and Nicoll, 1982a). Taken together these experiments provided strong evidence for the hypothesis that mitral cell dendrites release an excitatory transmitter which in turn mediates GABA release from granule cell dendrites to generate mitral cell inhibition. A crucial piece of additional evidence for this interpretation was offered by intracellular recordings from granule cells. During mitral cell inhibition, granule cells hyperpolarized concomitantly with a reduced frequency of spontaneous excitatory synaptic potentials (Jahr and Nicoll, 1982a).

In vitro experiments in the olfactory bulb of the turtle have uncovered an additional example of dendrodendritic synaptic interaction. After block of GABAergic inhibition, regenerative responses, generated by a depolarizing current pulse in a mitral cell, were followed by a depolarizing afterpotential (DAP) whether or not TTX was present. Unlike TTX-resistant DAPs in other cells, this one was strictly Ca^{2+}-dependent and was abolished by substituting Ca^{2+} with Ba^{2+}.

Finally, the DAP was found to be blocked by α-aminoadipate, which antagonizes receptor responses mediated by aspartate – the presumed transmitter in mitral cells. Based on their experiments, Nicoll and Jahr (1982) concluded that Ca^{2+} influx in mitral cell dendrites releases a transmitter that not only excites postsynaptic dendrites of granule cells but also mediates autoexcitation of the presynaptic mitral cell. Since, after block of inhibition, the DAP also occurs following antidromic stimuli that fail to generate impulses in the cell recorded from, mutual excitatory interactions between dendrites of different mitral cells are strongly suggested. Such mechanisms may account for the secondary excitation in mitral cell following olfactory nerve stimulation under conditions of blocked inhibition (Nowycky *et al.*, 1981b).

Activation of the olfactory nerve and the lateral olfactory tract has revealed three phases of inhibition. The GABA-mediated inhibition from granule cells could account for two early waves of mitral cell hyperpolarization following a stimulus of either pathway (Mori *et al.*, 1981b). A third wave of inhibition with a duration approaching 5 s (Mori *et al.*, 1981c) had a number of novel properties. It was not mediated by Cl^- and was not associated with a clear conductance change (Nowycky *et al.*, 1981a). The amplitude of this slow inhibition was greatest when the mitral cell was close to its resting potential but decreased when elicited from more hyperpolarized or depolarized levels of membrane potential. A reversal potential was absent (Mori *et al.*, 1981c; Nowycky *et al.*, 1981a).

The origin of slow inhibition is not known but one possibility seems to be that it reflects removal of autoexcitation during mitral cell inhibition. This could explain its gradual disappearance with removal of GABA-mediated inhibition (Nowycky *et al.*, 1981a,b).

The experiments on the isolated olfactory bulb of the turtle have provided new insights into the functional flexibility of local neuronal networks. The intrinsic processing of afferent input to the glomeruli of mitral cells is subjected to considerable regulation. Fast prepotentials in the primary dendrites help to boost excitatory synaptic responses (Mori *et al.*, 1982). Signal transfer to the cell body is undoubtedly affected by dendrodendritic interaction with interneurons. The fact that recurrent inhibition from granule cells is regulated over a wide range by norepinephrine, enkephalin and glutamate (Jahr and Nicoll, 1982b) suggests that the intrinsic transfer of synaptic potentials in mitral cells intimately depends on the state of the local interneurons that they contact (Mori *et al.*, 1982).

Not only synaptic transfer is regulated. Axons in separate bundles of the olfactory nerve converge on individual mitral cells. The generation of excitatory synaptic potentials in mitral cell dendrites in glomeruli following a test stimulus was shown to be inhibited locally by a preceding conditioning stimulus over a period of more than 1 s. These experiments (Waldow *et al.*, 1981; Mori *et al.*, 1984) clearly indicate that complex synaptic processing involving olfactory nerve terminals, dendrites of mitral and periglomerular cells and terminals of glomerular cells, takes place in the glomerulus in relative isolation from the rest of the mitral cell.

At the network level the 2-deoxyglucose technique has helped to define the overall picture of sensory processing in the olfactory bulb (Greer *et al.*, 1981). The experiments show increased metabolic activity after olfactory nerve stimulation confined to the synaptic laminae, indicating that all known intrinsic synapses participate. After blocking GABAergic inhibition, metabolic activity is also enhanced in the unstimulated half of the bulb. Such experiments suggest that even very focused afferent inputs may involve distributed processing. It is also of considerable interest to note that the functional network for processing afferent activity is dynamically sculptured by the interneuronal activity. It remains to be seen if isolated preparations that include the sensory epithelium might help to elucidate how natural stimuli are processed.

II.2 The Telencephalon

The turtle telencephalon *in vitro* offers the experimenter the prospect of being able to study cortical processing from the subcellular level to the level of neuronal networks in one and the same preparation. The configuration of the preparations is also tolerant to a wide range of flexible modifications to suit different technical demands. The turtle telencephalon is structurally simpler and includes fewer elements than the mammalian counterpart. This all suggests a potential shortcut to general principles of cortical function. Not surprisingly, therefore, a good deal of work on *in vitro* telencephalic preparations from the turtle has been aimed at clarifying similarities with and differences from results from mammals. Most extensively studied to date is the visual, dorsal cortex.

In the dorsal cortex of the turtle the three known cell types – the pyramidal cells, the stellate cells and the ependymal cells – are unambiguously identifiable *in vitro* by one of several independent electrophysiological criteria. In addition, all three cell types are readily distinguished morphologically after dye injection and histological processing.

Records from pyramidal cells are obtained in the cell body layer; the cells have action potentials with two distinct levels of amplitudes. During trains of action potentials evoked by depolarizing current pulses, there is marked time-dependent adaptation of spike frequency (Connors and Kriegstein, 1986). In response to an afferent stimulus, pyramidal cells generate excitation followed by prominent inhibition (Kriegstein and Connors, 1986).

Stellate cells are recorded from their cell bodies in the subcellular or the molecular layer. They have fast action potentials with uniform amplitudes. Time-dependent frequency adaptation during spike trains is minimal (Connors and Kriegstein, 1986). The response to afferent stimuli is almost exclusively excitatory (Kriegstein and Connors, 1986).

The intrinsic response properties of the two types of neurons in the visual cortex are remarkably different. Action potentials in pyramidal cells have several sites of origin. Particularly intriguing is the suggestion that the low-amplitude spike can be supported by dendrites as well as by the axon (Connors and Kriegstein, 1986; Kriegstein and Connors, 1986). The experiments also showed that the low-amplitude spikes had a lower threshold than the high-amplitude spikes, when evoked both by depolarizing current injected at the soma and by excitatory synaptic responses generated in distal dendrites (Connors and Kriegstein, 1986; Kriegstein and Connors, 1986). The low-amplitude spikes closely resemble fast prepotentials in hippocampal pyramidal cells. The visual cortex of the turtle seems to offer an excellent opportunity to analyze the mechanism and function of fast prepotentials under simplified and more readily controlled experimental conditions.

The ionic conductances so far identified in turtle pyramidal cells include a time-dependent inward rectifier and TTX-sensitive Na^+ conductances responsible for both types of action potentials.

Ca^{2+} conductance is masked by K^+ conductances under normal conditions but generates a prominent long-lasting Ca^{2+} spike in the presence of extracellular TEA or intracellular Cs^+. There is also evidence for a Ca^{2+}-dependent K^+ conductance mediating a slow spike afterhyperpolarization (Connors and Kriegstein, 1986; Kriegstein and Connors, 1986). The more compact stellate cells appear to have less complex intrinsic properties, but detailed experimental analysis has not yet been performed.

Kriegstein and Connors (1986) made an extensive study of the functional organization of synaptic interactions in the dorsal cortex *in vitro*. Activation of superficial cortical afferents evoked a long-lasting excitatory response in stellate cells and a phase of brief excitation followed by long-lasting inhibition in pyramidal cells.

Like the CA1 region in the mammalian hippocampus, the excitatory inputs to pyramidal cells showed a striking laminar organization with depth from the pial surface. The inhibitory response in pyramidal cells was mediated by both feedforward and feedback connections. The early part of the inhibitory response had all the classical characteristics of interneuronal inhibition by GABA receptors. It was mediated by a Cl^- conductance, could be blocked by bicuculline and was stimulated by local iontophoretic application of GABA. In contrast, the late phase of inhibition seemed to be mediated by a K^+ conductance coupled to a different, and as yet unidentified, transmitter receptor (possibly a $GABA_B$ receptor). Part of this late response was clearly dissociable from the Ca^{2+}-dependent afterhyerpolarization.

More complex synaptic responses in pyramidal cells in the dorsal cortex include temporal facilitation of excitatory synaptic responses and spatially and temporally labile reversal potential for GABA-mediated inhibition.

The analysis suggests a rather clear basic synaptic circuitry in the dorsal cortex in which extrinsic afferents provide primary excitation to stellate and pyramidal cells in parallel. Pyramidal cell collaterals mediate feedback excitation of both pyramidal and stellate cells while stellate cells mediate feedback and feedforward inhibition of pyramidal cells.

Electrotonic intercellular connections are notably absent. It remains to be seen how the basic circuitry is regulated by noradrenergic, cholinergic and other subcortical afferent systems known to project to the cortex.

Aided by the analysis of the cellular elements and their synaptic interactions, Kriegstein (1986) has performed the first intracellular analysis of cortical processing of visual information *in vitro*. In a preparation that included the eyes, optic nerves and the brain, pyramidal cells in the visual cortex responded to brief flashes of light in the contralateral eye with an excitatory–inhibitory sequence. A light stimulus lasting several seconds evoked this sequence at the onset and at the offset of the flash. The main part of the inhibitory response was due to a GABA-mediated increased Cl^- conductance. In cases of burst responses during the excitatory phase there was evidence of a Ca^{2+}-dependent K^+ conductance contributing to the subsequent inhibition. Simultaneous intracellular recording from a pyramidal cell and extracellular recording from a nearby stellate cell suggested that the light-evoked inhibition of pyramidal cells was mediated by co-activated stellate cells. Since the retina of the turtle has been the subject of numerous investigations (which will not be reviewed here), the preparation devised by Kriegstein offers unique opportunities for elucidating fundamental questions on cortical processing with a precision not previously possible. A new derivative of this eye-attached preparation shows further promise for the study of the visual system (Rosenberg and Ariel, 1988).

The mammalian hippocampus is by far the most studied slice preparation. One of the main lines of research has focused on epileptogenesis and epileptiform activity with this preparation as a rewarding *in vitro* model. *In vitro* studies in the equivalent region of the turtle brain – the medial cortex – give good indications of this being an even simpler model. Shen and Kriegstein (1986) have shown that the anatomical distinction between a dorsal and ventral portion of the medial cortex is reflected in the physiological properties of pyramidal cells and synaptic connectivity of stellate cells in the two regions.

The ventromedial cortex seems to share most of the intrinsic electrophysiological properties of the visual cortex. This is not so for the dorsal medial cortex, however. In this region most pyramidal cells engage in regular burst activity. In some pyramidal cells this is an intrinsic property, since the burst frequency changes with the bias current through the recording electrode. In other cells, bursting is insensitive to bias current and hyperpolarization reveals ‘giant’ synaptic potentials with the same interval at bursting as at rest. Shen and Kriegstein also encountered pyramidal cells in which bursting was a mixed endogenous/exogenous phenomenon. In all bursting pyramidal cells, however, the intrinsic properties clearly contributed to the configuration of the burst. The hyperpolarization following each burst was predominantly due to GABAergic inhibition mediated by a Cl^- conductance, most likely mediated by local stellate cells.

Unlike stellate cells in other cortical areas, those in the dorsal medial cortex respond to afferent stimuli with only brief excitation followed by a prominent phase of inhibition. This seems to indicate that stellate cells in this part of the hippocampus of the turtle are synaptically interconnected.

Epileptiform activity in preparations including dorsal and medial cortex was observed in the presence of bicuculline. Burst responses occurred with very similar frequency in different areas of the same preparation. After interrupting the connections between the subfields of the preparation, however, epileptiform bursts only occurred in the dorsal medial cortex. These findings show that the dorsal medial cortex in the turtle, like the CA3 region in the mammalian hippocampus, can act as a pacemaker for seizure activity in other cortical areas. The specific characteristics of the dorsal medial cortex may give new clues as to the minimal requirements for epileptiform activity. The lack of electrotonic connections between pyramidal cells in the turtle excludes this as a basic requirement for synchronized activity. The results obtained by Shen and Kriegstein suggest that the endogenous burst characteristics of pyramidal cells, mutual inhibition between stellate cells and mutual excitatory connections between pyramidal cells could be the essentials. More recent findings by Larson-Prior and Slater (1988a) indicate reduced inhibition to be the fundamental initiator of epileptiform activity. They also find that blocking receptors for excitatory amino acids using kynurenic acid or 2-amino-5-phosphovalerate (2-APV) reduces the amplitude and duration but does not extinguish the epileptiform burst discharge in pyramidal cells. This could indicate that endogenous burst properties play a more prominent role than connectivity.

It should be noted that the different susceptibility for epileptiform activity in different areas of the telencephalon may be relative and changeable. In the isolated visual cortex, repeated iontophoretic application of glutamate sometimes transformed a subregion of the preparation to a local generator of synchronized burst discharge in which both stellate cells and pyramidal cells took part (Kriegstein and Connors, 1986).

An extremely promising new direction in *in vitro* studies of the turtle brain has been opened up by the introduction of explant and dissociated cultures of embryonic cortex (Shen *et al*., 1988; Kriegstein *et al*., 1988; Wilson *et al*., 1988). The experiments clearly show that embryonic neurons continue to develop in culture along paths that closely resemble normal development *in vivo*. The whole-cell patch-clamp technique (Blanton *et al*., 1989) has enabled electrophysiological analysis despite the small size of the cells. In this way the functional maturation of neurons and neuronal circuits during development can be followed both in acute experiments and after a period in culture. As an example, the first experiments have shown that $GABA_A$ receptors develop prior to synaptogenesis (Shen *et al*., 1988). Whether this mainly serves to protect against uncontrolled electrophysiological activity or has a wider developmental significance remains to be seen.

One of the more interesting features of the turtle brain is the presence of well-defined ependymal cells that usually have their cell bodies on the ventricular surface and send their processes all the way to the pial surface. In both the dorsal cortex and cerebellum these cells are apparently the only representatives of astrocytic glia and thus provide a welcome simplification. Connors and Ransom (1987) studied the electrophysiological properties of the ependyma in the dorsal cortex. In common with other glia they have a very negative resting potential of some −90 mV, have fast time constants and lack excitability. The membrane potential responds to changes in $[K^+]_o$ in accord with the Nernst equation. An important observation was that, by combining simultaneous recording from nearby ependyma and intracellular Lucifer yellow dye injection, coupling between ependyma could be shown. By increasing ambient CO_2 in the isolated preparation, the coupling could be abolished, presumably by intracellular acidification.

II.3 The Cerebellum

The cerebellum of the turtle, either isolated by severing it from the brain at the peduncles, or in the form of conventional brain slices, has been studied *in vitro*. We have used these preparations to examine cerebellar physiology as such and also as well-defined models in which new techniques could be rigorously evaluated. The major asset of the cerebellum in this work is the regular and defined cellular organization.

Studies of cellular electrophysiology have so far focussed on the Purkinje cell, which is the principal neuron of the cerebellar cortex, and its response to afferent excitation. The intrinsic response properties of Purkinje cells in the turtle and in mammals are basically identical. The axon and the cell body generate fast action potentials (1 ms duration) and depolarizing plateau potentials mediated by TTX-sensitive Na^+ channels, while dendrites support slow action potentials and depolarizing plateau potentials mediated by Ca^{2+}. Ca^{2+}-dependent K^+ channels are also mainly located in the dendrites. Voltage-dependent K^+ channels, which are responsible for repolarization, and non-selective channels responsible for time-dependent inward rectification, are probably ubiquitous.

The remarkable firing pattern of Purkinje cells during long-lasting depolarizing current pulses has been analyzed in detail. After a brief burst at the onset of a moderately intense current pulse, the frequency of Na^+ spikes quickly declined to a minimum and then accelerated slowly to reach a higher steady-state level after 5–10 s (Hounsgaard and Midtgaard, 1988). At higher current amplitudes the phase of acceleration was interrupted by Ca^{2+} spikes after a period of some hundred milliseconds to seconds, depending on the stimulus intensity. The phase of acceleration and the delayed occurrence of Ca^{2+} spikes were shown to be due to inactivation of a transient K^+ conductance activated by the initial depolarization. This conductance seemed to play a major role in determining the safety factor, configuration and threshold of Ca^{2+} spikes in Purkinje cells. It was partly inactivated at rest and its activation, inactivation and de-inactivation overlapped in a voltage range from well below the resting membrane potential to well above the threshold for Na^+ spikes.

Repeated Ca^{2+} responses in Purkinje cell dendrites, whether induced by depolarizing current pulses or climbing fiber activation, were shown to be followed by a powerful long-lasting hyperpolarization (Hounsgaard and Midtgaard, 1989). The amplitude and duration of this poststimulus hyperpolarization depended on the number and frequency of Ca^{2+} responses with maxima of 20 mV and more than 1 min. These results showed that climbing fiber activity (the only source of Ca^{2+} responses under normal conditions) regulated the excitability of Purkinje cells on a long-term basis.

The Ca^{2+} component of the climbing fiber responses was also shown to be regulated by the transient outward current. When this current was inactivated, the Ca^{2+} component of climbing fiber responses was large while the Ca^{2+} component was minimal at times of complete de-inactivation of the transient outward current (Chan *et al*., 1989). The effectiveness of climbing fiber responses in generating the long-lasting hyperpolarization (LHP) varied with the magnitude of their Ca^{2+} component (Hounsgaard and Midtgaard, 1989).

Since the state of inactivation of the transient outward current is determined in a voltage range that includes the synaptic responses of all excitatory and inhibitory afferents, they all contribute to the control of Ca^{2+} influx during climbing fiber responses. In this sense, synaptic integration in Purkinje cells intrinsically involves heterosynaptic interaction. It was also argued that intrinsic response properties, operating on a time-scale from milliseconds to minutes, make this the natural time window for synaptic integration (Hounsgaard and Midtgaard, 1989).

An important basis for these conclusions was the development of reliable techniques for differential polarization of Purkinje cells (Chan and Nicholson, 1986; Tranchina and Nicholson, 1986). This approach was facilitated by the flattened ellipsoidal shape of the cerebellum in which the cell bodies of Purkinje cells are located midway between the pial and ventricular surfaces with their dendrites predominantly vertically oriented towards the pial surface. With the isolated cerebellum placed between two plate electrodes, current could be passed along the main axis of the dendrites. This produced an extracellular field of 13–43 mV/mm across the cerebellum as compared with 4–9 mV/mm when a similar current was used along the same axis in the medium without the cerebellum present. Perpendicular to this axis the electric field induced by current flow was constant. In response to a low-frequency sinusoidal current, Purkinje cells were most strongly excited when the extracellular potential at the soma was at its maximum negative value as compared with the potential close to the pial surface. The threshold for modulating Purkinje cell activity was 10–20 mV/mm. Combining the results obtained from similar experiments with three classes of stellate cells (based on location and dendritic morphology revealed by intracellular horseradish peroxidase staining), it was concluded that the effectiveness of neuronal modulation by imposed electric fields strictly depended on cellular processes being oriented in the direction of the field (Chan and Nicholson, 1986).

The analysis of differential polarization of neurons by applied external electric fields evoked by slow sinusoidal currents between the plate electrodes was taken one step further by recording the transmembrane potential in Purkinje cells (Chan *et al*., 1988a). This was accomplished by subtracting the extracellular potential from the intracellular potential, as measured simultaneously with two electrodes at the same depth from the pial surface. Such records were obtained in different cells from the tip of the dendrites to the cell body and confirmed that Purkinje cells were differentially polarized by applied external fields. The transmembrane potential recorded in the molecular layer midway between the pial surface and the cell body layer was unaffected by the field. Deeper and more superficial to the neutral depth (i.e. the level where the induced transmembrane current flow in the cellular cable structures changes sign) the recorded transmembrane potential was modulated with increasing amplitude. The modulation of the transmembrane potential on either side of the neutral depth was symmetrical but in opposite directions, i.e. superficially Purkinje cells depolarized while being hyperpolarized in depth and vice versa. At higher field intensities the modulation became asymmetrical as a consequence of the voltage-dependent non-linear membrane properties. Field-induced depolarization of the proximal and somatic regions generated Na^+ spikes and Na^+ plateaux, while field-induced depolarization of the distal dendritic region induced Ca^{2+} spikes and Ca^{2+} plateaux. It had been suggested that Ca^{2+} spikes in turtle Purkinje cells were generated in spiny dendrites and that individual spiny dendrites rather than branch points were 'hot spots' (Hounsgaard and Midtgaard, 1988). The results of differential polarization not only provided crucial evidence for this idea but also showed the transient hyperpolarization to be generated by the terminal dendrites (Chan *et al*., 1988a). These conclusions hinged on the unique branching pattern of turtle Purkinje cells. Originating from the cell body, smooth proximal dendrites project no further than the deepest one-third of the molecular layer. The spiny dendritic branchlets that they give rise to mostly branch before reaching the mid-molecular layer to ascend unbranched to the pial surface. In turtle Purkinje cells, therefore, the cell body and smooth dendrites are located deep to the level where the induced membrane polarization reverses sign and only spiny dendrites are superficial to this depth in the mid-molecular layer. The basic behavior of Purkinje cells during differential polarization is illustrated in Figure 3.

The exceptionally well-defined distribution of the intrinsic active response properties was used to analyze the distribution and voltage dependence of the regenerative component of climbing fiber responses (Chan *et al*., 1989). Using the differential polarization technique it was possible to show that climbing fiber terminals, although located on the smooth proximal dendrites, evoked a regenerative Ca^{2+} response in the spiny dendrites. The amplitude and time-course of the regenerative climbing fiber response was smoothly graded with the membrane potential and had the characteristics of a local response rather than a propagated all-or-none spike. The intricate regulation of climbing fiber responses and Purkinje cell excitability was shown to reside with the slowly inactivating outward current. The amplitude and duration of a climbing fiber response directly depended on the amount of transient outward current available for activation – which, of course,

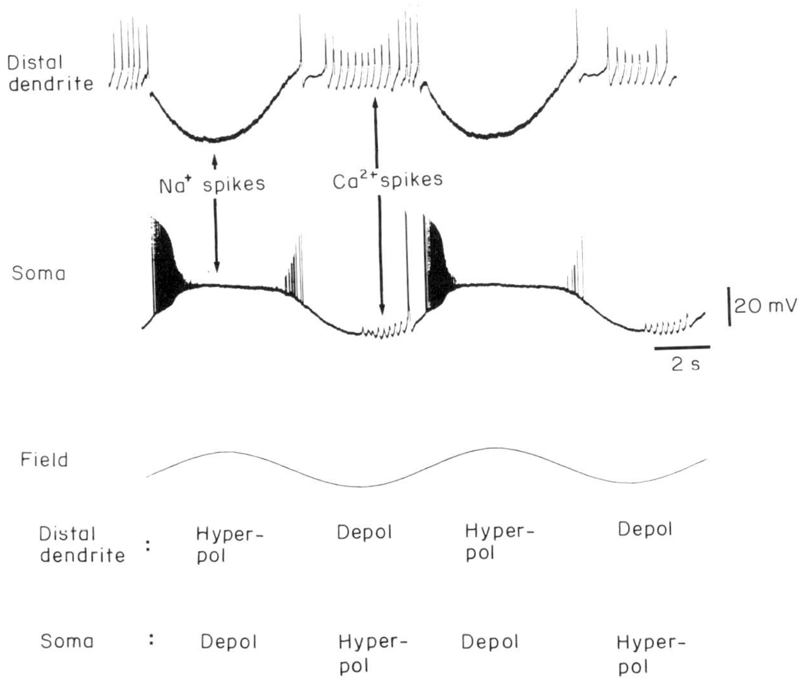

Figure 3 Differential polarization of distal dendrites and the cell body of Purkinje cell by an applied electric field (records from two different cells). The upper record is of transmembrane potential from a distal dendrite, and the middle record is from a cell body during passage of a sinusoidal current across the isolated cerebellum between plate electrodes at the pial and ventricular surfaces. The differential effect on the transmembrane potential is labeled below. Note that the depolarization of the soma was so large that the Na^+ spikes inactivated. (Hounsgaard and Chan, unpublished data)

also determined the degree of inactivation of outward current during the response. The use of differential polarization not only helped to uncover the richly faceted synaptic interactions in Purkinje cells (Hounsgaard and Midtgaard, 1989), but also provided the rigorous verification that allows the technique to be used in cells in which only transmembrane potential measured at the cell body is available.

Further intracellular studies of Purkinje cells, reported only in abstract form at this time (Chan *et al*., 1988b; Larson-Prior and Slater, 1988b), suggest that the isolated turtle cerebellum may become a valuable preparation for the exploration of functional plasticity.

A curious population response, probably pathophysiological in nature, but found in many cortices, is spreading depression (SD) (Leao, 1944; Nicholson, 1984). SD continues to be more than a curiosity, however, since there is mounting evidence that it is a major component of migraine headache (Lauritzen and Nicholson, 1988).

Briefly, SD consists of an extinction of neuronal activity that spreads across brain regions at a velocity of a few millimeters per minute. This extinction is accompanied by a negative extracellular potential of 10 mV or more and a rise in $[K^+]_o$ to 30–40 mM from its normal baseline of about 3 mM, while $[Ca^{2+}]_o$, $[Na^+]_o$ and $[Cl^-]_o$ all drop by large amounts. After a period measured in tens of seconds or minutes, the brain recovers normal activity. Glutamate has long been considered as a possible mediator of SD and recently interest has centered on the NMDA receptor. Using the isolated turtle cerebellum, Lauritzen *et al*. (1988) showed that, while NMDA could indeed initiate SD, so could quisqualate and kainate. Moreover, NMDA receptors appeared to be absent from the molecular layer (Rice and Nicholson, 1990) but SD could occur there.

The preservation of extended neuronal networks *in vitro* remains one of the most inviting uses of turtle preparations. The first step has been taken in the study of the integrative interaction of the cerebellorubrospinal system (Houk and Keifer, 1988; Keifer and Houk, 1989). In a preparation that included the cerebellum, the brainstem and the upper spinal cord, stimuli to the contralateral cord activated cells in the magnocellular nucleus ruber synaptically and antidromically. The level of spontaneous activity and the magnitude and duration of synaptically evoked burst responses in rubral units were found to be curtailed by the cerebellum. Elimination of Purkinje cell inhibition, either by partial transsection of the cerebellar peduncles or by local injection of lidocaine, greatly facilitated both patterns of activity in rubral units. These experiments provide a basis for analyzing basic cerebellar functions and critically testing aspects of global modules.

In addition to the studies of basic cerebellar electrophysiology described above, the turtle cerebellum has been used to evaluate and develop a number of new technical approaches to brain function. Thus the diffusion properties of the molecular layer have been explored by the micro-ejection of tetramethylammonium ions (TMA^+) and their measurement at a known distance away with an appropriate ISM (Nicholson and Rice, 1986). These studies revealed similar diffusion characteristics to those of other brain regions with a volume fraction of 0.21 (i.e. the volume of the extracellular space was 21%) and a tortuosity of 1.65 (i.e the free diffusion coefficient was diminished by the square of this number, reflecting the obstruction of free diffusion by the presence of cellular processes). Rice and Nicholson (1987) employed voltametric microsensors based on carbon fibers to study ascorbate in the isolated turtle cerebellum. They discovered that ascorbate was concentrated during SD but that this was primarily due to the transient reduction in extracellular space volume that is known to accompany the ion movements associated with SD. More recently voltametric microsensors have been used to study dopamine diffusion in the isolated turtle cerebellum (Rice and Nicholson, 1988).

A further recent study (Chesler and Chan, 1988) using pH-sensitive micropipettes confirmed that, as in other tissues, local stimulation of a beam of parallel fibers in the cerebellum leads to a fast alkaline-going transient followed by a slow acid-going phase (Figure 2). While the latter is probably due to the production of metabolic acids, the former alkaline-going component remains mysterious in origin. Chesler and Chan were able to show that it may be generated by a channel-mediated transmembrane flux of proton equivalents. This might imply the existence of either proton or bicarbonate selective channels.

In another series of technical studies, the isolated turtle cerebellum has been used to provide an understanding of the basis of magnetic signals from the brain. Using a magnetometer consisting of a superconducting quantum interference device (SQUID), Okada *et al.* (1987) and Okada and Nicholson (1988) conclusively demonstrated that local stimulation of the parallel fiber–Purkinje cell system or excitation via the mossy fiber–climbing fiber afferents produced a magnetic field of as much as 1 pT (picotesla) at a distance of 17 mm from the cerebellar surface. At this distance the source of the magnetic field was accurately modeled by a dipole and was consistent with the picture given by simultaneously mapping the evoked field potentials with a conventional extracellular microelectrode.

When SD was initiated in the isolated turtle cerebellum a magnetic field was elicited with several distinct components and a magnitude as large as 10 pT at 17 mm from the tissue surface. These results suggest that if SD is a component of migraine headache then there should be a magnetic field associated with it that could readily be detected with a SQUID magnetometer.

Taken together these magnetic studies, which are in progress at this time, confirm that the intrinsic current loops produced by populations of active neurons can generate magnetic fields of sufficient magnitude to be detected with advanced magnetometers. These investigations are also revealing which elements of neuronal activity give rise to specific magnetic components and are aiding the development of magnetic recording techniques for the non-invasive monitoring of brain function.

II.4 The Spinal Cord

The spinal motor system is an attractive neuronal network to study. The large size of spinal motoneurons facilitates intracellular recordings. The result of their activity – muscle contraction – is uniquely intelligible. In addition the spinal neuronal network that controls motoneurons can be influenced in well-differentiated ways via peripheral sensory nerves. *In vitro* preparations of the turtle spinal cord promise new ways to exploit these experimental advantages.

The simplest access to intracellular recording from turtle motoneurons is in transverse sections of the spinal cord (Hounsgaard *et al*., 1988). Cervical or lumbar sections 2–4 mm thick mounted on-end allow microelectrodes to be positioned in the lateral motor column under direct visual guidance. During intracellular recordings, motoneurons were identified by antidromic activation from the ventral root, by orthodromic conduction to the ventral root of action potentials evoked by intracellular current pulses and by their intrinsic response properties. Turtle motoneurons in this preparation reproduced most of the response patterns that characterize mammalian motoneurons *in vivo*. This included time-dependent inward rectification, initial segment (IS) and soma–dendritic components of the action potential and fast and slow afterhyperpolarizations. The firing pattern during step depolarizations also revealed biphasic adaptation of the firing frequency. Based on these findings it was not surprising to find that the underlying ionic conductances in turtle motoneurons were qualitatively identical to those demonstrated in cat motoneurons. Inward rectification was shown to be produced by a non-selective conductance blocked by Cs^+, the rising phase of the action potential by a TTX-sensitive Na^+ conductance, the falling phase and the fast afterhyperpolarization (AHP) by a TEA-sensitive K^+ conductance. The hump-like prolongation of the repolarizing phase of the action potential was mediated by a Ca^{2+} conductance and the slow AHP in part by an apamin-sensitive Ca^{2+}-dependent K^+ conductance. Finally, the experiments revealed an additional nifedipine-sensitive Ca^{2+} conductance with minimal influence on voltage generation under neutral conditions (Hounsgaard and Mintz, 1988).

Although the response properties and ionic conductances of motoneurons in cats and turtles are strikingly similar, some deviations were noted. The temporal separation of the fast and slow AHPs is much more distinct in the turtle than in mammals and the M spike is not present in the turtle. These differences aside, the experiments seemed to justify the use of the *in vitro* preparation as a model system for studies of not only turtle motoneurons, but also of general properties of spinal motoneurons as such.

The ability to affect specific ion conductances with some selectivity during *in vitro* experiments inspired an effort to identify the conductances responsible for the separate components of the intrinsic firing patterns in turtle motoneurons. These experiments immediately demonstrated that any change in ion conductance affected the frequency–current (*f–I*) relation and the degree and time-course of adaptation of firing frequency but no single conductance was found to be responsible for any particular element of the firing patterns (primary–secondary range firing, early and late adaptation). The experiments also showed that the firing patterns of motoneurons were particularly sensitive to even small changes in K^+ conductance. In fact any reduction of the K^+ current near resting membrane potential tended to uncover a voltage-dependent depolarizing plateau potential. The plateau potential was mediated by a non-inactivating, nifedipine-sensitive Ca^{2+} conductance, distinctly different from the conductance that mediates the transient Ca^{2+} influx during action potentials. In summary it was found that Ca^{2+} influx in motoneurons was mediated by two different types of Ca^{2+} channels. The responses mediated by each of the two Ca^{2+} conductances were bounded by different K^+ conductances so that their relative contributions to the firing pattern in motoneurons could be regulated separately by changing specific K^+ conductances. The normal firing pattern in turtle motoneurons and the enhanced Ca^{2+}-mediated responses after reduced K^+ conductance are illustrated in Figure 4.

The outcome of this initial work shows that many basic properties of

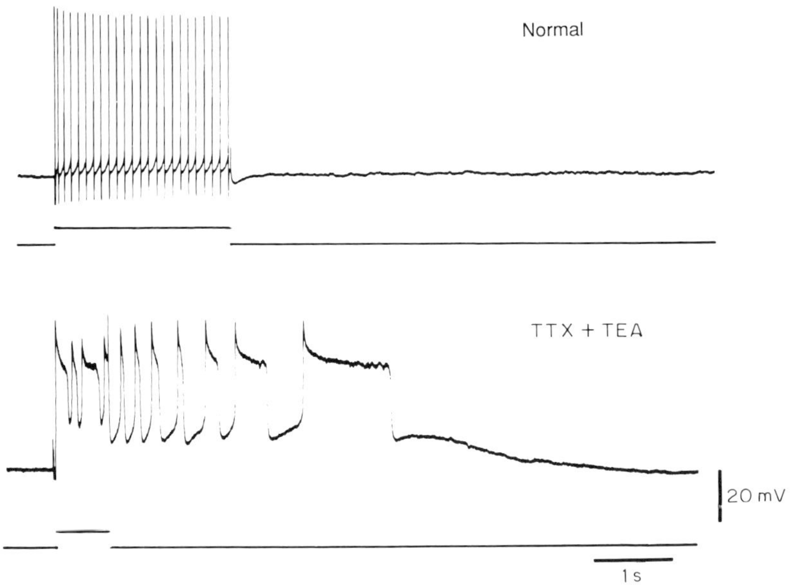

Figure 4 Pharmacologically modifiable response properties of spinal motoneuron *in vitro*. The upper record shows the response to superthreshold current pulse in normal medium. After application of TTX (10^{-6} M), TEA (25×10^{-3} M) uncovers a Ca^{2+}-mediated spike and plateau responses (lower record). (Hounsgaard and Mintz, unpublished data)

motoneurons can be studied legitimately and effectively in the turtle spinal cord *in vitro*. The stable recording conditions and the flexible access to pharmacological and ionic manipulations have improved the separation and characterization of ion conductances and response properties. The most significant finding is that the basic response properties, previously thought to be a stable parameter for synaptic integration in motoneurons, can be regulated over a wide range by small changes of specific ion conductances. Experiments still under way have shown that regulation of this kind can be mediated by transmitters. In turtle motoneurons serotonin induces nifedipine-sensitive plateau potentials by a specific reduction of the Ca^{2+}-dependent part of the slow AHP (Hounsgaard and Kiehn, 1985, 1989). NMDA and acetylcholine induce voltage-dependent bursting pacemaker activity by separate receptors and entirely different mechanisms (Mintz, 1987). Dynamic regulation of the response properties of motoneurons by transmitters is almost certainly part of normal motor behavior, but exactly how they contribute remains to be seen. The pursuit of this and related questions will require *in vitro* preparations in which coordinated and functionally meaningful neuronal activity is preserved. Such an integrated preparation was devised by the first reported *in vitro* study on the spinal cord of the turtle (Keifer and Stein, 1983). They isolated the postcervical spinal cord with a portion of the carapace so that one side included the lateral shell bridges with intact sensory nerves for eliciting the scratch reflex. The more caudally located nerves that mediate the motor component of the scratch reflex were also preserved. In this preparation tactile stimuli applied to the lateral shell bridges generated a phasic motor response in the three efferent nerves recorded from which were almost indistinguishable from recordings *in vivo* during the scratch reflex. While these experiments were designed to address the issue of central pattern generation from a global point of view, the same model provides an excellent platform for investigating the detailed mechanisms at the cellular level. Further development of integrated *in vitro* preparations of the spinal motor system is in progress. Lennard *et al*. (1989) have shown that aseptically dissected sections of the spinal cord with peripheral nerves and muscles attached maintain stable reflex patterns for 3–5 days in culture, with little change in histochemical parameters for muscle fibers.

III CONCLUSIONS

The mechanism of metabolic control and anoxia resistance is far from resolved at this time and promises further rewarding investigations. Even without this detailed understanding the isolated brain preparation is yielding important new insights into the membrane properties of neurons and glia in different brain regions. The really significant studies, however, will exploit the ability of this preparation to provide complete neuronal circuits both within a given brain region and between different regions. Such studies are now under way in several laboratories. This approach will be complemented by the ability to culture this robust brain over long periods and so study developmental issues. Finally the isolated brain is proving an invaluable platform for new technical developments which in turn will lead to novel insights into the physiology. Having made a convincing argument for the merit of this preparation it may be prudent to end on a note of caution. The isolated brain is indeed different from the *in vivo* one. The blood circulation, in addition to providing the very best means of supplying metabolic substrates and removing metabolic wastes, almost certainly contains many circulating hormones and neuromodulators, originating in organs outside the brain, that cannot presently be mimicked in our bathing solutions. Over the long periods of study now possible with this preparation, such deficiencies may come to significantly affect results. It will therefore be necessary sometimes to return to the intact animal, blood, guts and all.

REFERENCES

Belkin, D.A. (1963). 'Anoxia: tolerance in reptiles', *Science*, **139**, 492–493.
Berkson, H. (1966). 'Physiological adjustments to prolonged diving in the Pacific green turtle (Chelonia agassizii)', *Comp. Biochem. Physiol.*, **18**, 101–119.

Blanton, M.G., Turco, J.J.W., and Kriegstein, A.R. (1989). 'Whole-cell recording from neurons in slices of reptilian and mammalian cerebral cortex', *J. Neurosci. Methods*, **30**, 203–210.

Burton, R.F. (1975). *Ringer Solutions and Physiological Salines*, Wright-Scientecnica, Bristol.

Chan, C.Y., and Nicholson, C. (1986). 'Modulation by applied electric fields of Purkinje and stellate cell activity in the isolated turtle cerebellum', *J. Physiol. (Lond).*, **371**, 89–114.

Chan, C.Y., Hounsgaard, J., and Nicholson, C. (1988a). 'Effects of electric fields on transmembrane potential and excitability of turtle cerebellar Purkinje cells *in vitro*', *J. Physiol. (Lond).*, **402**, 751–771.

Chan, C.Y., Lee, A., and Wong, S.M. (1988b). 'Synaptic plasticity in the parallel fiber–Purkinje cell junction in isolated turtle cerebellum', *Soc. Neurosci. Abstr.*, **14**, 170.

Chan, C.Y., Hounsgaard, J., and Midtgaard, J. (1989). 'Excitatory synaptic responses in turtle cerebellar Purkinje cells', *J. Physiol. (Lond).*, **409**, 143–156.

Chesler, M., and Chan, C.Y. (1988). 'Stimulus-induced extracellular pH transients in the *in vitro* turtle cerebellum', *Neuroscience* **27**, 941–948.

Connors, B.W., and Kriegstein, A.R. (1986). 'Cellular physiology of the turtle visual cortex: distinctive properties of pyramidal and stellate neurons', *J. Neurosci.*, **6**, 164–177.

Connors, B.W., and Ransom, B.R. (1987). 'Electrophysiological properties of ependymal cells (radial glia) in dorsal cortex of the turtle, Pseudemys scripta', *J. Physiol. (Lond).*, **385**, 287–306.

Cserr, H. F., DePasquale, M., and Jackson, D. C. (1988). 'Brain and cerebrospinal fluid ion composition after long-term anoxia in diving turtles', *Am. J. Physiol.*, **255**, R338–R343.

Feng, F.C., Rosenthal, M., and Sick, T.J. (1988a). 'Suppression of evoked potentials with continued ion transport during anoxia in turtle brain', *Am. J. Physiol.*, **255**, R478–R484.

Feng, F.C., Sick, T.J., and Rosenthal, M. (1988b). 'Orthodromic field potentials and recurrent inhibition during anoxia in turtle brain', *Am. J. Physiol.*, **255**, R485–R491.

Greer, C.A., Mori, K., and Shepherd, G.M. (1981). 'Localization of synaptic responses in the *in vitro* turtle olfactory bulb using the (^{14}C)2-deoxyglucose method', *Brain Res.*, **217**, 295–303.

Harless, M., and Morlock, H. (1979). *Turtles. Perspectives and Research*, Wiley, New York.

Heisy, S.R. (1968). 'Cerebrospinal fluid (CSF) composition and volume of the ventricular system of the turtle brain', *Fed. Proc.*, **27**, 287.

Hitzig, B.M., and Jackson, D.C. (1978). 'Central chemical control of ventilation in the unanesthetized turtle'. *Am. J. Physiol.*, **235**, R257–R264.

Hochachka, P.W. (1986). 'Defense strategies against hypoxia and hypothermia', *Science*, **231**, 234–241.

Houk, J.C., and Keifer, J. (1988). 'Model of cerebellum as an array of adjustable motor pattern generators: microelectrode studies using the isolated turtle brainstem-cerebellum', in *Neural Networks from Models to Applications* (Eds L. Personnaz and G. Dreifus), pp. 121–130, IDSET, Paris.

Hounsgaard, J., and Kiehn, O. (1985). Ca^{++} dependent bistability induced by seretonin in spinal motoneurons', *Exp. Brain Res.*, **57**, 422–425.

Hounsgaard, J., and Kiehn, O. (1989). 'Serotonin-induced bistability of turtle motoneurones caused by a nifedipine-sensitive calcium plateau potential', *J. Physiol. (Lond.)*, **414**, 265–282.

Hounsgaard, J., and Midtgaard, J. (1988). 'Intrinsic determinants of firing pattern in Purkinje cells of the turtle cerebellum *in vitro*', *J. Physiol. (Lond.)*, **402**, 731–749.

Hounsgaard, J., and Midtgaard, J. (1989). 'Synaptic control of excitability in turtle cerebellar Purkinje cells', *J. Physiol. (Lond.)*, **409**, 157–170.

Hounsgaard, J., and Mintz, I. (1988). 'Calcium conductance and firing properties of spinal motoneurones in the turtle', *J. Physiol. (Lond.)*, **398**, 591–603.
Hounsgaard, J., Kiehn, O., and Mintz, I. (1988). Response properties of motoneurones in a slice preparation of the turtle spinal cord. *J. Physiol., Lond.*, **398**, 575–589.
Jackson, D.C., and Heisler, N. (1982). 'Plasma ion balance of submerged anoxic turtles at 3°C: The role of calcium lactate formation', *Resp. Physiol.*, **49**, 159–174.
Jackson, D.C., and Heisler, N. (1983). 'Intracellular and extracellular acid–base and electrolyte status of submerged anoxic turtles at 3°C', *Resp. Physiol.*, **53**, 187–201.
Jackson, D.C., and Silverblatt, H. (1974). 'Respiration and acid–base status of turtles following experimental dives', *Am. J. Physiol.*, **226**, 903–909.
Jackson, D.C., and Ultsch, G. R. (1982). 'Long-term submergence at 3°C of the turtle, *Chrysemys picta bellii*, in normoxic and severely hypoxic water: II. Extracellular ionic responses to extreme lactic acidosis', *J. Exp. Biol.*, **96**, 29–43.
Jahr, C.E., and Nicoll, R.A. (1980). 'Dendrodendritic inhibition: Demonstration with intracellular recording', *Science*, **207**, 1473–1475.
Jahr, C.E., and Nicoll, R.A. (1982a). 'An intracellular analysis of dendrodendritic inhibition in the turtle *in vitro* olfactory bulb', *J. Physiol. (Lond.)*, **326**, 213–234.
Jahr, C.E., and Nicoll, R.A. (1982b). 'Noradrenergic modulation of dendrodendritic inhibition in the olfactory bulb', *Nature*, **297**, 227.
Keifer, J., and Houk, J.C. (1989). 'An *in vitro* preparation for studying motor pattern generation in the cerebellorubrospinal circuit of the turtle', *Neurosci. Lett.*, **97**, 123–128.
Keifer, J., and Stein, P.S.G. (1983). 'In vitro motor program for the rostral scratch reflex generated by the turtle spinal cord'. *Brain Res.*, **266**, 148–151.
Kriegstein, A.R. (1987). 'Synaptic responses of cortical pyramidal neurons to light stimulation in the isolated turtle visual system', *J. Neurosci.*, **7**, 2488–2492.
Kriegstein, A.R., and Connors, B.W. (1986). 'Cellular physiology of the turtle visual cortex: synaptic properties and intrinsic circuitry', *J. Neurosci.*, **6**, 178–191.
Kriegstein, A.R., Shen, J.M., Lee, G.M., and Blanton, M.G. (1988). 'Development of inhibitory function during corticogenesis in the turtle', in *The Forebrain of Reptiles. Recent Advances in Understanding the Structure and Function of the Forebrain in Reptiles* (Eds W.K. Schwerdtfeger, and W.J.A.J. Smeets), pp. 131–141, Karger, Basel.
Larson-Prior, L.J., and Slater, N.T. (1988a). 'GABAergic inhibition and epileptiform discharges in the turtle hippocampus *in vitro*', *Brain Res.*, **460**, 369–375.
Larson-Prior, L.J., and Slater, N.T. (1988b). 'Novel slow EPSPs evoked by climbing and parallel fiber activation in turtle cerebellar Purkinje cells', *Biophys. J. Abstr.*, **55**, 177a.
Lauritzen, M., and Nicholson, C. (1988). 'The pathophysiology of classical migraine, in *Headache: Problems in Diagnosis and Management* (Ed. A. Hopkins), pp. 77–109, W.B. Saunders, London.
Lauritzen, M., Rice, M.E., Okada, Y., and Nicholson, C. (1988). 'Quisqualate, kainate and NMDA can initiate spreading depression in the turtle cerebellum', *Brain Res.*, **475**, 317–327.
Leao, A.A.P. (1944). 'Spreading depression of activity in the cerebral cortex', *J. Neurophysiol.*, **7**, 359–390.
Lennard, P.R., Boss, V., and Shirley, K.A.S. (1989). 'Long-term *in vitro* turtle preparation for the study of spinal organization', *J. Neurosci. Methods*, **29**, 165–169.
Lutz, P.L., McMahon, P.M., Rosenthal, M., and Sick, T.J. (1984). 'Relationships between aerobic and anaerobic energy production in turtle brain in situ', *Am. J. Physiol.*, **247**, R740–R744.

Lutz, P.L., Rosenthal, M., and Sick, T.J. (1985). 'Living without oxygen: turtle brain as a model of anaerobic metabolism', *Mol. Physiol.*, **8**, 411–525.
Mintz, I. (1987). 'Analyse *in vitro* des propriétés membranaires actives des motoneurones de tortue et de leur contribution aux oscillations induites par l'acide *N*-méthyl-D-aspartique et les agonistes muscariniques', *Thèse de Doctorat de l'Université Paris 6*.
Mori, K., and Shepherd, G.M. (1979). 'Synaptic excitation and long-lasting inhibition of mitral cells in the *in vitro* turtle olfactory bulb', *Brain Res.*, **172**, 155–159.
Mori, K., Nowycky, M.C., and Shepherd, G.M. (1981a). 'Electrophysiological analysis of mitral cells in the isolated turtle olfactory bulb', *J. Physiol. (Lond.)*, **314**, 281–294.
Mori, K., Nowycky, M.D., and Shepherd, G.M. (1981b). 'Analysis of synaptic potentials in mitral cells in the isolated turtle olfactory bulb', *J. Physiol. (Lond.)*, **314**, 295–309.
Mori, K., Nowycky, M.D., and Shepherd, G.M. (1981c). 'Analysis of long-duration inhibitory potential in mitral cells in the isolated turtle olfactory bulb', *J. Physiol. (Lond.)*, **314**, 311–320.
Mori, K., Nowycky, M.C., and Shepherd, G.M. (1982). 'Impulse activity in presynaptic dendrites: analysis of mitral cells in the isolated turtle olfactory bulb', *J. Neurosci.*, **2**, 497–502.
Mori, K., Nowycky, M.C., and Shepherd, G.M. (1984). 'Synaptic excitatory and inhibitory interactions at distal dendritic sites on mitral cells in the isolated turtle olfactory bulb', *J. Neurosci.*, **4**, 2291–2296.
Nicholson, C. (1984). 'Comparative neurophysiology of spreading depression in the cerebellum', *An. Acad. brasil. Cienc.*, **56**, 481–494.
Nicholson, C., and Hounsgaard, J. (1983). 'Diffusion in the slice microenvironment and implications for physiological studies', *Fed. Proc.*, **42**, 2865–2868.
Nicholson, C., and Rice, M.E. (1986). 'The migration of substances in the neuronal microenvironment', *Ann. N.Y. Acad. Sci.*, **481**, 55–71.
Nicoll, R.A., and Jahr, C.E. (1982). 'Self-excitation of olfactory bulb neurones', *Nature*, **296**, 441–444.
Nowycky, M.C., Mori, K., and Shepherd, G.M. (1981a). 'Blockade of synaptic inhibition reveals long-lasting synaptic excitation in isolated turtle olfactory bulb', *J. Neurophysiol.*, **46**, 649–658.
Nowycky, M.C., Mori, K., and Shepherd, G.M. (1981b). 'GABAergic mechanisms of dendrodendritic synapses in isolated turtle olfactory bulb', *J. Neurophysiol.*, **46**, 639–648.
Okada, Y.C., and Nicholson, C. (1988). 'Magnetic evoked field associated with transcortical currents in turtle cerebellum', *Biophys. J.*, **53**, 723–731.
Okada, Y.C., Lauritzen, M., and Nicholson, C. (1987). 'Magnetic field associated with neural activities in an isolated cerebellum', *Brain Res.*, **412**, 151–155.
Okada, Y.C., Lauritzen, M., and Nicholson, C. (1988). 'Magnetic field associated with spreading depression: a model for the detection of migraine', *Brain Res.*, **442**, 185–190.
Penney, D.G. (1974). 'Effects of prolonged diving anoxia on the turtle, *Pseudemys scripta elegans*', *Comp. Biochem. Physiol.*, **47A**, 933–941.
Rice, M.E., and Nicholson, C. (1987). 'Interstitial ascorbate in turtle brain is modulated by release and extracellular volume change', *J. Neurochem.*, **49**, 1096–1104.
Rice, M.E., and Nicholson, C. (1988). 'Dopamine diffusion in turtle brain measured with high-speed cyclic voltammetry', *Soc. Neurosci. Abstr.*, **14**, 742.
Rice, M.E., and Nicholson, C. (1990). 'Glutamate- and aspartate-induced extracellular potassium and calcium shifts and their relation to those of kainate, quisqualate and NMDA in the isolated turtle cerebellum. *Neuroscience* (in press).

Robin, E.D., Vester, J.W., Murdaugh, H.V., Jr., and Millen, J.E. (1964). 'Prolonged anaerobiosis in a vertebrate: Anaerobic metabolism in the freshwater turtle', *J. Cell. Comp. Physiol.*, **63**, 287–291.

Robin, E.D., Lewiston, N., Newman, A., Simon, L.M., and Theodore, J. (1979). 'Bioenergetic pattern of turtle brain and resistance to profound loss of mitochondrial ATP generation'. *Proc. Natl Acad. Sci. USA*, **76**, 3922–3926.

Robin, E.D., Robin, D.A., Ackerman, R., Lewiston, N., Hance, A.J., Caligiuri, M., and Theodore, J. (1981). 'Prolonged diving and recovery in the freshwater turtle, *Pseudemys scripta*', *Comp. Biochem. Physiol.*, **70A**, 359–364.

Rosenberg, A.F., and Ariel, M. (1988). 'Cells in the turtle's basal optic nucleus are direction selective over a broad velocity range', *Soc. Neurosci. Abstr.*, **14**, 990.

Shen, J.M., and Kriegstein, A.R. (1986). 'Turtle hippocampal cortex contains distinct cell types, burst-firing neurons, and an epileptogenic subfield', *J. Neurophysiol.*, **56**, 1626–1649.

Shen, J.M., Huguenard, J.R., and Kriegstein, A.R. (1988). 'Development of GABA responsiveness in embryonic turtle cortical neurons', *Neurosci. Lett.*, **89**, 335–341.

Shepherd, G.M. (1979). *The Synaptic Organization of the Brain*, Oxford University Press.

Sick, T.J., Rosenthal, M., LaManna, J.C., and Lutz, P.L. (1982). 'Brain potassium ion homeostasis, anoxia, and metabolic inhibition in turtles and rats', *Am. J. Physiol.*, **243**, R281–R288.

Suarez, R.K. (1988). 'Thinking with and without oxygen: energy metabolism in vertebrate brains', *Can. J. Zool.*, **66**, 1041–1045.

Tranchina, D., and Nicholson, C. (1986). 'A model for the polarization of neurons by extrinsically applied electric fields', *Biophys. J.*, **50**, 1139–1156.

Waldow, U., Nowycky, M.C., and Shepherd, G.M. (1981). 'Evoked potential and single unit responses to olfactory nerve volleys in the isolated turtle olfactory bulb', *Brain Res.*, **211**, 267–283.

Wilson, A.M., Desan, P.H., and Kriegstein, A.R. (1988). 'Development of the turtle cerebral cortex *in vitro*', *Soc. Neurosci. Abstr.*, **14**, 92.

Preparations of Vertebrate Central Nervous System *In Vitro*
Edited by H. Jahnsen
Published by John Wiley & Sons Ltd

CHAPTER 7

The Rabbit Retina *In Vitro*

ADELBERT AMES III and RICHARD H. MASLAND
Massachusetts General Hospital,
Boston, Massachusetts 02114, USA

I INTRODUCTION

The retina is part of the central nervous system with respect to its embryology, anatomy, chemistry and function, and it has two features that make it especially attractive for experimental study. (1) It is relatively easy to isolate and to maintain *in vitro* in a near-physiological state, as will be discussed in more detail below. (2) It is composed of relatively simple neural circuits, through which the primary flow of information is unidirectional. Retinal neurons process the information received by the photoreceptor cells and encode it as trains of action potentials that travel down the optic nerve. At the level of input–output analysis this encoding has been fairly well described. Some of the information processing is quite sophisticated; for example., one group of retinal ganglion cells responds to motion *per se*, in a particular direction along a particular axis, irrespective of whether it is a white object moving against a dark background or a dark object moving against a white background. Except perhaps for the spinal cord, there is no central nervous system structure whose function is so well understood.

Retinas are thin sheets of tissue that can be separated from the adjacent non-neural tissue along natural cleavage planes without damage to cellular elements. The retinas of the rabbit and some other animals are essentially avascular. They are nurtured *in vivo* by diffusion from capillary beds on their surfaces, so that nurturing the retina *in vitro* with medium moving over its surfaces simulates fairly closely the normal condition. The retina can be maintained *in vitro* with nearly normal function for many hours or days. The isolated retina remains responsive to its natural physiological input, light, so that it can be activated in a way that is essentially identical to its activation in its normal position in the brain. Since retinas are so uniform and well defined in structure, it is relatively easy to compare results that are obtained on different preparations.

The isolated retina is unusually accessible to experimental manipulation. The problem of a blood–brain (or blood–retina) barrier is bypassed. The retina is so thin that agents introduced into the medium reach the cells promptly, and substances released by the cells appear in the medium promptly. It can be photically stimulated, either diffusely or with a high degree of spatial resolution. It is readily accessible for electrical recordings, either of large cell groups, using gross electrodes on the optic nerve or across the retina, or of single cells, using microelectrodes within the retina. It can be monitored optically while intact, either in the living state or following partial or complete fixation. These features permit a wide variety of experimental approaches. A few of them, with which we have had personal experience, will be surveyed here. The anatomy and physiology of the retina have been well reviewed by others (Rodieck, 1973; Hubel, 1988; Dowling, 1988). In this chapter we hope to give a view of the isolated retina as an experimental preparation, indicating the types of experiments that have been done and the directions in which new undertakings may be possible.

II GENERAL METHODS

II.1 Isolation of Retina and Maintenance *In Vitro*

Our studies have been performed on retinas obtained from 2–4-kg New Zealand White rabbits. Their retinas are 3–4 cm^2 in area and about 0.1 mm in thickness, with dry weights of 9–12 mg and wet weights of 60–85 mg. The rabbits are usually sedated with intravenous sodium pentobarbitone (15 mg/kg) and brought to a surgical anesthetic level with ether by mask. There has been no evidence of persisting effects of the anesthesia in the isolated retina. We have also used pentobarbitone in combination with valium, and other workers use combinations of ketamine and zylazine; the combinations produce a greater margin of anesthetic safety than can be obtained with barbiturate alone. If a full complement of rhodopsin is desired in the isolated tissue, the rabbit is dark adapted and the isolation performed in dim red light. Little if any rhodopsin is regenerated following bleaching after the photoreceptors have been separated from the pigment epithelium.

Isolation of the retina involves enucleation, hemisection of the globe near the equator, removal of the vitreous, eversion of the posterior eye cup, and separation of the retina to the optic disc followed by excision of the pigment epithelium, choroid and sclera together from around the disc. The segment of nerve that remains attached to the retina forms a convenient handle for transferring the retina, but is removed before chemical analysis or before studies in which chemical exchange between retina and medium is being monitored. To obtain retinas with longer segments of nerve (up to 1 cm) for electrical recordings, the extraocular muscles are ligated near the globe and cut, and the ligatures are used to extract the eye while cutting the nerve with narrow-tipped scissors inserted deeply into the optic foramen. The globe can be hemisected with minimal deformation if it is retained in a snug-fitting cup and cut with a razor blade (Figure 1A). This facilitates removal of the vitreous *en masse* with large forceps. As the vitreous body is lifted away, medium is poured into the eyecup to help displace it. The rest of the isolation is performed under medium. The eyecup is everted over the rounded end of a Teflon rod and its edges impaled on two projecting needles to hold it in place. The retina then can be separated from the pigment epithelium with a small glass rod (Figure 1B). If the retina is being removed dark-adapted, retinal separation is facilitated by having a source of dim red light placed beneath the dissecting vessel so that the retina is viewed in silhouette.

The medium is designed to simulate the fluid distal to the blood–retina barrier that bathes retinal cells *in vivo*, and we have used published analyses of spinal fluid as the best guide available to its composition. A simplified medium containing only the inorganic electrolytes (120 mM NaCl, 23 mM $NaHCO_3$, 3.1 mM KCl, 0.5 mM KH_2PO_4, 1.2 mM $MgSO_4$, 1.15 mM $CaCl_2$) and 6 mM glucose maintains retinas for several hours with near-normal electrophysiological function. A more complete medium containing 35 additional organic constituents, including amino acids, vitamins and nucleosides, has been designed (Ames and Nesbett, 1981) and is now (February, 1989) being sold by Sigma as Ames Medium. The more complete medium is required for the maintenance of some metabolic processes, e.g. a balanced protein turnover (Parks *et al*., 1976). It has been found to maintain retinas for more than 2 days with quite normal glucose utilization, protein synthesis, and electrophysiological function (Ames and Nesbett, 1981). Bacteria have not been a problem if the medium is made up on the day of the experiment, if it is kept cold until shortly before use, and if it is completely exchanged at intervals of less than 4 h.

Several precautions are required in establishing suitable conditions *in vitro*. The medium should not be allowed to become alkaline in the course of its preparation

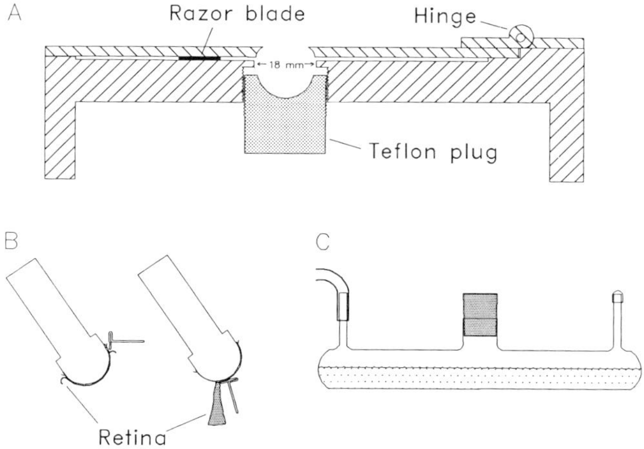

Figure 1 (A) Apparatus for hemisecting the rabbit eye. The eye is removed from the orbit and placed in the cup whose vertical dimension can be adjusted with threaded Teflon plug. A hinged lucite lid is swung down (as shown in the figure) to restrain the eyeball. The eyeball is them hemisected by a razor blade (Personna, hairshaping blade) slid between the supporting surface and the lid. (B) Early and late stage of separation of the retina from pigment epithelium. The eye has been hemisected, the vitreous removed, and the resulting eyecup has been everted over a rounded (13 mm diameter) Teflon rod containing two protruding needles that are pushed through the sclera to retain it on the rod. The left portion of the figure shows a glass rod developing a plane of cleavage between the retina and pigment epithelium; the right portion shows dissection completed with the retina attached only at the optic disc. (C) Incubation boat containing 20 ml of medium. Gas enters at one port and leaves at the other, which is covered by a loose-fitting cap. The boat is rocked on its long axis through 3° at about 1 Hz to provide movement between the retina and medium and between the medium and gas phase

in order to prevent the precipitation of $Ca_3(PO_4)_2$ or $CaCO_3$. This can be done by gassing with 100% CO_2 before adding the $CaCl_2$ and storing in high CO_2 until use. Special precautions must be taken to avoid the introduction of toxins. Incidents of toxic damage that we have encountered have included Ag from Ag–AgCl electrodes, NH_3 from tubing cleaned with NH_4OH, and unknown (presumably bacterial-derived) substances in glass-redistilled water. Accordingly, we restrict the solids that come in contact with the medium to glass, Teflon, lucite and silicone rubber. They are cleaned with a tissue-culture detergent and rinsed extensively; plastic components are exposed to detergent only briefly to avoid adsorption. Water is obtained from a source being used successfully by tissue culturists. Bench tops are covered with ash-free filter paper to avoid pickup of contaminants by instruments or pipette tips. The effects of toxins may be delayed, with the retina appearing healthy at first and several hours elapsing before the decline in function is evident. The medium as finally used can be tested for its ability to support growth of cells in culture.

Oxygen may become a limiting factor in the diffusional exchange between medium and tissue. Its availability depends on how well the retina is deployed within the medium and on the depth of the unstirred layer surrounding the tissue. Movement vigorous enough to disrupt the unstirred layer at the interface between tissue and medium is necessary to overcome diffusion limitations, even when the bulk medium has been equilibrated with 95% O_2. In our experience, the gas equilibrated with the medium can contain as little as 40% O_2, but not much less, when the retina is in a vigorously rocked vessel or is being aggressively superfused.

The pH of the retina is determined by a bicarbonate buffer *in vitro*, as it is *in vivo*. It is therefore important that the medium always remain equilibrated with 5% CO_2. If the retina is removed to a gas phase for more than a few seconds, the gas should contain 5% CO_2. The cells are not easily damaged by the fall in pH associated with an increase in CO_2, but appear more vulnerable to the rise in pH associated with a reduction in CO_2. We have usually kept the temperature of the isolated retina between 30 and 37°C. Increasing temperature through this range generally increases cellular activity, with a 6% per degree increase in glycolysis and oxygen consumption. The lower temperatures are more forgiving with respect to any unphysiological demands imposed by the experimental protocol. Increasing the temperature above 37°C is quite hazardous. The b-wave of the ERG or the light-evoked action potential from the nerve stump are good indicators of the health of the *in vitro* tissue. A more detailed account of the medium and general methods of incubation has been published (Ames and Nesbett, 1981).

II.2 Preparing Retina for Histological Examination and for Chemical Measurements

The mechanical strength (when grasped by the optic nerve stump) and the unusual thinness of the retina may be used to advantage in preparing the tissue for analysis. For example, the distribution of a diffusible [^{3}H]-labelled solute could be preserved for autoradiography by quick-freezing the tissue on thin tin foil and then freeze-drying it (Masland and Mills, 1979). The success of this technique depended critically on the fact that the tissue is thin enough to freeze very rapidly. Labile metabolites such as ATP and GTP have been retained by quenching the tissue at 0°C in 0.5 M perchloric acid (Goldberg *et al.*, 1983). Monovalent electrolytes can be quantitatively recovered by extraction with 0.75 M HNO_3. A method has been developed for wet ashing the retina at temperatures of less than 150°C that is useful for measuring total Ca, Mg and P as well as Na, K and Cl (Ames *et al.*, 1967). For studies involving isotope counting, a clear colorless solution of the retina, with minimal quenching of the radioactivity, can be prepared in about 1 h by adding 225 μl of water to the lyophilized tissue followed after 45 min by 1 ml of Nuclear-Chicago Solubilizer (NCS) with mixing by magnetic stirrer (Ames and Nesbett, 1983).

The tissue can be fixed for light microscopy using conventional fixatives, the choice of fixative depending on the type of histological or histochemical procedure to be carried out. In general, we flatten the retina in the living state and apply fixative to it in its flattened configuration (Tauchi and Masland, 1984; Sandell and Masland, 1986). Four cuts are made in the tissue so that it can take the shape of a Maltese cross. A convenient procedure in cases where the speed of fixation is not crucial is to float the living tissue onto a piece of filter paper, hold it in the flattened shape by sandwiching with a second filter paper, and apply fixative to the sandwich. Where it is desired to keep the tissue immersed in medium throughout fixation, the retina can be floated in its native shape over a glass sphere (a marble). The medium is gradually replaced by fixative so that the tissue is fixed in its normal hemispherical shape (Steinman and Ames, 1974). Fixation for electron microscopy can be accomplished using conventional fixative mixtures. Experimentally induced volume changes were best preserved by immersing the retina at 4°C in a solution identical to the medium except for the addition of isotonic osmium tetroxide (Webster *et al*., 1969). The speed with which the isolated retina can be fixed is one of its desirable features, especially when intracellular solutes must be immobilized for histochemistry or autoradiography.

II.3 Electrical Recordings

Several types of apparatus have been devised for recording electrically from the isolated retina. An apparatus for recording the light-evoked compound action potential from the optic nerve stump (Ames and Gurian, 1960) used platinum electrodes, one of which encircled the nerve near the disc while the other impaled the crushed end of the nerve and acted as the reference. (Platinum electrodes were used because they are not toxic; when measurements of slow electrical events are to be made, Ag–AgCl electrodes are used, isolated from the tissue by saline or agar saline bridges.) A syringe was attached to the chamber to lower the level of medium so that the nerve and electrode were in the gas phase at the time of a recording. Recording was thus between the optic nerve at its exit from the retina and the crushed end of the nerve, with the extracellular space as a resistive barrier. These recordings of the optic nerve response *in vitro* have most of the characteristics of the optic nerve response as seen *in vivo*, with the exception (also seen in recordings from single cells) that the nerve response fatigues more quickly to repeated stimulation than in the intact animal. The reason for this is unclear. Recording by this method is useful as a survey of the activity of a large number of optic nerve fibers.

In order to make extracellular recordings of the action potentials of single ganglion cells, the retina was stabilized, vitreous surface upward, on an inflatable silicone membrane in a chamber mounted on a microscope stage (Figure 2F). A Teflon-coated platinum–iridium electrode was advanced to the surface of the tissue under direct vision (Ames and Pollen, 1969; Masland and Ames, 1976). Recordings may also be made with tungsten in glass or varnish-coated electrodes. The spacing between the ganglion cells is wider in the rabbit's retina than in many others and this allows rather large electrodes to be used. The position of the electrode on the retinal surface can be readily adjusted, under direct vision, to obtain the loudest possible spikes on an audio monitor. The retina may be stimulated during recording by a variety of means. We have usually directed light through the reversed optics of the microscope, often using the two eyepieces of a conventional binocular miscoscope to provide two optical channels. An advantage of this strategy is that it ensures very accurate focussing of the stimuli. Probably the best arrangement is to use the a trinocular head on the microscope, with the vertical port receiving light from a conventional optical system. An alternative is to focus light on the tissue via the microscope's condenser.

For d.c. recordings of the transretinal potential and its response to light (the electroretinogram), the retina need only be deployed between two electrodes. In one arrangement it was spread over a hollow plastic mandrel with perforations at the top. One Ag–AgCl electrode was sealed in a short silicone tube that penetrated the base of the mandrel and the other was immersed in the fluid outside (Figure 2D). In another, the retina was simply deployed over a domed spiral of platinum wire and a second piece of wire was placed in the medium above it (Masland and Ames, 1975; Pepperberg and Masland, 1978). In a third case very small electroretinographic signals were expected from the isolated retinas of infant rabbits. The medium was briefly drained from the surface of the everted eyecup and the uppermost surface of the retina was contacted by a saline-moistened cotton thread. Recordings were made between the thread and a reference placed below the tissue. The resulting partial 'air gap' gave larger signals than when the tissue was entirely immersed (Masland, 1977). In general, though, the resistance of the bulk medium is high enough that shunting is not a problem and the retina need not be mounted as an uninterrupted membrane.

II.4 Photic Stimulation

Depending on the experimental objective, photic stimuli may be very bright or very dim, continuous or intermittent, diffuse or focused, and (if focused) stationary or moving. Light intensity can be controlled by controlling the output at the source or by interposing neutral density filters. It is possible to characterize the light dose in terms of photons $\mu m^{-1}\ s^{-1}$ capable of bleaching rhodopsin and thereby to estimate the number of photoisomerizations of rhodopsin per photoreceptor (e.g. Ames *et al*., 1986). Short stimuli with rapid rise and fall times can be achieved using a shutter in the light path or, in the case of diffuse stimuli, using lamps containing a rapidly reacting phosphor. Small, discrete patterns of light, either stationary or moving, can be generated by introducing the stimulus 'backwards' through microscope optics as described above or by introducing it through a mandrel made from a fiber optic bundle (O'Malley and Masland, 1989).

Since the secondary and tertiary (in sequence of response) cells in the retina

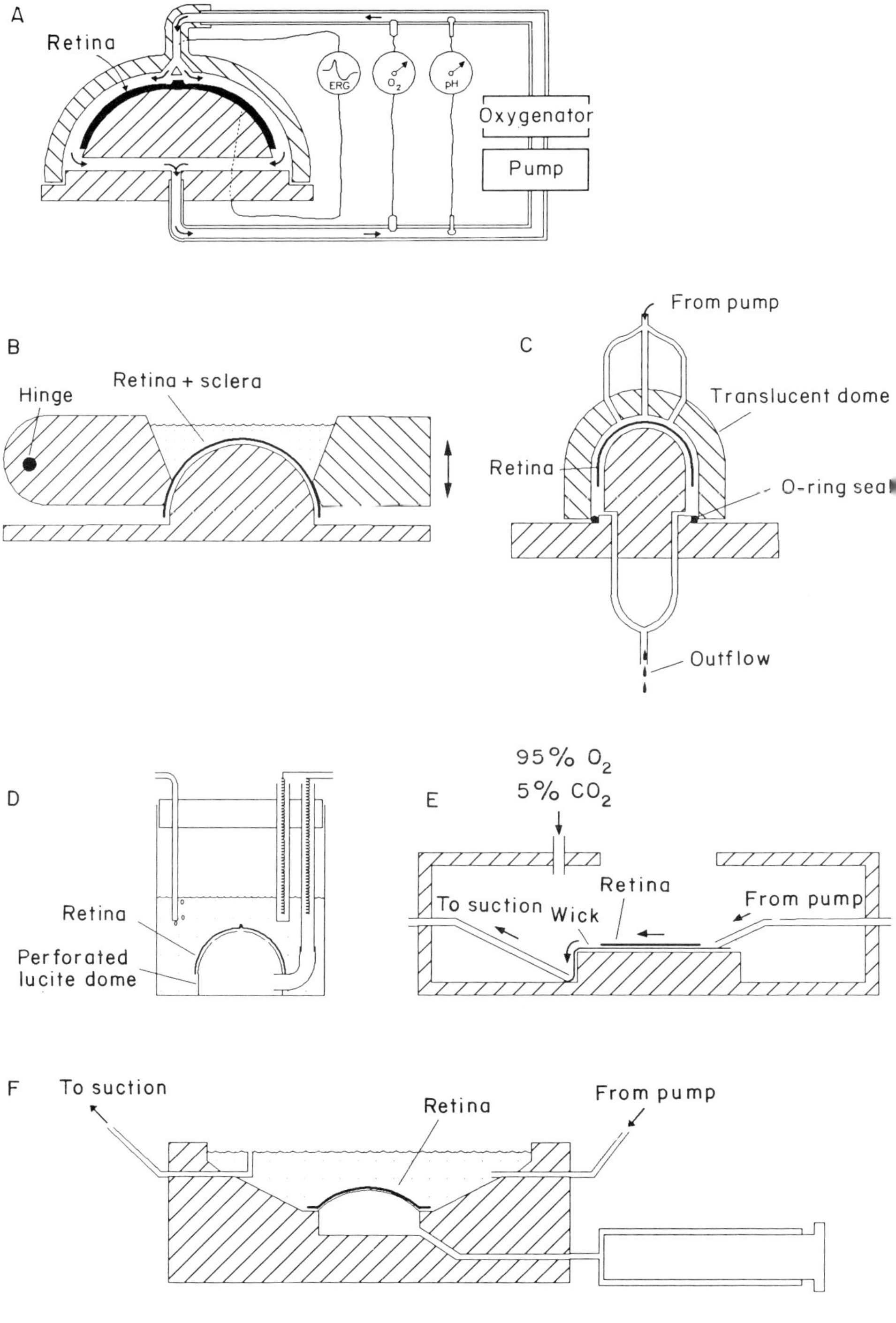
A
Retina
ERG
O_2
pH
Oxygenator
Pump
B
Hinge
Retina + sclera
C
From pump
Translucent dome
Retina
O-ring seal
Outflow
D
Retina
Perforated lucite dome
E
95% O_2
5% CO_2
Retina
To suction
Wick
From pump
F
To suction
Retina
From pump

Figure 2 (A) Apparatus for monitoring retinal energy metabolism and light-evoked transretinal electrical responses. The retina is deployed over a lucite mandrel (stabilized by pin at top, not shown) and superfused by a rapid flow of medium entering and leaving through radially arranged ports. Medium is circulated by peristaltic pump through a bubble chamber (the lung) that equilibrates it with 5% CO_2 and 40% O_2. Oxygen and pH electrodes measure the 'arterio-venous' differences or, when the lung is short-circuited (in another mode), the rates at which O_2 and pH fall in the closed system. (B) Chamber in which the retina is incubated in the presence of the sclera, choroid, and pigmented epithelium. The eye is removed and the vitreous is lifted away. The resulting eyecup is everted and spread over a lucite dome. A second piece of lucite supported on a pivot is swung down to rest lightly on the retinal surface. The second piece becomes the walls of a small dish (the retinal surface forms the floor), which is filled with incubation medium. In this chamber the retina remains in contact with the pigment epithelium and can reisomerize rhodopsin. It has been used with both the rabbit and the cat retinas. (C) Chamber for rapid superfusion of the retina. The retina is spread over a dome and a second symmetrical dome is clipped over the first. The translucent upper dome forms a seal at an O-ring. The internal volume of the chamber is approximately 400 μl. The medium flows over the retina at 8 ml/min, providing rapid exchange of the chamber's internal contents. (D) ERG chamber. The retina is deployed over a hollow plastic mandrel with perforations at the top, and held in place with a silicone rubber O-ring (not shown). Recordings are made between two Ag–AgCl electrodes in glass tubes containing a half-saturated KCl solution in agar. One electrode is inserted (forming a seal) into a short silicone tube that penetrates the base of the mandrel. Movement of medium is maintained by bubbling 95% O_2, 5% CO_2 through two fine polyethylene tubes (only one shown). (E) System in which the retina is incubated in a thin film of incubation medium. The retina is placed on top of a wick made of gauze. Medium flows onto one end of the wick and is removed from the other end of the wick by suction. A thin film of medium flows through the wick and the retina remains immersed in it. The thickness of the film of medium over the retina is approximately 200 μm. The atmosphere around the retina and wick is maintained at 95% O_2, 5% CO_2. The movement of medium and thickness of the layer of medium over the retina are controlled by adjusting the rate of flow of the medium and the suction that removes medium from the wick. This chamber has been used for electrophysiological experiments. In those experiments the electrophysiological activity of the retinal ganglion cells appeared to be more vigorous than in chambers where the retina is immersed in medium. This chamber adopts features from chambers used to incubate brain slices, notably that a thin film of medium on top of the tissue is in direct contact with the oxygenated atmosphere. A somewhat similar chamber was used by Baughman and Gilbert for incubating slices of visual cortex (personal communication). (F) Chamber for recording with microelectrodes. The retina is spread over a thin silicone membrane, and is held down by the weight of a platinum ring slightly larger than the membrane (not shown). The silicone membrane is inflated into a dome by a screw-operated syringe, which tenses the retina slightly to improve its immobilization. A chamber similar to this has also been used for studies of the release of neurotransmitters into the superfusing medium, but the exchange of solutes is relatively inefficient are in general phasic in their responses, their activity is increased much more by flashes of light than by steady light. In various studies in which we have sought to achieve maximal activation of these cells, we have used three different criteria for their level of activity – 2-deoxyglucose uptake (Ames *et al*., 1980), acetylcholine release (Masland and Livingstone, 1976), and the rate of glycolysis (Ames and Zager, 1987). By all three criteria, maximal activation was achieved with short (10–50 ms) flashes delivered at 3 or 4 Hz, with an intensity sufficient to elicit a maximal b-wave.

II.5 Measurements of the Volumes of the Intracellular and Extracellular Compartments and of the Rate of Exchange by Diffusion between the Extracellular Fluid and the Medium

For some experiments (e.g. measurement of intracellular Na^+ concentration) it is important to know how tissue water is partitioned between the intracellular and extracellular compartments. Inulin, mannitol, and the readily elutable Na^+ and Cl^-, were all found to have similar volumes of distribution and to characterize an extracellular phase that constituted about 30% (w/w) of the retina (Ames and Nesbett, 1966). It is interesting that this value is somewhat larger than the size of the extracellular space as it appears in electron microscopy. The reason for the disagreement remains unclear.

For some experiments, the rate of diffusion of substances between the cells and the medium may be an important consideration (e.g. in studies of the kinetics of the release of endogenous substances from the cells, or in studies of the rate of response of the cells to an agent delivered via the medium). Diffusion rates through the interstitial fluid have been assessed by labeling the interstitial fluid with [^{14}C]mannitol and then measuring its rate of elution into unlabeled medium. The results were consistent with theoretical calculations from Crank's (1975) sorption curve and showed that for solutes the size of mannitol the exchange between the interstitial fluid of the retina proper and the medium is more than 90% complete in 30 s (Ames *et al*., 1976).

III SPECIFIC TYPES OF EXPERIMENT

III.1 Measurements of Membrane Transport

Because the retina is so thin, rates of diffusion between medium and interstitial fluid are rapid relative to rates of transport across plasma membranes, which makes it possible to use this tissue to study transport in and out of CNS cells. The rate of leucine influx into retinal cells has been determined from the rate of disappearance of labeled leucine from the extracellular fluid of a retina temporarily removed from the medium, and the rate of leucine efflux from retinal cells has been determined from the rate at which labeled leucine appeared in the medium following labeling of the intracellular fluid to steady state. Similar measurements have been made of Na^+ efflux. Since these measurements required a restricted volume of medium and involved short time periods, the retina was held continuously by forceps at the optic disc and moved in time to the beat of a metronome through a series of beakers containing as little as 3 ml of medium, with the residence time in each beaker as short as 2 s. In order to distinguish the leucine and Na^+ coming from the extracellular fluid from that coming from the cells, it was necessary (with double labeling) to use mannitol as a measure of the extracellular fluid and hence of the extracellular component of the substances of interest. These types of measurements are described in more detail by Ames *et al*. (1976). A superfusion system for elution studies with a high temporal resolution is described below.

III.2 Measurements of Retinal Metabolism

Changes in the levels of precursors and products in the fluid bathing a retina can be used as a measure of its metabolism. This approach has the advantage of permitting serial measurements to be made on a single retina so that the retina can serve as its own control, e.g. in experiments testing the effects of a chemical variable or of a change in photic stimulation. Rates of glucose utilization have been measured by serial sampling of the 6 ml of medium in a small incubation boat containing two retinas (Ames *et al*., 1980). In a more recently developed experimental procedure for measuring O_2 consumption and lactic acid production, the retina is mounted on a mandrel incorporated in a miniature heart–lung apparatus (Figure 2A), and is superfused by a l-mm-deep layer of medium moving at 4.6 ml/min. The medium is recirculated by a peristaltic pump through a chamber in which it is equilibrated by bubbling with a gas phase of 5% CO_2, 40% O_2 and 55% N_2. Oxygen and pH electrodes sample the medium before and after it has passed over the retina, providing, in one mode, continuous recordings of the 'arterio-venous' differences in O_2 and pH. In the other mode, which has been used more often, the bubble chamber which serves as the lung is temporarily bypassed so that the O_2 consumed by the retina cannot be replaced and the CO_2 and H_2CO_3 generated cannot be removed. In this mode, the volume of recirculating medium is only 2.3 ml, so the O_2 consumption of an average retina reduces the O_2 content of the medium by about 12% in the 1-min period over which the measurement is made – an accurately measurable change – and the acid generated in this period reduces the pH by about 0.040 units, which is also quite accurately measured. Rates of lactic acid production are calculated assuming that the measured fall in pH is attributable to the CO_2 produced by oxidative metabolism and to the lactic acid from glycolysis, and using the measurements of O_2 consumption to determine the contribution of the former. The measurements of O_2 consumption and lactic acid production under control conditions in darkness (0.12 and 0.15 μmol/min respectively) have been in good agreement with those previously reported and have shown the changes expected in response to changes in lighting conditions. Thus, steady light, of intensity sufficient to markedly reduce the photoreceptors' dark current, reduced O_2 consumption by $12 \pm 0.9\%$ (SE) ($n = 7$; $P \ll 0.001$); and 10-ms light flashes at the frequency of 4 Hz, which strongly activate neurons of the inner retina, increased glycolysis by $26 \pm 2.6\%$ (SE) ($n = 4$; $P \ll 0.001$). As indicated by these figures, the variability in results obtained on replicate retinas is relatively small, while the response of the retina to physiological stimuli (and also to pharmacologic agents) is often quite large. Electrodes incorporated in the heart–lung apparatus, one on either side of the retina, are used to record the light-evoked changes in transretinal potential (the *in vitro* ERG), making it possible to correlate the changes in energy metabolism with changes in function measured simultaneously (Ames and Zager, 1987; Zager and Ames, 1988).

III.3 Studies of Neurotransmitter Release

Because the retina is so thin, compounds released by retinal cells appear promptly in the superfusing medium, and neurotransmitter precursors or pharmacologic agents diffuse quickly from the medium to the cells. Since the retina can be stimulated by light, the synapses can be studied during their normal functions.

Most of our studies of neurotransmitter function have used radiochemical methods. Although it is quite possible to study the synthesis and release of endogenous neurotransmitters, radiochemical methods allow measurement of very dilute concentrations of neurotransmitters. This has the advantage that high rates of superfusion can be used, with the accompanying gain in temporal resolution (see below). Typically, the retina's neurotransmitter pools are isotopically labeled while the retina is in an incubation boat (Figure 1C); the radioactive precursor is cleared from the retina's extracellular spaces by a brief (3–6 min) rinse in medium containing no radioactive compounds; and the retina is transferred to a superfusion system for studies of neurotransmitter release.

When retinas were exposed to [^{3}H]choline, for example, they accumulated it and synthesized [^{3}H]acetylcholine. Stimulation with light – originally with the retina mounted in the simple chamber shown in Figure 2F – caused a clear increase in the rate of release of acetylcholine into the superfusing medium. Other experiments showed that the increased secretion was accompanied by an increased rate of acetylcholine turnover within the tissue (Masland and Livingstone, 1976).

There was no increase in acetylcholine secretion when the retinas were stimulated with steady light: flashing light was required. This implied that the acetylcholine-releasing cells were responding to the photic transients. To investigate this possibility further, experiments were done using the rapid-flow chamber shown in Figure 2C. The retina is draped over a dome and a second, slightly larger, dome is clipped over it. The outer dome (uppermost in the figure) seals around the lower one by means of an O-ring. The enclosed volume within the chamber is about 400 μl, of which the retina occupies about 70 μl. Incubation medium is pumped through the chamber at a rate of 8 ml/min, so that the internal volume of the chamber is calculated to be replaced every 2.5 s, and this was confirmed in experiments that measured the rate of clearance of radioactive acetylcholine that had been added to the medium in the chamber. Thus the rate of clearance of the chamber was more rapid than the rate at which small molecules diffuse out of the retina's extracellular space (as described above, the half-time is about 15 s) so the chamber's kinetics were not limiting for purposes of this experiment. The earliest release of acetylcholine was detected about 5 s after the photic stimulation. This is slow compared to the release of neurotransmitters in synapses, but was fast enough, for a photic stimulus that lasted 30 s, to reveal both the on and the off transient of acetylcholine release in response to light (Masland *et al.*, 1984a).

The use of radiochemical methods also allows direct observation of the neurotransmitter content of the cells whose neurotransmitter secretion is under study. This is done by conventional autoradiography. For example, one can stimulate the release of a neurotransmitter (or other compound that is released or consumed) and later use autoradiography to see how the cellular content of the compound in question has changed. An example was the original identification of the cholinergic amacrine cells (Masland and Mills, 1979). Tissues were labeled with radioactive acetylcholine. A comparison was then made, using autoradiography, between tissues that had been stimulated to release their acetylcholine and tissues that had been labeled but not stimulated. The cholinergic cells were identified as those cells that became depleted of silver grains when the retina was stimulated. A slightly more more sophisticated experiment used a pharmacologic agent 2-amino-4-phosphonobutyric acid (APB) to silence the retinal cells (ON cells) that increase their activity in response to light (Slaughter and Miller, 1981). Retinas that contained radioactive acetylcholine were stimulated with light in the presence or absence of APB. Those cells whose acetylcholine release was retarded by APB were identified autoradiographically as the ON cholinergic amacrine cells (Masland *et al.*, 1984a). A third example was a study of the co-release of acetylcholine and GABA. Histochemical studies had shown that a group of amacrine cells contains both acetylcholine and GABA (Brecha *et al.*, 1988; Kosaka *et al.*, 1988; Vaney and Young, 1988). The question was whether or not the cells released GABA. To test it the cells were labeled by exposure to [^{3}H]GABA. One group of retinas was stimulated to release GABA and a second group was not. Autoradiographs of the retinas that had been stimulated to release GABA showed depletion of GABA from all of the retina's GABA-containing neurons, indicating that at least some of the GABA released by the retina originated in cells that also contained acetylcholine (O'Malley and Masland, 1989).

In all three of these cases the rapid exchange between medium and cells provided by the isolated retina preparation was essential: it made it possible to obtain a high, and uniform, level of isotopic labeling of the neurotransmitters of interest; it permitted measurement of neurotransmitter release with fast temporal resolution; and the quick clearing of radioactive compounds from the tissue diminished complications introduced by the subsequent recapture and reincorporation of the radioactive compounds that were released. Furthermore, the ability to interrupt the experiment at a desired moment in time made it possible to 'freeze' a chain of physiological events for histochemical study. The examples cited above will perhaps suggest the variety of experiments that can be carried out when radiochemical and autoradiographic techniques are combined.

III.4 Visualizing Retinal Cells

The thinness and repetitive structure of the retina suggest the possibility of observing it optically in the living state, and special methods allow one to visualize some of the retinal neurons either as individual entities or as populations. The repeating pattern of retinal circuits in the tangential plane makes it easy to survey large numbers of cells of identical type. It is sometimes possible to observe retinal cells during electrophysiological recording.

An obvious approach to visualizing cells in a tissue that is only 100 μm thick is the use of differential interference contrast optics (DIC). In our experience, however, DIC does not work as well as would be expected. The reason appears to be that the photoreceptor outer segments are very refractile (this is to be expected from structures whose function is to trap light). When the outer segments are absent, as can occur if one intentionally rubs off the photoreceptor outer segments, the images are better. In retinas from newborn rabbits, in which the inner and outer segments have not yet developed, the images are also better than in those from adult rabbits, even though retinas from newborn rabbits are more than twice as thick as retinas from adult rabbits and thus would be expected to yield worse DIC images (McArdle *et al*., 1977). One way to circumvent the difficulty is by the use of retinal slices, which have been made from the retinas of cold-blooded vertebrates (Werblin, 1978). For retinal slicing the tissue is laid flat and then cut into slices a few hundred micrometers thick using a simple tissue chopper. The slices are turned on their sides and viewed with DIC or Hoffman optics. Cells can be seen quite clearly in such slices and electrophysiological recordings from visually chosen cells have been successful. This technique is particularly useful for the study of cells along the retina's through-pathway, i.e. photoreceptors, bipolar cells, and narrow-field amacrine cells. In retinal slices the long processes of the laterally conducting cells are damaged.

An alternative strategy, which works for intact retinas, is to label the cells with fluorescent markers. A great variety of fluorescent molecules for labeling living cells are available. These bind to different parts of the cell, or are accumulated by subsets of the cells. There is some risk that illuminating a tissue labeled with fluorescent molecules will damage it, but this appears to be more the exception than the rule if the intensity of the fluorescent illumination and the concentration of the fluorochrome used for labeling are kept within reasonable limits. Living rabbit retinas can be stained with acridine orange, for example, without noticeable damage to the tissue's electrophysiological responsiveness. For visualizing cells during recording it is desirable to incubate the tissue without isolating it from the retinal pigment epithelium – e.g. as shown in Figure 2B. The retinal pigment epithelium is required for the resynthesis of rhodopsin, which is bleached by the illumination used to excite the fluorescence of acridine orange.

Fluorescent marking of the cells may be used as a guide to aiming micropipettes for intracellular recording, especially in retinas with large neurons. In principle, the ability to aim the micropipette increases the probability of a direct hit on the identified cell. It also allows one to select the class of cell to be impaled. In practice, however, this may not offer as large an advantage as might at first seem. Aiming micropipettes is time-consuming, and most attempts to impale mammalian neurons fail even with visual guidance. As a strategic matter, a small number of carefully aimed attempts at penetration may yield fewer successful recordings than the much larger number of penetrations that may be attempted without visual guidance. Newer recording methods may ameliorate this difficulty (Miller and Coleman, 1987).

On the other hand, fluorescent labeling is clearly useful for anatomical studies. The overall strategy is to use one fluorescent molecule, introduced into the medium or injected intraocularly before the experiment, to stain the cell body or nucleus and to then guide a micropipette to the fluorescent cell and inject it with a second diffusible marker. The injected marker fills the cell, revealing its processes. It is convenient for the injected marker also to be fluorescent, as is Lucifer yellow, because the experimenter can then monitor the process of cell filling. This type of injection can be carried out in living tissue maintained by the methods described above; in tissue isolated and maintained in a physiological solution but kept at room temperature, in which case the conditions of incubation are more forgiving; or in tissue that has been lightly fixed with paraformaldehyde. Light fixation with aldehyde does not destroy the ability of the phospholipid bilayer to contain molecules such as Lucifer yellow; the use of fixed tissue has advantages of convenience and of allowing certain kinds of histochemistry (see below). The primary fluorescent marker used to visualize the cell bodies can be a universal stain such as acridine orange. The family of nuclear stains exemplified by Hoechst 33258 are also useful. Although such molecules stain all the cells, the size, shape, and location within the tissue of the cells give clues as to the final identity of the cell. For example, the large multipolar cell bodies located in the ganglion cell layer of mammalian retinas usually belong to the class of α ganglion cells (Tauchi and Masland, 1985; Peichl *et al*., 1987). One quickly becomes able to predict which type of cell will be revealed upon injection. Injection of cells by these means thus resembles a controllable Golgi stain; many examples of defined cell classes may be accumulated (Masland, 1986). Fluorescence-guided injection is probably less capricious than the Golgi, because one can choose to inject any cell whose nucleus contains stainable DNA. There is less risk of whole cell populations remaining undetected and one can be more or less assured of an accurate sample of the major cell classes. As with the Golgi technique, though, there is some danger of incomplete filling of the cells. This is not an issue for cells with small dendritic fields (less than about 1 mm). It appears to be a greater problem for cells with very long and very thin processes, especially when the injections are done in fixed tissue (Vaney, 1986; Dacey, 1988). An example of a retinal ganglion cell injected in a tissue originally stained with acridine is shown in Figure 3. At the top is shown a whole mount of a cat retina stained with acridine orange. The stained neuron indicated in the upper micrograph was injected with Lucifer yellow, to reveal the characteristic shape of a β-ganglion cell.

If the primary fluorescent stain is specific to a certain cell class, one can use this method to collect examples of members of that class. An example is the use of 4,6-diamidino-2-phenylindole (DAPI) to stain the cholinergic amacrine cells (Masland *et al.*, 1984b; Tauchi and Masland, 1984). A second example is the use of 5,7-dihydroxytryptamine to identify indoleamine-accumulating cells for injection (Sandell and Masland, 1986). A third example – one made possible by the injection of cells in fixed tissue – is the injection of the dopaminergic cells. The catecholamines contained in the dopaminergic cells form a fluorescent reaction product upon fixation with a mixture of paraformaldehyde and glutaraldehyde. The fluorescent cell bodies were impaled and injected to show the whole shape of the cell (Voigt and Wässle, 1987; Tauchi and Masland, 1988). A final example is the injection of cells that have been filled by retrograde transport of fluorescent markers, of which a great number are now available. The fluorescent marker is applied to the nerve terminals *in vivo*, and the axons then transport the fluorescent marker retrogradely to the cell body. At a later time the retina is isolated and the cell bodies are visualized by their content of the fluorescent marker. The cells are then injected to reveal the whole shape of the retrogradely labeled neuron (see Buhl and Peichl, 1986).

IV DISCUSSION

Some of the characteristics of the isolated retina appear almost paradoxical. It is extremely fragile in most respects, but it is remarkably resistant to mechanical damage when it is manipulated with forceps applied to the optic nerve stump. This is a consequence, presumably, of the convergence at the optic disc of the axons of the hundreds of thousands of ganglion cells dispersed over the retinal surface. The function of the retina is extremely sensitive to changes in the concentrations of medium constituents and to trace amounts of heavy metals or of organic toxins; but, once appropriate experimental conditions have been established, preparations that appear to function normally for at least 8 h can be obtained with great regularity. The vulnerability to toxins may be explained by the fact that only 5 mg of protein (the retinal content) is being exposed during the course of isolation and experimentation to large volumes of medium that have been in contact with large areas of glass and plastic. Though it is possible to isolate retinas freehand with only a scalpel and forceps, it is virtually impossible to obtain reasonably intact

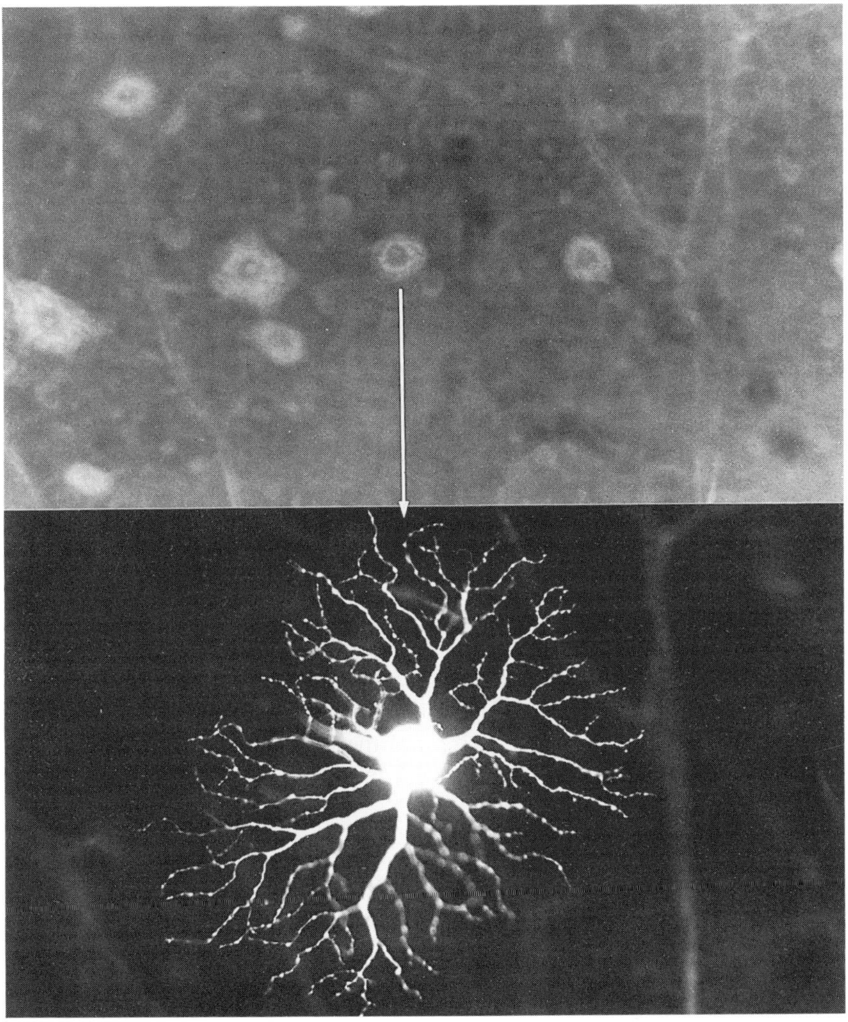

Figure 3 Injection of retinal ganglion cells with Lucifer yellow. The top panel shows a flat-mount view of a cat's retina. The retinal ganglion cells are visible by their staining with acridine. The cell indicated was impaled with a micropipette filled with Lucifer yellow, under visual control. Injection with Lucifer yellow revealed the dendritic morphology shown below. (Micrographs by Dr Masaki Tauchi (see Tauchi and Masland, 1985))

retinas this way even with optimal illumination; whereas with simple mechanical devices it is relatively easy to obtain intact retinas even in dim red light.

It has always seemed of importance to us to try to establish control conditions that simulate as closely as possible the physiological conditions to which the retina had been exposed *in vivo*. This may add appreciably to the work of the investigator, examples being provision of the many organic solutes upon which cells ultimately depend (now much simplified by the Sigma medium) and the CO_2 equilibration required for a bicarbonate buffer. Though a simplified medium using a non-volatile buffer may seem adequate for some types of experiment and is easier to use, we feel there are compelling reasons for attempting to duplicate the *in vivo* situation. Of these, the most important is that such preparations achieve something resembling a steady state. As we have already remarked, preparations poisoned by toxins, starved by inadequate perfusion, or stressed by improper extracellular ion concentrations may retain some of their physiological activity for several hours. When the *in vivo* condition is duplicated carefully – even for experiments much shorter than the potential life of the preparations – the investigator can be much more confident that the process under study will be quantitatively interpretable in terms of the *in vivo* situation and in terms of the work of other investigators.

The ultimate challenge presented by the *in vitro* retina preparation is not the establishment of control conditions that preserve normal function but the development of protocols that realize more fully its experimental potential. The preparation provides the investigator with an almost unique opportunity to study intact neural circuits that are operating under circumstances that permit microscopic observation, physiological stimulation, measurements of electrophysiological responses, introduction of experimental agents, and recovery of substances released by the cells.

ACKNOWLEDGEMENTS

This work was supported by NIH grants EY01075, EY05747, and NS25753. We thank Michael Weiner for help with the illustrations.

REFERENCES

Ames, A., III, and Gurian, B.S. (1960). 'Measurement of function in an *in vitro* preparation of mammalian central nervous tissue', *J. Neurophysiol.*, **23**, 676–691.

Ames, A., III, and Nesbett, F.B. (1966). 'Intracellular and extracellular compartments of mammalian central nervous tissue', *J. Physiol. (Lond.).*, **184**, 215–238.

Ames, A., III, and Nesbett, F.B. (1981). '*In vitro* retina as an experimental model of the central nervous system', *J. Neurochem.*, **37**, 867–877.

Ames, A., III, and Nesbett, F.B. (1983). 'Pathophysiology of ischemic cell death: I. Times of onset of irreversible damage; importance of the different components of the ischemic insult', *Stroke*, **14**, 219–226.

Ames, A., III, and Pollen, D.A. (1969). 'Neurotransmission in central nervous tissue: A study of isolated rabbit retina', *J. Neurophysiol.*, **32**, 424–442.

Ames, A., III, and Zager, E.L. (1987). 'Reducing cellular energy requirements as protection against CNS ischemia', in *Cerebral Ischemia and Hemorheology* (Eds A. Hartmann and W. Kuschinsky), pp. 241–248, Springer, Berlin.

Ames, A., III, Tsukada, Y., and Nesbett F.B. (1967). 'Intracellular Cl, Na, K, Ca, Mg, and P in nervous tissue: response to glutamate and to changes in extracellular Ca', *J. Neurochem.*, **14**, 145–159.

Ames, A., III, Parks, J.M., and Nesbett, F.B. (1976). 'Transport of leucine and sodium in central nervous tissue: Studies on *in vitro* retina', *J. Neurochem.*, **27**, 999–1015.

Ames, A., III, Parks, J.M., and Nesbett, F.B. (1980). 'Synthesis and degradation of retinal proteins in darkness and during photic stimulation', *J. Neurochem.*, **35**, 142–147.

Ames, A., III, Walseth, T.F., Heyman, R.A., Barad, M., Graeff, R.M., and Goldberg, N.D. (1986). 'Light-induced increases in cGMP metabolic flux correspond with electrical responses of photoreceptors', *J. Biol. Chem.*, **261**, 13034–13042.

Brecha, N., Johnson, D., Peichl, L., and Wässle, H. (1988). 'Cholinergic amacrine cells of the rabbit retina express GABA and GAD immunoreactivity', *Proc. Natl Acad. Sci. USA*, **85**, 6187–6191.

Buhl, E.H., and Peichl, L. (1986). 'Morphology of rabbit retinal ganglion cells projecting to the medial terminal nucleus of the accessory optic system', *J. Comp. Neurol.*, **253**, 163–174.

Crank, J. (1975). *The Mathematics of Diffusion*, 2nd edn, Clarendon Press, Oxford.

Dacey, D.M. (1988). 'Dopamine-accumulating retinal neurons revealed by *in vitro* fluorescence display a unique morphology', *Science*, **240**, 1196–1198.

Dowling, J.E. (1988). *The Retina: An Approachable Part of the Brain*, Belknap, Cambridge.

Goldberg, N.D., Ames, A., III, Gander, J.E., and Walseth, T.F. (1983). 'Magnitude of increase in retinal cGMP metabolic flux determined by ^{18}O incorporation into nucleotide alpha-phosphoryls corresponds with intensity of photic stimulation', *J. Biol. Chem.*, **258**, 9213–9219.

Hubel, D.H. (1988). *Eye, Brain, and Vision*, Freeman, New York.

Kosaka, T., Tauchi, M., and Dahl, J.L. (1988). 'Cholinergic neurons containing GABA-like and/or glutamic acid decarboxylase-like immunoreactivities in various brain regions of the rat', *Exp. Brain Res.*, **70**, 605–617.

Masland, R.H. (1977). 'Maturation of function in the developing rabbit retina', *J. Comp. Neurol.*, **175**, 275–286.

Masland, R.H. (1986). 'The functional architecture of the retina', *Sci. Am.*, **255**, 102–111.

Masland, R.H., and Ames, A., III (1975). 'Dissociation of field potential from neuronal activity in the isolated retina: failure of the b-wave with normal ganglion cell response', *J. Neurobiol.*, **6**, 305–312.

Masland, R.H., and Ames, A., III (1976). 'Responses to acetylcholine of ganglion cells in an isolated mammalian retina', *J. Neurophysiol.*, **39**, 1220–1235.

Masland, R.H., and Livingstone, C.J. (1976). 'Effect of stimulation with light on the synthesis and release of acetylcholine by an isolated mammalian retina', *J. Neurophysiol.*, **39**, 1210–1219.

Masland, R.H., and Mills, J.W. (1979). 'Autoradiographic identification of acetylcholine in the rabbit retina', *J. Cell Biol.*, **83**, 159–178.

Masland, R.H., Mills, J.W., and Cassidy, C. (1984a). 'The functions of acetylcholine in the rabbit retina', *Proc. R. Soc. Lond. B*, **223**, 101–119.

Masland, R.H., Mills, J.W., and Hayden, S.A. (1984b). 'Acetylcholine-synthesizing amacrine cells: identification and selective staining by using autoradiography and fluorescent markers', *Proc. R. Soc. Lond. B*, **223**, 79–100.

McArdle, C.G., Dowling, J.E., and Masland, R.H. (1977). 'Development of outer segments and synapses in the rabbit retina', *J. Comp. Neurol.*, **175**, 253–274.

Miller, R.F., and Coleman, P.A. (1987). 'Kainate receptors are the primary mediators of excitatory synaptic transmission among third-order neurons in the amphibian retina', *Invest. Ophthalmol. Vis. Sci.*, suppl. 28, 239.

O'Malley, D.M., and Masland, R.H. (1989). 'Co-release of acetylcholine and GABA by a retinal neuron', *Proc. Natl. Acad. Sci. USA*, **86**, 3414–3418.

Parks, J.M., Ames, A., III, and Nesbett, F.B. (1976). 'Protein synthesis in central nervous tissue: Studies on retina *in vitro*', *J. Neurochem.*, **27**, 987–997.

Peichl, L., Ott, H., and Boycott, B.B. (1987). 'Alpha ganglion cells in mammalian retinae', *Proc. R. Soc. Lond. B*, **231**, 169–197.

Pepperberg, D.R., and Masland, R.H. (1978). 'Retinal-induced sensitization of light-adapted rabbit photoreceptors', *Brain Res.*, **151**, 194–200.

Rodieck, R.W. (1973). *The Vertebrate Retina*, Freeman, San Francisco.

Sandell, J.H., and Masland, R.H. (1986). 'A system of indoleamine-accumulating neurons in the rabbit retina', *J. Neurosci.*, **6**, 3331–3347.

Slaughter, M., and Miller, R.F. (1981). '2-amino-4-phosphonobutyric acid: a new pharmacological tool for retina research', *Science*, **211**, 182–185.

Steinman, L., and Ames, A., III (1974). 'The sites of synthesis and the subsequent migration of newly synthesized protein in retina', *Tissue Cell*, **6**, 137–152.

Tauchi, M., and Masland, R.H. (1984). 'The shape and arrangement of the cholinergic neurons in the rabbit retina', *Proc. R. Soc. Lond. B*, **223**, 121–139.

Tauchi, M., and Masland, R.H. (1985). 'Local order among the dendrites of an amacrine cell population', *J. Neurosci.*, **5**, 2494–2501.

Tauchi, M., and Masland, R.H. (1988). 'Morphology of the catecholaminergic neurons in the rabbit retina', *Biomed. Res.*, **9** (suppl. 12), 135–138.

Vaney, I.D. (1986). 'Morphological identification of serotonin-accumulating neurons in the living retina', *Science*, **223**, 444–446.

Vaney, D.I., and Young, H.M. (1988). 'GABA-like immunoreactivity in cholinergic amacrine cells of the rabbit retina', *Brain Res.*, **438**, 369–373.

Voigt, T., and Wässle, H. (1987). 'Dopaminergic innervation of AII cells in mammalian retina', *J. Neurosci.*, **7**, 4115–4128.

Webster, J. deF., Ames, A., III, and Nesbett, F.B. (1969). 'A quantitative morphological study of osmotically induced swelling and shrinkage in nervous tissue', *Tissue Cell*, **1**, 201–216.

Werblin, F.S. (1978). 'Transmission along and between rods in the retina of the tiger salamander', *J. Physiol. (Lond.)*., **294**, 613–626.

Zager, E.L., and Ames, A., III (1988). 'Reduction of cellular energy requirements. Screening for agents that may protect against CNS ischemia', *J. Neurosurg.*, **69**, 568–579.

Preparations of Vertebrate Central Nervous System *In Vitro*
Edited by H. Jahnsen
Published by John Wiley & Sons Ltd

CHAPTER 8

The Isolated and Perfused Mammalian Hypothalamus

CHARLES W. BOURQUE
Center for Research in Neuroscience,
McGill University and Division of Neurology, Montreal General Hospital,
1650 Cedar Avenue, Montreal, Quebec H3G 1A4, Canada

I INTRODUCTION: FROM BRAIN SLICES TO PERFUSED EXPLANTS

I.1 Importance and Contribution of *In Vitro* Methods

The application of *in vitro* techniques to the study of the vertebrate brain has led to the formulation of important new concepts in our understanding of central nervous system (CNS) function. These methods permit a control of the extracellular ionic environment and a stability for intracellular recording that cannot be achieved *in vivo*. Experimentally, the use of *in vitro* preparations has revealed that a multitude of ionic conductances modulate the electrical behaviour and responsiveness of individual cells. Consequently, we now appreciate that the complexity of the vertebrate brain results not only from the intricate connectivity of neuronal circuits, but also from the computational power of individual neurons with specific intrinsic activities (Llinás, 1988). The output of a neuron to a given synaptic signal is indeed strongly influenced by the types, density and topographic distribution of the ionic channels which are expressed in its membrane. It is therefore crucial to determine the exact complement of ionic mechanisms present in a particular system if we are to understand its intrinsic behaviour as well as the means by which relevant exogenous or synaptically released substances can modulate or control its electrical activity. The experiments which need to be performed to obtain this type of information can best be accomplished *in vitro*.

In the twenty years or so that followed the introduction of brain slices for electrophysiological studies (Yamamoto and McIlwain, 1966), the accumulation of published technical information has led to a consensus of procedures required for their preparation and maintenance *in vitro* (e.g. Kerkut and Wheal, 1981; Dingledine, 1984; Schurr *et al*., 1987). As a result, the establishment of new preparations dedicated to the analysis of defined intrinsic and/or synaptic mechanisms has become a matter of routine. By following a few simple rules and using commercially available equipment, most laboratories can now adapt or devise a viable preparation of isolated CNS tissue which suits their specific experimental needs. In many cases a variety of preparations encompassing the same area are used to target different questions in the most direct and productive manner.

As frequently mentioned in this book, the specific advantages of each isolated preparation must also be weighed against significant problems. The successful exploitation of these techniques therefore depends on our ability to make the best possible use of the sound features of a preparation while avoiding those which may present problems of interpretation. As a rule it is always preferable to use *in vitro* analysis to study the *details* of a mechanism which has already been identified and is known to occur *in vivo*, than to give primary *in vitro* observations a speculative physiological meaning which may have no basis in the intact animal.

I.2 Limitations of the Slice Preparation

The attractive possibility of obtaining stable intracellular recordings from 'intact' neurons bathed in defined solutions has incited the continued refinement of the slice technique and has allowed the biophysical analysis of many intrinsic and synaptic mechanisms. In general, slices (300–700 μm thick) obtained from most brain regions can be kept viable for at least 6–8 h in a chamber where they are either submerged in an oxygenated artificial cerebrospinal fluid (ACSF), or placed at an ACSF–O_2 interface. Ample detail concerning the preparation, maintenance and use of brain slices can be found elsewhere in this book. One feature common to all slice preparations, however, is that the metabolic support of cellular viability depends on the diffusion of small molecules (e.g. O_2, CO_2 and glucose) between the surface and depth of the slice. Within a slice, PO_2 falls rapidly with distance from the surface (Bingmann and Kolde, 1982) and the slope of this drop becomes steeper with increasing metabolic rate. Under the usual recording conditions (33–35°C), the adequacy of diffusional processes seems to be restricted to a distance of about 200 μm from either surface, limiting the maximal thickness of the tissue sample to about 400 μm (see Richards, 1981). Thus, while brain slices are extremely useful for studying the intrinsic cellular properties of neurons and synaptic interactions existing within the plane of section, they do not allow the study of nerve circuits which depend on axons coursing more than a few hundred micrometres in each dimension.

I.3 Intra-arterially Perfused CNS Explants

An important advance in the development of *in vitro* techniques and an attempt to overcome the limited range of synaptic connectivity in slices was the introduction of intra-arterial perfusion (Yarom and Llinás, 1979; Llinás *et al*., 1981; Llinás and Mühlethaler, 1988a,b; Walton and Llinás, 1982; Bourque and Renaud, 1982, 1983, 1984a,b; Mühlethaler and Llinás, 1984, 1985). This method takes advantage of the existing arterial system to provide oxygen and nutrients to relatively large (>700 μm in each axis) portions of isolated brain. Because this technique is particularly well suited to brain regions which are extensively vascularized, it also offers the advantage of being applicable to brain regions which may not withstand common slicing procedures.

This chapter describes a preparation of perfused rat basal forebrain which was originally developed to analyse the physiological and biophysical mechanisms underlying the electrical behaviour of mammalian hypothalamic neurosecretory neurons (Bourque, 1984; Bourque and Renaud, 1982, 1983, 1984a,b). This preparation can be adapted to study a wide variety of other neurons, in terms of both their intrinsic cellular behaviour and their synaptic connectivity. The technical description of this method has been made as general and as complete as possible to provide some assistance to investigators wishing to use or develop similar preparations.

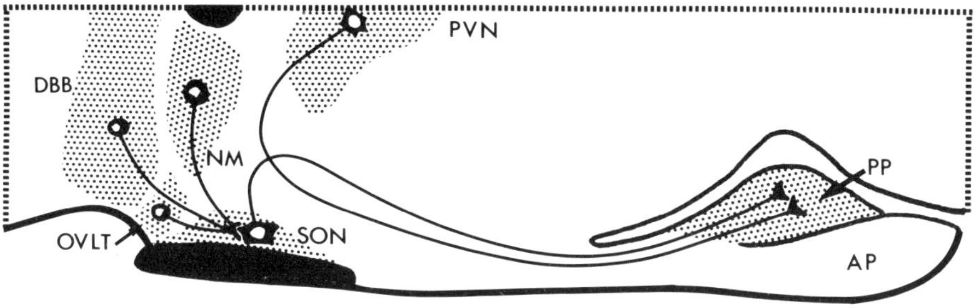

Figure 1 Schematic sagittal diagram of the rat basal hypothalamus showing the location of the paraventricular (PVN) and supraoptic (SON) nuclei. These nuclei contain the magnocellular neurosecretory cells whose axons project to the posterior pituitary (PP) where they release oxytocin or vasopressin. Also illustrated are the organum vasculosum lamina terminalis (OVLT), the diagonal band of Broca (DBB) and the nucleus medianus (NM). These areas are known to project to SON neurons and to regulate their electrical activity. Other axonal projections to and from these areas are not shown. The solid black area at the bottom represents the optic chiasm. The magnocellular subdivision of the PVN is not usually included in perfused hypothalamic explants

I.4 Development of *In Vitro* Hypothalamic Preparations

The hypothalamus modulates the functions of the autonomic nervous system through descending pathways that eventually impinge on preganglionic sympathetic and parasympathetic neurons. Hypothalamic neurons also regulate endocrine functions, both through humoral connections with the anterior pituitary gland (via the portal vessels) and through neural connections with the posterior pituitary. During the 1970s, rising interest in the function of neuropeptides as transmitters and hormones led several groups of investigators to focus their attention on mammalian hypothalamic neurosecretory neurons (e.g. Arnauld *et al*., 1974; Koizumi and Yamashita, 1972; Renaud, 1978; Wakerley and Lincoln, 1973). Two useful features of these cells are that they (1) can be identified electrophysiologically and (2) can serve as models of peptide-secreting neurons. In particular, the magnocellular neurosecretory cells (MNCs) which secrete oxytocin or vasopressin are probably the most intensely studied group of neurosecretory neurons of any animal. In mammals, these cells are located in the supraoptic (SON) and paraventricular (PVN) nuclei from where they send axonal processes through the neurohypophysial tract and into the posterior lobe of the pituitary gland (Figure 1). There, each axon branches into thousands of terminal dilatations from which peptides are released into the general circulation. Because of the specificity of this anatomical projection SON and PVN neurons can be identified electrophysiologically as MNCs by showing an antidromic response following electrical stimulation of the neural lobe (Yagi *et al*., 1966). Extracellular recordings obtained from MNCs *in vivo* have therefore provided a wealth of information concerning their activity patterns and general physiology (for review see Poulain and Wakerley, 1982). However, the observation of complex response patterns, including the generation of stereotyped bursting activity, led to a growing need for information concerning the intrinsic properties of these neurosecretory cells. Unfortunately, intracellular recordings from mammalian hypothalamic MNCs are notoriously difficult to obtain *in vivo* because they are small in size (12–25 μm) and they are located in a mechanically unstable area of the brain (for example, the SON receives its vascular supply as direct branches of the anterior cerebral artery). There was therefore much incentive for the development of the hypothalamic slice (Brimble *et al*., 1978; Haller *et al*., 1978; Hatton *et al*., 1978), from which stable intracellular recordings could be obtained (e.g. Dudek *et al*., 1980; MacVicar *et al*., 1982; Andrew and Dudek, 1984). An inevitable problem with hypothalamic slices, however, is that the axonal output of SON and PVN neurons is severed. This makes the identification of impaled neurons as neurosecretory cells difficult without the application of post-recording immunocytochemical techniques (Kayser *et al*., 1982; Yamashita *et al*., 1983; Smithson *et al*., 1984). Another specific disadvantage of slice preparations for the analysis of the magnocellular neurosecretory system is the limited synaptic connectivity they retain, particularly with forebrain structures thought to be involved in the regulation of body fluid homeostasis (see Figure 1).

In an attempt to devise a preparation that retained relevant inputs as well as the output connection of MNCs (i.e. the neurohypophysis), Bourque and Renaud (1982, 1983) adapted the technique of intravascular perfusion described by Yarom and Llinas (1979) to an explant of rat basal forebrain that extended rostrally from the interpeduncular nucleus to the diagonal band of Broca, and laterally to the medial amygdala (see Figure 2). As described in the following section, this portion of the brain is particularly well suited to maintenance by intravascular perfusion because much of its arterial supply arises directly from the readily accessible internal carotid arteries.

I.5 Hypothalamic Vasculature and *In Vitro* Perfusion

The arterial supply of the rat hypothalamus has been well described by Ambach and Palkovits (1979) using the double ink perfusion technique (Ambach and Palkovits, 1974). The hypothalamus is vascularized from the basal surface, by arteries which originate from the circle of Willis. The latter, which comprises the posterior communicating, posterior cerebral, anterior cerebral, middle cerebral, and the anterior communicating arteries, is itself supplied by the basilar and internal carotid arteries (see Figure 2). The arteries which emerge from the circle of Willis

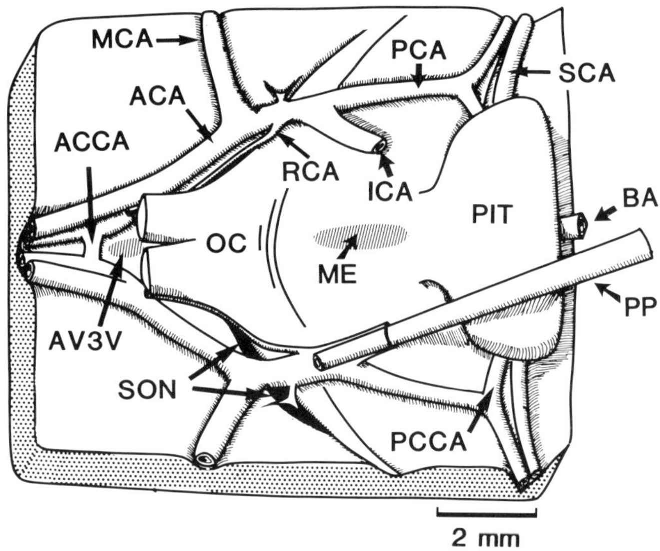

Figure 2 Diagram of a perfused hypothalamic explant cannulated through the right internal carotid artery (ICA). MCA: middle cerebral artery; ACA, anterior cerebral artery; PCA, posterior cerebral artery; RCA, retrochiasmatic artery; SCA, superior cerebellar artery; BA, basilar artery; PIT, pituitary; PP, perfusion pipette; PCCA, posterior communicating artery; ACCA, anterior communicating artery; AV3V, anterior ventral third ventricle; SON, supraoptic nucleus; ME, median eminence; OC, optic chiasm

and enter the brain through the anterior and posterior perforated substance do not contribute to the blood supply of the hypothalamus, but they form arches over it in the rostrocaudal direction. It is also a characteristic of the angioarchitectonics of the hypothalamus that the regions supplied by different arteries 'cover each other like shells' (Ambach and Palkovits, 1979). While an exhaustive review of hypothalamic vascular ramification is beyond the scope of this chapter, a minimum of information is required to appreciate the effects of cannulating selected vessels for *in vitro* perfusion. The general hypothalamic vascular contributions of the circle of Willis will therefore be outlined, proceeding in a rostrocaudal order.

Vessels arising from the anterior communicating artery supply the preoptic region and the suprachiasmatic nucleus. Branches of the anterior cerebral artery supply the rostral part of the SON, the anterior hypothalamus, and parts of the preoptic region. Only a few branches of the middle cerebral artery go to the hypothalamus. These, however, include two short lateral supraoptic arteries (which supply much of the SON) and the lateral thalamostriatal arteries (which also send branches to the SON). The majority of the arteries supplying the hypothalamus arise as three groups of branches from the internal carotid artery: (1) the retrochiasmatic artery, which supplies the medial SON and the anterior hypothalamus via a large number of branches; (2) the tuberal arteries supplying parts of the ventromedial and dorsomedial nuclei; and (3) the hypophysial arteries. Specifically, there are three hypophysial arteries that leave the internal carotid prior to its entry into the circle of Willis. The anterior and middle hypophysial arteries supply the median eminence and form the primary plexus of the portal system. Branches of these two arteries also vascularize the arcuate nucleus. The posterior hypophysial artery turns caudally after its origin and reaches the dorsal aspect of the neurohypophysial stalk. Its branches form a plexus on the caudal side of the median eminence behind the point at which the pituitary stalk becomes separated. Terminal branches of these vessels (posterior infundibular arteries) supply the posterior pituitary. Branches of the posterior cerebral artery also supply the hypothalamus. One such branch, the posterior tuberal artery, is the main arterial supply of the dorsomedial nucleus and of the posterior hypothalamus. Finally, arteries emerging from the posterior communicating and basilar arteries supply the lateral hypothalamus, the posterior hypothalamic nucleus and the mammillary body.

Clearly then, cannulation of the internal carotid arteries can provide extensive perfusion of an isolated explant of rat hypothalamus, provided that the vasculature of the circle of Willis is patent and undamaged.

II METHODS: PREPARATION AND MAINTENANCE OF HYPOTHALAMIC EXPLANTS

Although this chapter only provides specific information concerning the preparation and maintenance of perfused explants of rat basal forebrain, the approach and the methodology should be readily applicable to a large variety of '*en bloc*' CNS explants. Attempts to maintain explanted brain tissue by intravascular perfusion should, however, begin with a design of the geometry, dimensions, and orientation of the desired area, based on a sound understanding of its vascular supply. Any explant having an intact and extensive vascularization that can be accessed by cannulating a limited number of vessels can, in principle, be kept alive *in vitro* using the techniques described below.

II.1 Perfusion System and Recording Chamber

Explants of hypothalamus are cannulated and perfused after having been attached to the base of a recording chamber. This chamber (Figure 3A) consists of a circular well drilled into a Perspex disc. Except for the top 2–3 mm, the well is filled with Sylgard elastomer (Corning Ltd), giving the chamber a total volume of less than 300 μl (<200 μl with the tissue in place). The transparent rubbery base offered by the Sylgard allows the tissue to be fixed in place using small 'insect' pins and to be transilluminated (from below) for improved visibility. The recording chamber is housed in a depression at the top of a Perspex bath (Figure 3C). The bath itself contains a 250-ml water jacket which is regulated thermostatically using an insulated thermocouple as feedback source. Inside this water jacket a 30-cm segment of coiled polyethylene tubing (PE-240) acts as a heat exchanger and carries oxygenated perfusion medium. This tube is connected to the proximal end of a perfusion pipette via a short length (3–6 cm) of flexible Tygon or silicone tubing. Hypodermic needles can be inserted into the flexible connector to infuse small quantities of drugs (Figure 3C).

Cannulation of the explant is performed by placing the perfusion pipette into an internal carotid artery with the aid of a micromanipulator. Perfusion pipettes are manufactured by hand by heating glass capillary tubes (1.5 mm o.d.) over a small flame (see Figure 3B). Pipette tips are usually broken back to an external

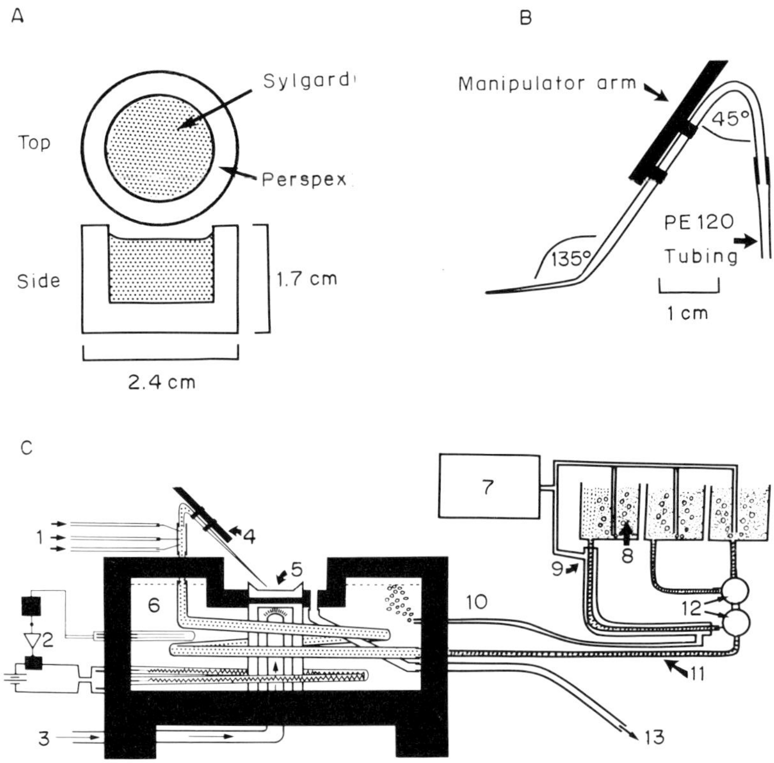

Figure 3 Diagram of the perfusion apparatus. The perfusion chamber consists of a well drilled into a thick Perspex disc (A). The bottom of the chamber is filled with Sylgard elastomer into which small pins can be inserted to attach the explant. Internal carotid arteries are cannulated by inserting a glass perfusion pipette (B) into the cut end of the vascular stump using a micromanipulator. The complete perfusion system is schematized in (C): (1) infusion lines, (2) temperature regulator, (3) fibre optic light source, (4) arm of manipulator and perfusion pipette, (5) perfusion chamber, (6) water jacket, (7) O_2/CO_2 source, (8) ACSF reservoir, (9) jacketed perfusion line, (10) outflow of the perfusion line jackets, (11) selected perfusion line, (12) valve assembly, (13) drain. (From Bourque, 1984)

diameter of 150–300 μm to achieve perfusion rates of 0.9–1.6 ml/min. When two cannulae are required, the end of the heat exchanger is inserted into a T-piece and two flexible connectors are attached to separately manipulated pipettes.

Because the flow of ACSF is relatively slow, the volume of fluid which has to course outside the water jacket should be minimized to prevent it from cooling more than one or two degrees Celsius. Drops in temperature at this point are unfavourable because: (1) they increase the oxygen-carrying capacity of a solution which has already been saturated at a higher temperature, thereby decreasing effective PO_2, (2) they affect the desired recording temperature, and (3) compensation of this temperature drop by raising that of the water jacket can cause the formation of oxygen bubbles. The shape of pipettes and their attachment to manipulators, therefore, should be designed to minimize the length of the proximal connector(s).

In our laboratory, perfusates are simply gravity-fed from 60–2000-ml reservoirs located 0.8–1.3 m above the perfusion chamber. The solutions inside these containers are continuously gassed with a 95% O_2/5% CO_2 mixture and delivered through individual polyethylene perfusion lines (Figure 3C) to Teflon-coated, electronically operated solenoid valves (General Valve Inc.). A disposable millipore filter (0.2 μm pore size) can be placed in series at the inlet of each valve to intercept debris and avoid the frustration of discovering a clogged cannula in the middle of a recording session. The configuration of the valves and mode of perfusate selection are depicted in Figure 3C. The valve assembly offers a single common outlet which communicates directly with the initial portion of the polyethylene heat exchanger located inside the bath. The delivery lines linking the reservoirs to the valve assembly are themselves jacketed with a larger calibre Tygon tubing through which the O_2/CO_2 mixture is forced to flow. The gas outflow from this perfusion line jacket is located just prior to the valve assembly from where it connects directly to the water jacket via a port drilled along the side of the bath (Figure 3C). This arrangement forces the gas through the heated water, thereby mixing it, and provides a humidified, oxygen-rich environment to the explant via a series of holes directed towards the recording chamber (Figure 3C). These precautions avoid potential changes in pH and PO_2, particularly in solutions which remain stagnant for prolonged periods prior to use, and help keep non-submerged portions of the explant moist.

Smooth drainage is achieved with wicks of cotton positioned so as to oppose the surface tension forces which tend to build up excessive amounts of fluid over the explant. The wasted perfusate then leaves by gravity, through a hole in the bottom of the bath.

II.2 Illumination and Optics for Cannulation and Recording

Because the cannulation step and the removal of selected membranes (see section

II.8) are procedures involving the microdissection of small (<100 μm) transparent structures it is critical to optimize the visual optics of the system, especially for the untrained eye. This can best be achieved by (1) using a good-quality stereomicroscope without extended range attachments, (2) minimizing the level of fluid over the explant, (3) reducing or eliminating ambient light coming from behind the experimenter, (4) transilluminating the tissue with focussed light from a short distance (<1 cm) and (5) moving the position of the light source to maximize the contrast and shadows observed through the microscope. While illuminating the tissue from above can provide satisfactory visibility for static cannulation procedures (i.e. without ongoing flow), cannulation with flowing pipettes and accurate placement of recording electrodes are more difficult to perform under such lighting conditions. Transilluminated lighting also yields additional free space for micromanipulators.

II.3 Dissection of Hypothalamic Explants

Explants of basal hypothalamus are obtained from 150–350-g rats of either sex (various strains, including Sprague-Dawley, Wistar and Long-Evans, have all been used successfully). Following decapitation, bone rongeurs are used to expose the dorsal surface of the brain rostrally to the level of the olfactory bulbs and scissors are used to make a coronal cut through the olfactory tubercles, the optic nerves and the anterior cerebral arteries. Then, from a nose-up position, the rostral part of the brain is allowed to tilt dorsally outside the cranial vault. Using fine scissors, the internal carotid arteries are cut proximal to the floor of the cavity and the pituitary released from the sella turcica by making longitudinal cuts along the diaphragma sella. The brain is then removed from the vault by sectioning all attached cranial nerves. This part of the procedure usually requires less than 1 min.

The isolated brain is then placed dorsal face down onto the bottom of an inverted Petri dish coated with cyanoacrylate glue. After dropping a small amount of fresh oxygenated ACSF (at room temperature) onto the tissue, a razor blade is used to make four cuts normal to the ventral surface of the brain. First, coronal cuts are made (1) 2–4 mm rostral to the point where the optic nerves fuse into a chiasm and (2) through the rostral end of the pons. Second, two parasagittal sections are made about 4 mm on either side of the midline. Following each cut, the peripheral tissue is discarded, leaving an 'island' of basal forebrain glued to the dish. After spreading more fresh ACSF over the remaining tissue, the ventral 2–3 mm of this block is isolated by manipulating two razor blades in a scissor-like motion. The resulting 8 × 8 × 2-mm explant (Figure 2) is then transferred to the recording chamber where it is fixed in place with four stainless steel 'insect' pins (Fine Science Tools Ltd). This part of the procedure requires no more than 30–40 s. For right-handed users it is best to position the explant such that the rostral end faces the left, allowing the cannula(e) to be manipulated from the right side.

II.4 Cannulation Procedure

As will be detailed below, many short-term (3–12 h) recording applications only require that a single internal carotid artery be cannulated. Under stereoscopic visualization, the distal portion of the selected carotid artery is gently lifted with jeweller's forceps and a fresh cut of the vessel is made with fine scissors, upstream of its junction with the anterior and posterior cerebral arteries (Figure 4A,B). Whether the cannula should be flowing or not at the time of cannulation is largely a matter of preference, since both methods can be used successfully. The advantages

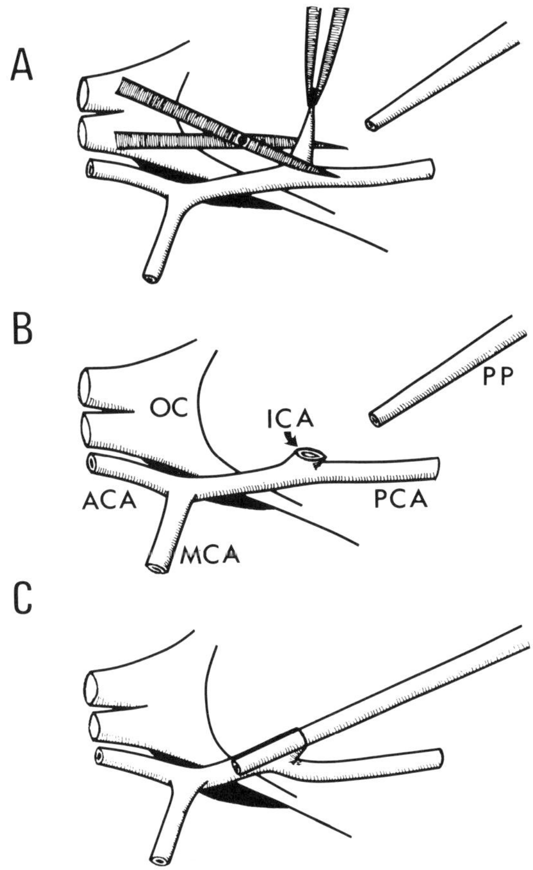

Figure 4 Cannulation of the internal carotid artery (ICA) in an explant of rat hypothalamus. As shown in (A), the arterial stump is held up with fine forceps and a fresh cut is made upstream of the bifurcation of the posterior cerebral artery (PCA). (B) The perfusion pipette (PP) is then inserted into the open stump of the ICA until (as shown in (C)) the tip is near the bifurcation of the middle (MCA) and anterior cerebral (ACA) arteries. OC, optic chiasm

of a flowing pipette are that (1) tissue perfusion begins as soon as the pipette and the vessel are lined up (i.e. even before it is actually cannulated) and (2) that the flow of ACSF maintains the vascular stump opened and increases the calibre of the artery, making cannulation easier. The disadvantages of ongoing flow are (1) that once a vessel has been flushed with ACSF it becomes difficult to see and (2) that active flow can sometimes cause the vascular stump to shift away from the approaching pipette. A micromanipulator is used to insert the perfusion pipette into the anterior cerebral artery, and to advance it to a point near the bifurcation of the anterior and middle cerebral arteries (Figure 4C). With some practice, unilateral cannulation can be completed within 4 min of decapitation.

Upon insertion of the perfusion pipette, vigorous flow is apparent from the rapid movement of red blood cells inside arteries and veins. Within seconds all blood cells clear rapidly from most surface vessels and a gradual blanching of the tissue can be observed. Residual strings of clotted blood can be pulled out from the open ends of affected venous vessels to improve drainage. Each vessel comprising the circle of Willis appears to be under considerable internal pressure from the perfusion fluid. The maintenance of a positive pressure inside this part of the vascular system evidently results from a partial but sustained constriction of the stump of most of the large arteries. The washout of red blood cells from the venous vasculature and blanching of the explant further suggest that the existing pressure is sufficient to force perfusate through significant portions of the capillary network. This interpretation is supported by the scarcity of blood-clotted capillaries which are usually found upon subsequent histology.

II.5 Artificial Cerebrospinal Fluid (ACSF)

The normal solution (ACSF) used for electrophysiological experiments consists of a modified Yamamoto's solution (e.g. Yamamoto and McIlwain, 1966) comprising (mM): NaCl 126, KCl 3, KH_2PO_4 1.2, $MgSO_4$ 1.3, $NaHCO_3$ 25.9, glucose 10 and $CaCl_2$ 1.0–4.0. Another solution which is also perfectly adequate and frequently used for electrophysiological experiments comprises simply (mM): NaCl 126, KCl 3, $MgCl_2$ 1.3, $NaHCO_3$ 25.9, glucose 10, $CaCl_2$ 2.0–4.0. The advantage of the second solution is that relatively large (1–5 mM) amounts of various divalent cations (e.g. Cd^{2+}, Co^{2+}, Mn^{2+}) can easily be added without precipitation. In all cases the solutions are kept saturated with 95% O_2 and 5% CO_2 which stabilizes the pH (at 35°C) to a value near 7.35. Finally, the osmolarity of the solution should be measured and, if necessary, adjusted to a value of 295 $\pm$ 1 mOsm/l.

II.6 Eliminating Oxygen Bubbles

Because of the high vascular pressure and flow rate inherent to perfused preparations, the presence of oxygen bubbles can be particularly problematic. Fortunately, because the arterial stumps are not tied off in the preparation presently described, a released oxygen bubble will preferentially travel through large-calibre vessels and leave the explant through one of the open ends. While this may not affect the explant's viability, the turbulence and vibration caused by the periodic release of bubbles usually prevents the maintenance of stable intracellular impalements. Care, therefore, must be taken to prevent the occurrence of oxygen bubbles in each preparation.

In most arrangements, solutions are oxygenated in a reservoir maintained at room temperature and then warmed to the desired recording temperature by having the fluid flow through some heat-exchanging device (in this case a water jacket). Because the amount of oxygen (ml/l) that can be dissolved in water decreases as temperature rises, warming a saturated solution under a constant atmospheric pressure will force some of the dissolved oxygen to return to the gas phase. In practice, once bubbles have started to form or 'seed' through this process, few manipulations are capable of resolving the problem if the explant is already in place. It is therefore important to prevent their occurrence at any point while setting up the apparatus. A solution to this problem is to warm up the ACSF at its point of oxygenation. However, care must be taken not to heat the ACSF reserve to a temperature which *exceeds* that of the recording chamber. Doing so would lower the maximal volume of oxygen which can be dissolved at the reservoir level and, consequently, reduce vascular PO_2 and the corresponding rate of oxygen diffusion within the cooler tissue. In practice, the formation of bubbles occurs when the bath temperature is about 10–12°C above room temperature. This observation may explain why problems with bubble formation often manifest themselves in a seasonal or geographical manner.

II.7 Preparation for Hormone Release Studies

An important adaptation of the preparation as originally described was recently made to allow measurements of hormone release to be obtained from perfused hypothalamic explants (Randle, 1985; Randle *et al*., 1986c). For this purpose explants are prepared as described above but with the following additional precautions.

First, the usual ACSF may be supplemented with bovine serum albumin (0.1%) to serve as carrier protein. Second, to minimize damage to the hypothalamo-neurohypophysial tract during removal of the brain from the skull, the diaphragma sella is gently torn using a hooked 24-gauge needle to free the pituitary from the sella turcica. Third, perfusion of the tissue is optimized by cannulating *both* internal carotid arteries and leaving *all* surface membranes intact. Fourth, instead of breaking the surface tension with cotton wicks, the ACSF is allowed to collect above the explant. An additional pipette (1.5 mm o.d.) is positioned caudal to the neurohypophysis to draw off the accumulating fluid with the help of a peristaltic pump. Fractions of effluent medium are collected into 16 × 100-mm polystyrene test tubes (on ice) and stored frozen (−20°C) until radioimmunoassays can be performed.

II.8 Preparation for Recording

To record from supraoptic MNCs, the arachnoid membranes are gently incised longitudinally between the anterior cerebral artery and the lateral aspect of the optic tract. The small tension provided by the remaining membranes usually causes a slight retraction of the anterior cerebral artery and provides sufficient exposure of the rostral portion of the SON to allow the insertion of a recording electrode. In most cases the anterior pituitary is removed by tearing open its ventral surface and peeling it away from the neuro-intermediate lobe. The neural stalk is then carefully desheathed and a bipolar stimulating electrode inserted into its fibres to antidromically activate MNCs (see Figure 5).

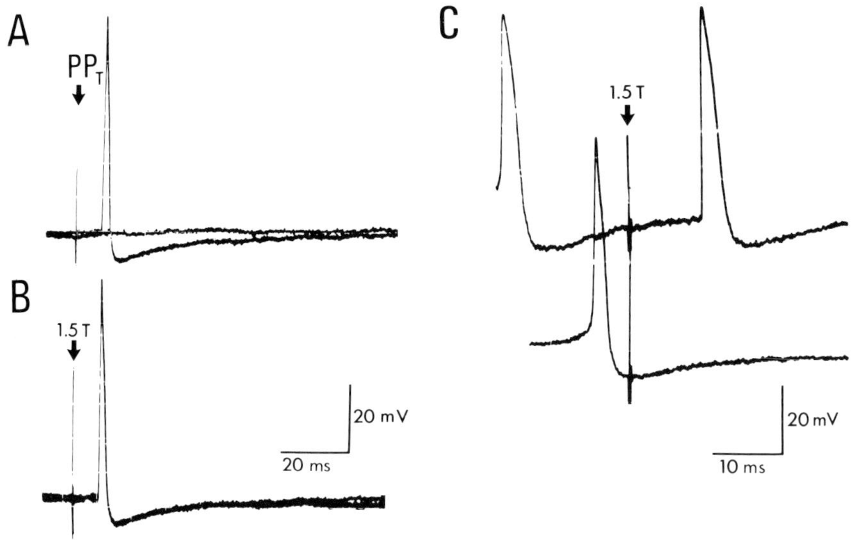

Figure 5 Antidromic activation of a supraoptic nucleus MNC following posterior pituitary (PP) stimulation in a perfused explant of rat hypothalamus. (A) All or none response at threshold (T). (B) Constant latency following suprathreshold stimulation. (C) Collision cancellation of the antidromic response when a spontaneous spike occurs within one latency interval of the stimulus (lower trace) but not if it occurs more than one interval before (upper trace). Note that no poststimulus synaptic response is revealed either during subthreshold stimuli (A) or after collision cancellation (C)

To obtain recordings from medial hypothalamic structures (e.g. the suprachiasmatic, ventromedial and arcuate nuclei), small additional tears in the pia-arachnoid membranes are made over the desired recording targets. If the cells of interest are tuberoinfundibular neurosecretory cells, a stimulating electrode can be placed into the median eminence for purposes of antidromic identification.

Finally, if desired, the optic nerves can be lifted by their rostral end and, using fine scissors, excised with a cut placed as near the optic chiasm as possible. This procedure exposes the organum vasculosum lamina terminalis, the diagonal band of Broca and the medial parolfactory area. More lateral (e.g. amygdaloid) structures can also be exposed for recording or stimulation purposes, by removing part of the covering membranes over the desired locations.

II.9 The 'Equilibration' Period

Following cannulation and the removal of membranes, the explant should be allowed at least 2 h of undisturbed perfusion before recording. A 3-h wait is usual practice in our laboratory. This period of 'equilibration' appears to be necessary, as attempts to record supraoptic neuronal activity within the first 10–20 min usually reveal very little spontaneous activity (Bourque, 1984; see also Schwartzkroin, 1981). This interval is also required to reach mechanical stability, as the tissue physically increases in size during the first 90 min (see section II.11).

II.10 Hydrogen Peroxide as a Source of Oxygen

Because simple saline solutions offer only a small fraction of the oxygen-carrying capacity of blood, early experiments on intravascularly perfused brain explants evaluated the potential benefits of adding H_2O_2 (0.003%) to the perfusion fluid (Llinás *et al.*, 1981; Bourque and Renaud, 1983). In one study (Llinás *et al.*, 1981), omission of the H_2O_2 caused cerebellar field responses to deteriorate rapidly during repetitive parallel fibre stimulation. This effect could be reversed by the addition of H_2O_2 to the perfusate. An elaborate study by Walton and Fulton (1983) confirmed that in concentrations as low as 0.001%, H_2O_2 can effectively supplement O_2/CO_2 bubbling to improve the long-term viability and response properties of certain CNS preparations. The mechanism by which H_2O_2 improves tissue oxygenation appears to involve a cellular catalase and the generation of molecular oxygen and water. It should be noted that the use of peroxide-containing solutions has been reported to cause oedema in isolated and perfused brainstem–cerebellum preparations (Llinás and Mühlethaler, 1988a). Furthermore, because the presence of free radicals may also be responsible for excessive swelling and other problems, it is probably wise to omit H_2O_2 from perfusing solutions unless its use leads to a demonstrable improvement in viability under the particular experimental conditions required. No apparent change in tissue viability seems to be gained by the addition of H_2O_2 to the ACSF perfusing hypothalamic explants (Bourque and Renaud, 1984a).

II.11 Tissue Swelling

As mentioned earlier, perfused explants of rat hypothalamus swell gradually during the 20–90 min which follow arterial cannulation (Bourque, 1984; Randle, 1985). This process ceases when the linear dimensions of the explant have increased some 5–10%. The shape and size of the tissue following this period stay largely unchanged as long as the explant remains viable (i.e. up to several days). Recordings from explants have shown that swelling does not, in the short term (8–12 h), affect the neuronal properties, viability or synaptic connectivity of MNCs when compared to recordings obtained in slices (e.g. Andrew and Dudek, 1984) or from intact animals (Renaud and Bourque, 1989). The occurrence of tissue swelling, therefore, has not been a major concern during studies on supraoptic MNCs. Solving this problem, however, may be beneficial or even necessary in order to study other hypothalamic networks.

The swelling of isolated tissue perfused with saline solutions results in part from the absence and washout of plasma proteins because the accompanying loss of oncotic pressure causes an accumulation of fluid in the interstitial space. Thus, oedema could be prevented or minimized by restoring the oncotic pressure of the ACSF which is required to balance the hydrostatic pressure exerted within the capillaries. This can be accomplished by the addition of high molecular weight colloids such as albumin or dextran. Because this technique is theoretically only completely successful if the oncotic and hydrostatic pressures are balanced, it would be useful, if not essential, to measure the hydrostatic pressure which exists within the vessels and regulate the flow rate via negative feedback to compensate for any changes in vascular pressure (see Llinas and Muhlethaler, 1988a).

II.12 Anaesthetics and the Preparation of Explants

Although this has not been examined quantitatively, it would appear that the synaptic and intrinsic properties of neurons within explants are completely identical whether these are prepared from animals decapitated with or without anaesthesia. The anaesthetics used thus far have included urethane, pentobarbital, ether and chloroform. Many of these substances are known to wash out rapidly from *in vitro* preparations (Bagust and Kerkut, 1981). The prolongation of GABA-mediated inhibitory postsynaptic potentials by pentobarbital is known to recover completely within minutes of the onset of a wash (Randle, 1985; Randle *et al.*, 1986a). The *density* of viable neurons encountered throughout the explant, however, appears to be reduced somewhat in explants prepared from anaesthetized rats. The effects of anaesthesia on the long-term viability of explants (i.e. beyond 12 h) have not been investigated.

II.13 The Special Case of 'Superfused' Explants

As a means of studying the regulation of vasopressin secretion, Sladek and Knigge (1977) introduced an 'organ-cultured' preparation of the rat hypothalamo-neurohypophysial system. This consists of a triangular block of hypothalamic tissue (2 mm thick) cut rostral to the optic chiasm, and lateral to either side of the median eminence. The resulting explant is about 5 mm wide at its most rostral portion, tapering back to the stalk separation at the median eminence. This block of tissue, which includes both supraoptic nuclei and their axonal projections to the neural lobe, can be kept under conventional tissue culture conditions for several days. Magnocellular neurosecretory neurons within cultured explants contain immunoreactive vasopressin (Sladek and Knigge, 1977), and display an ability to respond to a variety of chemicals by altered vasopressin secretion (Sladek and Joynt, 1980; Sladek and Armstrong, 1987; Gregg, 1986; Crandall and Gregg, 1986; Stern *et al.*, 1986). Acutely prepared and slowly superfused explants of the same geometry have also been used to record from antidromically identified MNCs (Armstrong and Sladek, 1982). Superfused hypothalamic explants therefore appear to be suitable for experiments on supraoptic neurosecretory neurons (e.g. Armstrong *et al.*, 1986; Bourque, 1986, 1988; Horikawa and Armstrong, 1988).

While the intrinsic electrical characteristics of supraoptic MNCs recorded in superfused explants are identical to those recorded in slices or perfused explants, little is known about other neurons contained within the block of tissue. It is probably wise to assume that cells impaled deeper than about 200 μm from any superfused surface may not display properties reflecting their behaviour *in vivo*.

III RESULTS: MORPHOLOGICAL STUDIES

III.1 Tissue Fixation

A number of histological procedures can be performed on perfused explants following immersion fixation in standard fixatives. A cold (4°C) solution of 4% paraformaldehyde in 0.1 M phosphate-buffered saline (pH 7.4) provides reasonable preservation and minimizes autofluorescence of the tissue. This feature is particularly useful when fluorescence is to be observed within tiny and distant portions of a neuron (e.g. Randle *et al.*, 1986b). Following 24 h of fixation whole explants can be embedded in agar gel and cut into 40–100-μm sections using a vibratome. Alternatively, fixed explants may be transferred to a solution of fixative (as above) containing 10% sucrose for 1–2 h in preparation for cutting using a cryostat. The tissue is then embedded in a suitable mounting medium (e.g. Tissue-Tek) and frozen in 50–100 ml of methylbutane previously chilled to −65°C in liquid nitrogen. Sections 6–40 μm thick can be cut subsequently using a cryostat at −25°C. Sections mounted on gelatin-coated slides can be dried and processed for a variety of purposes.

To maximize the quality of tissue preservation one might consider applying the fixative solution directly via the already cannulated vascular supply. However, the contamination of the apparatus with fixatives would require excessive cleaning procedures following each application. This practice is therefore not advisable.

III.2 Immunocytochemistry

Sections cut through the area of the SON and stained for the presence of vasopressin- or oxytocin-like immunoreactivity display intense reaction product, even following several hours of perfusion *in vitro* (Bourque, 1984). Immunocytochemical methods should similarly be applicable to a variety of basal forebrain neuronal systems following electrophysiological analysis and intracellular dye injection. This approach could be particularly useful for characterizing the peptidergic 'type' of various magnocellular or tuberoinfundibular neurosecretory cells.

III.3 Intracellular Dye Injection

Neurons recorded intracellularly can be injected with various enzymes (e.g. horseradish peroxidase), metals (e.g. Co^{2+}) or dyes (e.g. Lucifer yellow) which can subsequently be processed and visualized microscopically (e.g. Warner and Bate, 1987). This procedure is useful for the morphological analysis of physiologically identified neurons or for determining the nature of the neurotransmitter secreted by the neuron recorded. Perfused explants offer several advantages over slice preparations for the analysis of cellular morphology using intracellular dye labelling methods. The processes of many cells in perfused explants are not subject to the systematic ablation which characteristically occurs in brain slices. A more extensive or complete morphology of dye-injected cells can therefore be studied in this preparation (e.g. Randle *et al*., 1986b; Horikawa and Armstrong, 1988). One interesting and probably significant difference between hypothalamic slices and explants occurs when one examines supraoptic neurons following intracellular injection with the fluorescent dye Lucifer yellow (LY). The LY molecule has a diameter of 1.2–1.3 nm and can therefore diffuse freely through the 'electrotonic' or 'gap' junctions that can couple cells electrically. The occurrence of dye-coupling is therefore frequently interpreted as an indication of the existence of electrical coupling between neurons (Stewart, 1978). Interestingly, while dye-coupling is routinely observed between supraoptic neurosecretory cells impaled in hypothalamic slices (Andrew *et al*., 1981; Cobbett and Hatton, 1984; Yamashita *et al*., 1984), this phenomenon does not occur if the same group of neurons are impaled in perfused (Bourque, 1984; Reaves *et al*., 1983; Randle *et al*., 1986b) or superfused (Horikawa and Armstrong, 1988) explants. Dye-coupling in slices may therefore be caused by axotomy, which is known to induce dye-coupling between identified neurons in some invertebrates (Bulloch and Kater, 1982; Murphy *et al.*, 1983).

III.4 Ultrastructural Features of Perfused Hypothalamic Explants

As an index of tissue viability, the ultrastructure of doubly cannulated explants has been examined after 3–6 h of perfusion *in vitro* (unpublished results). Whole explants were fixed by immersion into a phosphate-buffered saline (PBS; pH 7.4) containing 3% glutaraldehyde. Following 12–24 h of fixation, the explant was cut into 50-μm coronal sections using a vibratome (Oxford). These were post-fixed overnight in 2% osmium tetroxide, dehydrated and embedded in epon. Areas including the suprachiasmatic, supraoptic, ventromedial and arcuate nuclei as well as the diagonal band of Broca, the organum vasculosum of the lamina terminalis and the preoptic area, were subsequently isolated, attached to epon blocks and trimmed prior to sectioning. Thin sections (70–80 nm) were then cut and processed for electron microscopy.

The ultrastructure of all areas examined (see above) was remarkably normal in appearance, with little sign of swelling or degeneration (Figure 6). The superficial aspects of the explant at its ventral surface showed no sign of a traumatic response to tissue preparation as is commonly seen over the cut face of brain slices (Lee *et al.*, 1981). In the coronal plane, occasional regions of gross disturbance occurred as bands bounded by layers of normal-looking tissue, suggesting that perfusion had failed through specific portions of the vascular system (see section I.5). At the light-microscope level a few arterioles were indeed found to contain clotted blood.

An increase in the extent of the perfused area may therefore be desirable for certain applications. Tying off the stumps of the anterior and middle cerebral arteries may help, since this would raise the local perfusion pressure and direct more of the flow towards the caudal vasculature of the circle of Willis. Small vessels can be tied off using fine strands of thread unravelled from surgical silk (Llinás and Mühlethaler, 1988a). Because a major source to the circle of Willis arises from the basilar artery, one may also consider cannulating this vessel.

IV RESULTS: PHYSIOLOGICAL STUDIES

IV.1 General Neuronal Activity

While the perfused hypothalamic explant was originally developed to carry out *in vitro* studies on identified neurosecretory neurons, unpublished recordings accumulated over the past few years indicate that neuronal viability is not limited to the magnocellular or tuberoinfundibular systems. Indeed, electrophysiological recordings made in explants have also been obtained from a variety of other basal forebrain areas, including the vertical and horizontal limbs of the diagonal band

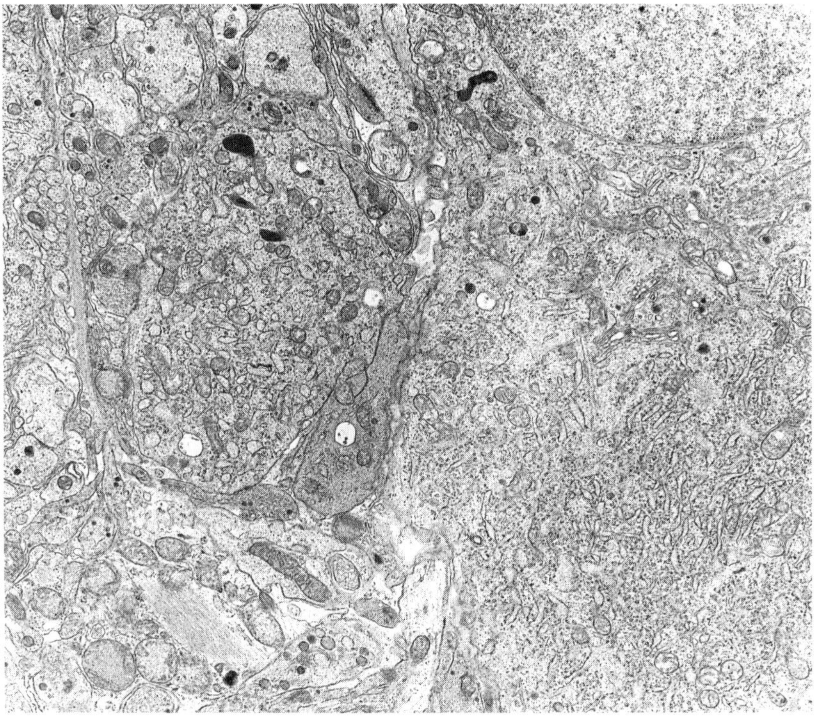

Figure 6 Low-power electronmicrograph showing a portion of the supraoptic nucleus following immersion-fixation of a hypothalamic explant after 4 h of perfusion *in vitro*

of Broca, the organum vasculosum of the lamina terminalis (Renaud *et al*., 1988), the median preoptic nucleus, the medial preoptic area, the lateral preoptic area, the suprachiasmatic nucleus, the supraoptic nucleus, the anterior hypothalamic area, the ventromedial hypothalamic nucleus, the arcuate nucleus, the medial amygdaloid nuclei and various mammillary nuclei. Clearly, these and many other neurons and their associated circuits are available for study using this preparation.

IV.2 Tuberoinfundibular Neurosecretory Neurons

Perfused hypothalamic explants offer a unique opportunity to perform experiments on identified tuberoinfundibular neurosecretory cells (TNCs), i.e. neurons which project to the median eminence where they secrete the release inhibiting or stimulating factors that regulate adenohypophysial secretion. Recording from characterized TNCs is virtually impossible to do in slice preparations unless the neurons recorded can be shown to contain a pertinent regulatory substance using a combination of dye labelling and immunocytochemistry (e.g. Kelly *et al*., 1984). Even with this information it is usually difficult or impossible to demonstrate that the neuron recorded projected to the median eminence. In contrast, such cells can be readily identified in perfused hypothalamic explants by their antidromic response to electrical stimuli applied at the median eminence. Preliminary observations obtained in explants have shown that identified TNCs can be recorded within the medial preoptic area, the ventromedial hypothalamus (Bourque and Renaud, 1984b) and the arcuate nucleus. Antidromic activation latencies ranging between 3 and 8 ms have been observed and spontaneous activity patterns resemble those seen *in vivo* (Renaud, 1981). Following antidromic activation some TNCs displayed a silent period lasting up to 150 ms which could also be induced by current intensities below the threshold for antidromic activation, suggesting the involvement of a recurrent inhibitory pathway. In addition, a population of non-antidromically activated ventromedial cells display orthodromic activation to the same stimulus, suggesting that they were interneurons activated by axon collaterals of tuberoinfundibular neurons. Initial experiments have shown also that stable (>60 min) intracellular recordings can be obtained from TNCs within these areas (Figure 7). The application of such techniques should therefore permit an extensive and productive analysis of the intrinsic, synaptic and pharmacological properties of identified TNCs.

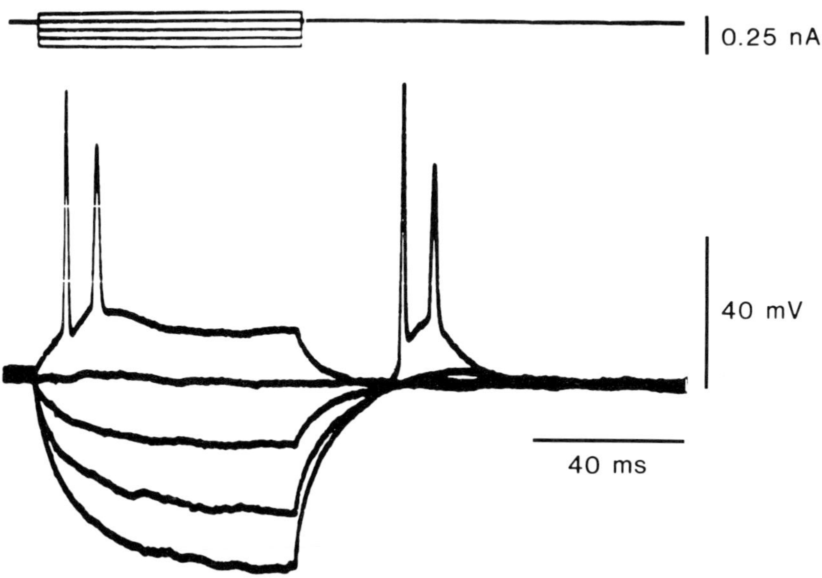

Figure 7 Intracellular recording obtained from a neuron located in the arcuate nucleus of a perfused explant of rat hypothalamus. Application of a depolarizing pulse triggers a doublet of fast action potentials superimposed upon a 'low-threshold' depolarizing envelope. Note that a similar response also occurs on the anode break of a large hyperpolarizing step

IV.3 Magnocellular Neurosecretory Neurons

Supraoptic MNCs have been extensively studied using perfused explants of rat hypothalamus, and a complete review of their intrinsic and synaptic physiology is beyond the scope of this chapter (see Renaud, 1988). Some aspects of this work, however, deserve mention since they illustrate the benefits provided by the use of explants.

Because dendritic and axonal processes are retained in this preparation, the passive and active properties displayed by MNCs may reflect the situation occurring *in vivo* more accurately. An interesting electrophysiological difference observed between slices and explants is that spontaneous phasic bursts recorded from vasopressin-secreting MNCs in slices (e.g. Andrew and Dudek, 1984) are typically much briefer than those recorded either *in vivo* or in explants (Armstrong and Sladek, 1982; Bourque and Renaud, 1983, 1984a). Based on current hypotheses (e.g. Bourque, 1987), this difference could result from a reduced capacity to buffer intracellular Ca^{2+} due to a diminished dendritic cytoplasmic volume, or from a reduced depolarizing drive if the underlying pacemaker current (Bourque, 1986) is generated within the dendrites.

Intracellular recordings obtained from MNCs in explants (Figure 5) have also shown that antidromic stimulation of their neurosecretory axons does not result in the activation of a synaptically mediated inhibitory postsynaptic potential (Bourque, 1984). The post-antidromic inhibition of supraoptic MNCs following stimulation of the posterior pituitary *in vivo* (e.g. Kelly and Dreifuss, 1970) is therefore presumably not mediated by a recurrent synaptic pathway, but by the activation of a transient Ca^{2+}-dependent K^+ current (Bourque *et al*., 1985; Bourque, 1988).

Electrical stimulation of the diagonal band of Broca inhibits the electrical activity of rat supraoptic MNCs *in vivo*, a response which is thought to result from the activation of peripheral baroreceptors (Jhamandas and Renaud, 1986). Intracellular recordings obtained *in vitro* using perfused explants have confirmed this response (Figure 8) and revealed that it is mediated by the activation of $GABA_A$ receptors located on the postsynaptic membrane of MNCs (Randle, 1985; Randle *et al*., 1986a; Randle and Renaud, 1987). Many other identified synaptic pathways are similarly available for study using hypothalamic explants.

IV.4 Neurosecretion from Perfused Explants

Explants cannulated and perfused through both carotid arteries have been used successfully to examine the release of oxytocin and vasopressin following the application of electrical (Nissen and Renaud, unpublished results) or chemical (Randle, 1985) stimuli. Immediately following preparation of the explant, both vasopressin and oxytocin are released at rates of 10–50 pg/min. This relatively intense initial secretory activity presumably results from a spike-independent

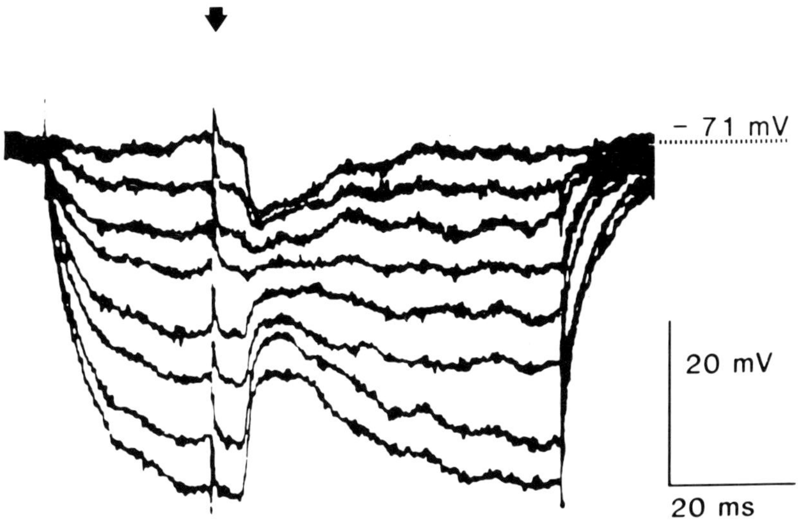

Figure 8 Current-clamp recording showing the response of a supraoptic MNC to electrical stimulation of the diagonal band of Broca (arrow) in a perfused explant. Hyperpolarizing current pulses of increasing magnitude progressively reduce the amplitude and ultimately reverse the polarity of the evoked IPSP. The dotted line indicates the resting potential

mechanism, since the SONs are electrically quiescent for the initial 15–30 min which follow cannulation. The rate of spontaneous release then gradually declines over some 30–50 min until they become stabilized at a value near or below 1 pg/min.

The release of both hormones can be evoked electrically (Figure 9) or by infusing α_1-adrenoreceptor agonists into the perfusate (Randle, 1985; Randle *et al*., 1986c). This response requires the presence of an intact hypothalamo-neurohypophysial pathway since damage to the neural stalk or addition of tetrodotoxin (1 μM) which blocks axonal conduction, prevented the release of vasopressin and oxytocin induced by noradrenaline in the same preparation (Randle, 1985). The intensity of the secretory response to the application of an agonist varies from one experiment to the next, peak rates of release ranging between 50 and 500 pg/min. The peak amplitude of multiple responses evoked within a single preparation, however, are very consistent, provided sufficient time is allowed between tests (i.e. about 1 h). These results suggest that the percentage of hypothalamo-neurohypophysial units which 'survive' the transfer to *in vitro* conditions determines the maximal rate of secretion that will be available, but that each 'surviving' unit is relatively robust and can continue to respond for some time (several hours). Measurements of neuropeptide release have yet to be made from explants maintained for extended periods of time (e.g. days).

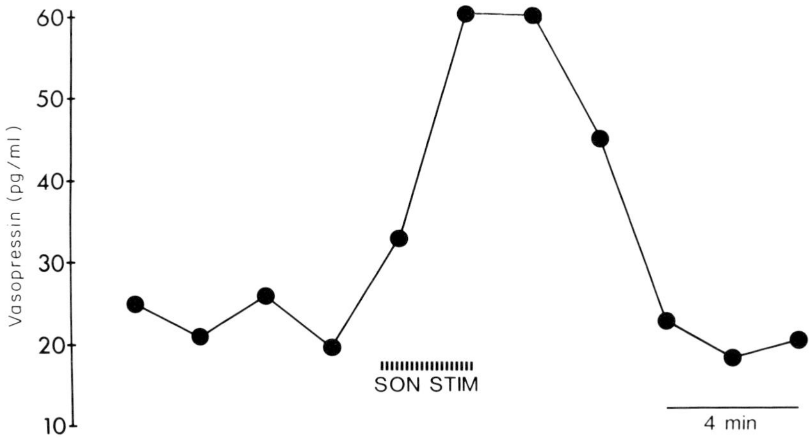

Figure 9 Effects of unilateral electrical stimulation of the supraoptic nucleus on vasopressin release from a perfused explant of rat hypothalamus. A bipolar stimulating electrode was inserted into the right SON of a perfused explant and electrical current pulses (500 μA × 150 μs) were applied at 10 Hz during the period indicated (bar). Note that the stimulus induced a large reversible rise in the concentration of assayable vasopressin within the sampled perfusate. (Courtesy of R. Nissen, L.P. Renaud and M. Mazurek)

IV.5 Life Expectancy of Perfused Hypothalamic Explants

Explants in which a single internal carotid artery is cannulated and perfused with oxygenated ACSF at 33–35°C display widespread (see section IV.1) and extensive viability for periods exceeding 12 h. For example, in a typical 300-μm track through the ipsilateral SON a dozen live neurons or so can be encountered. However, while spiking activity and synaptic potentials can also be recorded from neurons located in the contralateral SON, these do not usually fire as fast and cannot sustain firing as long as their ipsilateral counterparts, suggesting a reduced level of metabolic activity. These preparations are therefore best suited for a short-term analysis of the intrinsic and synaptic properties of structures located on the side ipsilateral to the cannulated artery.

Explants in which *both* carotids are cannulated and perfused with ACSF at 33–35°C display spontaneous synaptic potentials and spiking activity bilaterally, in any area examined. Normal activity and synaptic potentials can be recorded for more than 48 h in such preparations, with no apparent change in the density of live neurons (i.e. neurons/track). Following more than 4 days of perfusion, normal intrinsic activity can still be recorded from a large number of neurons (e.g. Figure 10). An examination of synaptic responses during long-term experiments has not been performed.

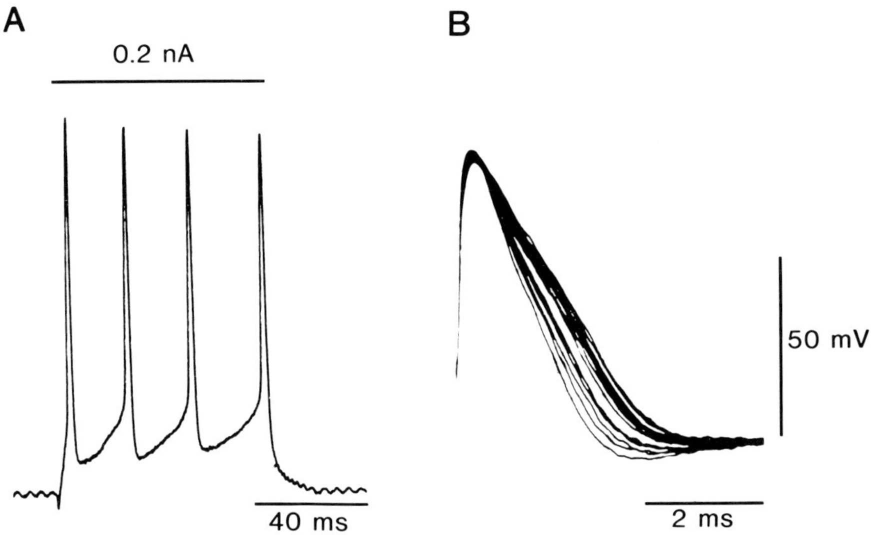

Figure 10 Current-clamp recordings of a supraoptic MNC obtained from an explant of hypothalamus after more than 90 h of perfusion *in vitro*. The explant was maintained at 32°C throughout except during daily recording sessions when the temperature was increased to 35°C. Note (A) that the cell can sustain continuous firing in response to a depolarizing current pulse and (B) that repetitive firing is associated with a pronounced broadening of the action potential. Both features are well-known characteristics of supraoptic MNCs

V CONCLUSIONS

While cell dissociation and tissue culture techniques offer the possibility of performing high-resolution biophysical measurements (e.g. single-channel recording $[Ca^{2+}]$ imaging), the functional analysis of intrinsic properties and their role in shaping the responsiveness of a specific cell upon activation by a particular pathway can only be performed in relatively intact preparations. Recording intracellularly from neurons in acutely prepared slices or explants of brain tissue therefore remains one of the most powerful methods in neurobiological research.

This chapter has described a preparation of rat basal forebrain which can be kept alive *in vitro* for several days by perfusion of ACSF through the existing vasculature. The isolated explant exceeds by far the dimensions of a conventional hypothalamic slice, being about 2–3 mm thick in its smallest dimension. Because of these dimensions, many neurons within explants remain completely intact, a feature which so far has proved to be particularly useful for the study of identified neurosecretory cells. Another important feature of explant preparations is that they retain several identified neural circuits. Forebrain inputs to supraoptic MNCs, for example, are available for detailed analysis using this preparation. Finally, perfused explants are ideally suited to functional studies on the role of synaptic connections onto MNCs since the output of MNCs, i.e. the secretion of oxytocin and vasopressin, can be measured by radioimmunoassay of collected effluent.

Perfused explants prepared from many other parts of the brain should offer similar advantages. The most significant advantage of these preparations is that they preserve complex synaptic circuits and open the way for a range of experiments that could not be performed with either slice or *in vivo* techniques. They offer a unique opportunity to study not only the intrinsic and synaptic properties of various neurons, but also how these interact to add a higher level of sophistication to signal processing within a synaptic network.

ACKNOWLEDGEMENTS

I thank R. Nissen, J. Randle and L. Renaud for data on the release of neuropeptides; L. Ohanessian and B. Oldfield for their expert advice with electron microscopy; and C. Jarvis and C. Yang for helpful comments. This work was supported by FCAR, FRSQ and the MRC of Canada.

REFERENCES

Ambach, G., and Palkovits, M. (1974). 'Blood supply of the rat hypothalamus. I. Nucleus supraopticus', *Acta Morphol. Acad. Sci. Hung.*, **22**, 291–310.

Ambach, G., and Palkovits, M. (1979). 'The blood supply of the hypothalamus in the rat', in *Handbook of the Hypothalamus*, Vol. 1 (Eds P.J. Morgane and J. Panksepp), pp. 267–377, M. Dekker, New York.

Andrew, R.D., and Dudek, F.E. (1984). 'Analysis of intracellularly recorded phasic bursting by mammalian neuroendocrine cells', *J. Neurophysiol.*, **51**, 552–566.

Andrew, R.D., MacVicar, B.A., Dudek, F.E., and Hatton, G.I. (1981). 'Dye transfer through gap junctions between neuroendocrine cells of rat hypothalamus', *Science*, **211**, 1187–1189.

Armstrong, W.E., and Sladek, C.D. (1982). 'Spontaneous phasic firing in supraoptic neurons recorded from hypothalamo-neurohypophysial explants in vitro', *Neuroendocrinology*, **34**, 405–409.

Armstrong, W.E., Gallagher, M.J., and Sladek, C.D. (1986). 'Noradrenergic stimulation of supraoptic neuronal activity and vasopressin release in vitro: mediation by an α_1-receptor', *Brain Res.*, **365**, 192–197.

Arnauld, E., Vincent, J.D., and Dreifuss, J.J. (1974). 'Firing patterns of hypothalamic supraoptic neurons during water deprivation in monkeys', *Science*, **185**, 535–537.

Bagust, J., and Kerkut, G.A. (1981). 'An in vitro preparation of the spinal cord of the mouse', in *Electrophysiology of Isolated Mammalian CNS Preparations* (Eds G.A. Kerkut and H.V. Wheal), pp. 338–365, Academic Press, London.

Bingmann, D., and Kolde, G. (1982). 'PO_2 profiles in hippocampal slices of the guinea pig', *Exp. Brain Res.*, **48**, 89–96.

Bourque, C.W. (1984). 'Membrane properties of rat supraoptic nucleus neurons in vitro', Ph.D. Thesis, McGill University.

Bourque, C.W. (1986). 'Calcium-dependent spike aftercurrent induces burst firing in magnocellular neurosecretory cells', *Neurosci. Lett.*, **709**, 204–209.

Bourque, C.W. (1987). 'Current and voltage-clamp studies of transient and pacemaker currents in neurosecretory neurons of the supraoptic nucleus', in *Inactivation of Hypersensitive Neurons* (Eds N. Chalazonitis and M. Gola), pp. 415–422, A.R. Liss Inc., New York.

Bourque, C.W. (1988). 'Transient calcium dependent potassium current in magnocellular neurosecretory cells of the rat supraoptic nucleus', *J. Physiol. (Lond.)*, **397**, 331–347.

Bourque, C.W., and Renaud, l.p. (1982). 'oSMOSENSITIVITY OF supraoptic nucleus (SON) neurons: studies utilizing intravascular perfusion in the isolated rat basal hypothalamus', *Soc. Neurosci. Abstr.*, **8**, 423p.

Bourque, C.W., and Renaud, L.P. (1983). 'A perfused in vitro preparation of hypothalamus for electrophysiological studies on neurosecretory neurons', *J. Neurosci. Methods*, **7**, 203–214.

Bourque, C.W., and Renaud, L.P. (1984a). 'Activity patterns and osmosensitivity of rat supraoptic neurones in perfused hypothalamic explants', *J. Physiol. (Lond.)*, **349**, 631–647.

Bourque, C.W., and Renaud, L.P. (1984b). 'In vitro neurophysiology of identified rat hypothalamic neuroendocrine neurons', *Neuroendocrinology*, **36**, 161–164.

Bourque, C.W., Randle, J.C.R., and Renaud, L.P. (1985). 'Calcium-dependent potassium conductance in rat supraoptic nucleus neurosecretory neurons', *J. Neurophysiol.*, **54**, 1375–1382.

Brimble, J.J., Haller, E.W., and Wakerley, J.B. (1978). 'Supraoptic and paraventricular units in slices incubated in iso- or hypertonic medium', *J. Physiol. (Lond.)*, **287**, 38p.

Bulloch, A.G.M., and Kater, S.B. (1982). 'Neurite outgrowth and selection of new electrical connections by adult helisoma neurons', *J. Neurophysiol. (Lond.)*, **48**, 569–583.

Cobbett, P., and Hatton, G.I. (1984). 'Dye coupling in hypothalamic slices: Dependence on in vivo hydration state and osmolality of incubation medium', *J. Neurosci.*, **4**, 3034–3038.

Crandall, M.E., and Gregg, C.M. (1986). 'In vitro evidence for an inhibitory effect of atrial natriuretic peptide on vasopressin release', *Neuroendocrinology*, **44**, 439–445.

Dingledine, R., (Ed.) (1984). *Brain Slices*, Plenum Press, New York.

Dudek, F.E., Hatton, G.I., and MacVicar, B.A. (1980). 'Intracellular recordings from the paraventricular nucleus in slices of rat hypothalamus', *J. Physiol. (Lond.)*, **301**, 101–114.

Gregg, C.M. (1986). 'Effect of cholinergic antagonists on basal and osmotically stimulated vasopressin release in compartmentalized hypothalamo-neurohypophysial explants', *Neuroendocrinology*, **44**, 378–383.

Haller, E.W., Brimble, M.J., and Wakerley, J.B. (1978). 'Phasic discharge in supraoptic neurones recorded from hypothalamic slices', *Exp. Brain Res.*, **33**, 131–134.

Hatton, G.I., Armstrong, W.E., and Gregory, W.A. (1978). 'Spontaneous and osmotically stimulated activity in slices of rat hypothalamus', *Brain Res. Bull.*, **3**, 497–508.

Horikawa, K., and Armstrong, W.E. (1988). 'A versatile means of intracellular labeling: injection of biocytin and its detection with avidin conjugates', *J. Neurosci. Methods*, **25**, 1–11.

Jhamandas, J.H., and Renaud, L.P. (1986). 'A γ-aminobutyric acid-mediated baroreceptor input to supraoptic vasopressin neurones in the rat', *J. Physiol. (Lond.)*, **381**, 595–606.

Kayser, B.E.J., Muhlethaler, M., and Dreifuss, J.J. (1982). 'Paraventricular neurones in the rat hypothalamic slice: Lucifer yellow injection and immunocytochemical identification', *Experientia*, **38**, 391-393.

Kelly, J.S., and Dreifuss, J.J. (1970). 'Antidromic inhibition of identified rat supraoptic neurones', *Brain Res.*, **22**, 406–409.

Kelly, M.J., Ronnekleiv, O.K., and Eskay, R.L. (1984). 'Identification of estrogen responsive LHRH neurons in the guinea pig hypothalamus', *Brain Res. Bull.*, **12**, 399–407.

Kerkut, G.A., and Wheal, H.V. (Eds) (1981), *Electrophysiology of Isolated Mammalian CNS Preparations*, Academic Press, London.

Koizumi, K., and Yamashita, H. (1972). 'Studies of antidromically identified neurosecretory cells of the hypothalamus by intracellular and extracellular recordings', *J. Physiol. (Lond.)*, **221**, 683–705.

Lee, K., Oliver, M., Schottler, F., and Lynch, G. (1981). 'Electron microscopic studies of brain slices: the effects of high frequency stimulation on dendritic ultrastructure', in *Electrophysiology of Isolated Mammalian CNS Preparations* (Eds G.A. Kerkut and H.V. Wheal), pp. 189–211, Academic Press, London.

Llinás, R. (1988). 'The intrinsic electrophysiological properties of mammalian neurons: insights into central nervous system function', *Science*, **242**, 1654–1664.

Llinás, R., and Mühlethaler, M. (1988a). 'An electrophysiological study of the in-vitro, perfused brain stem–cerebellum of adult guinea-pig', *J. Physiol. (Lond.)*, **404**, 215–240.

Llinás, R., and Mühlethaler, M. (1988b). 'Electrophysiology of guinea-pig cerebellar nuclear cells in the in-vitro brain stem–cerebellar preparation', *J. Physiol. (Lond.)*, **404**, 241–258.

Llinás, R., Yarom, Y., and Sugimori, M. (1981). 'Isolated mammalian brain in vitro: new technique for analysis of electrical activity of neuronal circuit function', *Fed. Proc.*, **40**, 2240–2245.

MacVicar, B.A., Andrew, R.D., Dudek, F.E., and Hatton, G.I. (1982). 'Synaptic inputs and action potentials of magnocellular neuropeptidergic cells: intracellular recording and staining in slices of rat hypothalamus', *Brain Res. Bull.*, **8**, 87–93.

Mühlethaler, M., and Llinás, R. (1984). 'Studies on the optimal conditions for the long-term survival of mammalian brain in vitro', *Soc. Neurosci. Abstr.*, **10**, 31p.

Mühlethaler, M., and Llinás, R. (1985). 'Burst firing properties of deep cerebellar nuclei cells studied in the isolated guinea pig brainstem cerebellum in vitro', *Soc. Neurosci. Abstr.*, **11**, 689p.

Murphy, A.D., Hadley, R.D., and Kater, S.B. (1983). 'Axotomy-induced parallel increases in electrical and dye coupling between identified neurons of helisoma', *J. Neurosci.*, **3**, 1422–1429.

Poulain, D.A., and Wakerley, J.B. (1982). 'Electrophysiology of hypothalamic magnocellular neurones secreting oxytocin and vasopressin', *Neuroscience*, **7**, 773–808.

Randle, J.C.R. (1985). 'Structure and function of rat supraoptic nucleus neurons in vitro', Ph.D. Thesis, McGill University.

Randle, J.C.R., and Renaud, L.P. (1987). 'Actions of γ-aminobutyric acid on rat supraoptic nucleus neurosecretory neurones in vitro', *J. Physiol. (Lond.)*, **387**, 629–647.

Randle, J.C.R., Bourque, C.W., and Renaud, L.P. (1986a). 'Characterization of spontaneous and evoked inhibitory postsynaptic potentials in rat supraoptic neurosecretory neurons in vitro', *J. Neurophysiol.*, **56**, 1703–1717.

Randle, J.C.R., Bourque, C.W., and Renaud, L.P. (1986b). 'Serial reconstruction of lucifer yellow-labeled supraoptic nucleus neurons in perfused rat hypothalamic explants', *Neuroscience*, **17**, 453–467.

Randle, J.C.R., Mazurek, M., Kneifel, D., Dufresne, J., and Renaud, L.P. (1986c). 'α_1-adrenergic receptor activation releases vasopressin and oxytocin from perfused rat hypothalamic explants', *Neurosci. Lett.*, **65**, 219–223.

Reaves, T.A., Jr, Hou-You, A., Zimmerman, E.A., and Hayward, J.N. (1983). 'Supraoptic neurons in the rat hypothalamo-neurohypophysial explant: double-labeling with Lucifer Yellow injection and immunocytochemical identification of vasopressin and neurophysin-containing neuroendocrine cells', *Neurosci. Lett.*, **13**, 137–142.

Renaud, L.P. (1978). 'Neurophysiological organization of the endocrine hypothalamus', in *The Hypothalamus* (Eds S. Reichlin, R.J. Baldessarini and J.B. Martin), pp. 269–301, Raven Press, New York.

Renaud, L.P. (1981). 'A neurophysiological approach to the identification, connections and pharmacology of the hypothalamic tuberoinfundibular system', *Neuroendocrinology*, **33**, 186–191.

Renaud, L.P. (1988). 'Magnocellular neuroendocrine neurons: update on intrinsic properties, synaptic inputs and neuropharmacology', *Trends Neurosci.*, **10**, 498–502.

Renaud, L.P., and Bourque, C.W. (1989). 'Intracellular recordings from rat supraoptic neurosecretory cells in vivo', *Can. J. Physiol. Pharmacol.*, **66**, Axxx.

Renaud, L.P., Nissen, R., and Bourque, C.W. (1988). 'Properties of organum vasculosum lamina terminalis (OVLT) neurons in the rat', *Soc. Neurosci. Abstr.*, **14**, 1179p.

Richards, C.D. (1981). 'The preparation of brain tissue slices for electrophysiological studies', in *Electrophysiology of Isolated Mammalian CNS Preparations* (Eds G.A. Kerkut and H.V Wheal), pp. 107–132, Academic Press, London.

Schurr, A., Tyler, T.J., and Tseng, M.T. (Eds) (1987). *Brain Slices: Fundamentals, Applications and Implications*, Karger, Basel.

Schwartzkroin, P.A. (1981). 'To slice or not to slice', in *Electrophysiology of Isolated Mammalian CNS Preparations* (Eds G.A. Kerkut and H.V. Wheal), pp. 15–50, Academic Press, London.

Sladek, C.D., and Armstrong, W.E. (1987). γ-Aminobutyric acid antagonists stimulate vasopressin release from organ-cultured hypothalamo-neurohypophyseal explants', *Endocrinology*, **120**, 1576–1580.

Sladek, C.D., and Joynt, R.J. (1980). 'Role of angiotensin in the osmotic control of vasopressin release by the organ cultured rat hypothalamo-neurohypophyseal system', *Endocrinology*, **106**, 173–178.

Sladek, C.D., and Knigge, K.M. (1977). 'Cholinergic stimulation of vasopressin release from the rat hypothalamo-neurohypophyseal system in organ culture', *Endocrinology*, **101**, 411–420.

Smithson, K.G., Cobbett, P., MacVicar, B.A., and Hatton, G.I. (1984). 'A reliable method for immunocytochemical identification of lucifer yellow, peptide containing mammalian central neurons', *J. Neurosci. Methods*, **10**, 59–69.

Stern, J.E., Mitchell, T., Herzberg, V.L., and North, W.G. (1986). 'Secretion of vasopressin, oxytocin and two neurophysins from rat hypothalamo-neurohypophyseal explants in organ culture', *Neuroendocrinology*, **43**, 252–258.

Stewart, W.W. (1978). 'Functional connections between cells as revealed by dye-coupling with a highly fluorescent naphthalimide tracer', *Cell*, **14**, 741–759.

Wakerley, J.B., and Lincoln, D.W. (1973). 'The milk ejection reflex of the rat: a 20 to 40 fold acceleration in the firing of paraventricular neurones during oxytocin release', *J. Endocrinol.*, **57**, 477–493.

Walton, K., and Fulton, B. (1983). 'Hydrogen peroxide as a source of molecular oxygen for in vitro mammalian CNS preparations', *Brain Res.*, **278**, 387–393.

Walton, K., and Llinás, R. (1982). 'Electrophysiological activity recorded from the cerebral cortex of an isolated mammalian brain maintained in vitro', *Soc. Neurosci. Abstr.*, **8**, 935p.

Warner, A.E., and Bate, C.M. (1987). 'Techniques for dye injection and cell labeling', in *Microelectrode Techniques: the Plymouth Workshop* (Eds N.B. Standen, P.T.A. Gray and M.J. Whitaker), pp. 169–185, The Company of Biologists Ltd, Cambridge.

Yagi, K., Azuma, T., and Matsuda, K. (1966). 'Neurosecretory cell: capable of conducting impulse in rats', *Science*, **154**, 778–779.

Yamamoto, C., and McIlwain, H. (1966). 'Electrical activities in thin sections from the mammalian brain maintained in chemically defined media in vitro', *J. Neurochem.*, **13**, 1333–1343.

Yamashita, H., Inenaga, K., Kawata, M., and Sano, Y. (1983). 'Phasically firing neurons in the supraoptic nucleus of the rat hypothalamus: immunocytochemical and electrophysiological studies', *Neurosci. Lett.*, **37**, 87–92.

Yamashita, H., Inenaga, K., Kawata, M., Sano, Y., and Kannan, H. (1984). 'Lucifer yellow-filled neurosecretory cells in the supraoptic and paraventricular nuclei', *Biomed. Res.*, **5**, 105–114.

Yarom, Y., and Llinás, R. (1979). 'Electrophysiological properties of mammalian inferior olive neurons in vitro brain stem slices and in vitro whole brain stem', *Soc. Neurosci. Abstr.*, **5**, 109p.

Index